D

Introduction

EXPLORING METALWORKING is a first course in the fundamentals of working with metal, using both hand and power tools.

EXPLORING METALWORKING is written in easy-to-understand language. It contains an abundance of illustrations. Extra color is used to clarify details, and to illustrate major processes in steelmaking.

EXPLORING METALWORKING provides constructional details on carefully selected projects. It also includes alternate designs and design variations which will help you design your own projects. The text tells and shows how to organize and operate a small manufacturing business in your school lab. You learn how to mass produce items with proven student-appeal.

Using selected material, EXPLORING METALWORKING may be used to present a six, nine, eighteen, or thirty-six week metalworking program.

EXPLORING METALWORKING emphasizes the important place metals occupy in our everyday lives. It explores metalworking career opportunities.

John R. Walker

Contents

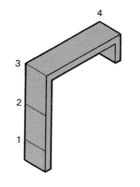

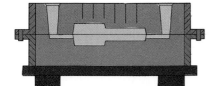

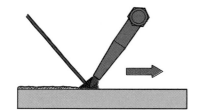

The field of metalworking technology offers many new opportunites. The concern for environmental protection and energy conservation, for example, has required the development of new metals and manufacturing techniques.
The experimental vehicle shown uses a gas turbine to drive a generator to charge the batteries. A gas turbine with continuous combustion and high operating temperatures has far lower polluting emissions than a piston engine. The high operating temperatures require metals able to withstand heat for long operating periods without failing.
The car has three modes of operation. The first mode is pure electric for pollution-free city driving with a range of 50 miles. In the second (hybrid) mode, the turbine is automatically activated to drive the generator when battery power declines to a certain level. In the third mode, the turbine alone is used when maximum power is required in emergency situations.
(Volvo Cars of North America, Inc.)

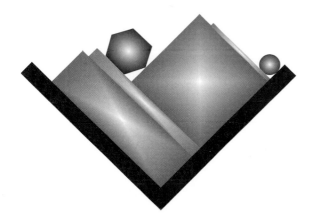

Unit 1
Careers in Metalworking Technology

After studying this unit, you will be able to identify the many job opportunities in the metalworking industries. You will also discover the importance of leadership skills in both school and career. And you will learn how to write a resume for use in your job search.

TECHNOLOGY is defined as the "know-how" that links science and the industrial arts. Its purpose is to solve problems and enhance (improve in value and/or quality) the natural and human-made environment. In the process, creativity, human skills, tools, machines, and resources are used.

HIGH TECHNOLOGY and STATE-OF-THE-ART TECHNOLOGY means that the area of technology to which reference is made employs the very latest ideas, techniques, research, tools, and machines available, Fig. 1-1.

Fig. 1-1. An artist's concept of the X-33 Advanced Technology Demonstrator test vehicle. The X-33 is a half-scale prototype being used in the development of a reusable, single-stage to orbit space vehicle. The construction of future space vehicles will require state-of-the-art metalworking technology. (NASA)

Fig. 1-2. Robotic work stations. The robots spot weld various body panels together. Productivity and quality is increased as the result of these robots.

Through METALWORKING TECHNOLOGY it is possible to change raw materials into finished products, Fig. 1-2. As an area of employment, it can provide interesting and challenging work.

Work has been found to be very important to life, well-being and security to the average person. Selecting an occupation is a serious and difficult task. It should not be done quickly. You must be ready to invest time and money to acquire the knowledge and skills needed to succeed in your chosen career.

There are many areas of specialization in metalworking technology. However, there are not many jobs for persons without the proper skills and training.

TYPES OF JOBS IN METALWORKING

Jobs in the metalworking industry fall into many categories. Yet all of them may be classified according to whether they are semiskilled, skilled, technical, or professional.

SEMISKILLED JOBS

Most semiskilled workers are employed in manufacturing where they:

1. Operate machines or equipment.
2. Assemble manufactured parts into complete units, Fig. 1-3.
3. Serve as helpers who assist skilled workers.
4. Work as inspectors or conduct tests to be sure manufactured parts are made to, and operate according to, specifications.

The majority of semiskilled workers work with their hands and require only a brief on-the-job training program. They are told what to do and how to do it. Frequently, semiskilled workers do the same job or operation over and over. Their work is closely supervised.

Some semiskilled workers are paid on the number of items they produce, and, therefore, make a good salary. However, most have annual earnings that are less than skilled workers.

8

Fig. 1-3. Many semiskilled workers assemble manufactured parts into complete units. They do not require a long training period. (Saturn)

Semiskilled workers are usually the first to lose their jobs when business slows down and are often the last to be rehired.

To qualify for better jobs, the semiskilled workers must enter an apprentice training program or attend classes at a community college, technical center, or night school.

SKILLED WORKERS

Skilled workers are considered the backbone of industry. They are the artisans who make the tools, machines, and equipment that transform raw materials, ideas, and designs into finished products.

Skilled workers may have acquired their skills through a formal apprentice training program, a lengthy on-the-job training program, or while in the Armed Forces, Fig. 1-4. Most such programs last 4 to 6 years. They are carefully planned to give the apprentice broad training in a specific area of

Fig. 1-4. The Armed Forces are one of the few places where it is possible to learn a trade. (U.S. Army)

9

interest: machinist, sheet metal specialist, welder, etc. The apprentice must also have a good background in math and be able to read and interpret drawings and prints.

Skilled workers are also able to work with their hands. They are expected to use independent judgment when using costly tools, machines, and raw materials.

Skilled workers continually update their skills as improvements and advances are made in technical processes. This is usually done through company sponsored educational programs.

There are many skilled trades in the metalworking industry. The MACHINIST operates and sets up machine tools such as lathes, drilling machines, milling machines, grinders, etc. Since they work to very close tolerances (the part must be precisely made), the machinist must be familiar with all types of precision measuring instruments.

The TOOL AND DIE MAKER is a highly skilled machinist who specializes in making tools and dies. These are needed to cut, shape, and form metal. They make jigs and fixtures used to guide cutting tools and position metal while it is being machined, formed, and assembled, Fig. 1-5.

Fig. 1-6. The instrument maker manufactures precise parts that make up the instruments and gauges used by industry, medicine, and science. (Master Lock Co.)

The INSTRUMENT MAKER, Fig. 1-6, makes precision parts for instruments used by industry, medicine, science, and government. They also repair, assemble, and test instruments.

A SET UP SPECIALIST, Fig. 1-7, sets up and adjusts machine tools so semiskilled workers can operate them.

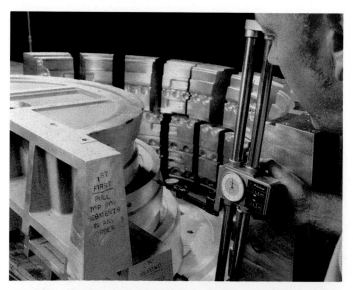

Fig. 1-5. This tool and die maker is checking the dies used for molding a plastic pattern used to cast a jet engine component. Master tooling assures that other sections of the jet, made elsewhere in the United States, Israel, and Europe will fit together perfectly.
(Precision Castparts Corp.)

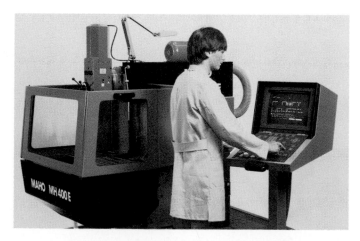

Fig. 1-7. The set up specialist prepares machine tools so that they can be operated by less skilled workers.
(Maho Machine Tool Corp.)

Fig. 1-8. Skilled steel erection workers had the dangerous job of building this bridge. Many of them travel all over the world to do this type of work. (Howard Bud Smith)

If you like working outside, you may want to be a skilled worker in the steel erection industry, Fig. 1-8. These experts have an adventurous (and dangerous) job.

The WELDER is highly skilled, Fig. 1-9. Types of work range from microwelding (welded pieces viewed through a microscope) to welding sections that weigh several hundred pounds.

Fig. 1-9. Welders ares skilled creatworkers. They must be familiar with metal characteristics and be able to read and understand drawings. They work indoors and outdoors, in all types of weather. (Lincoln Electric Co.)

Fig. 1-10. Many highly skilled men and women are involved in the manufacture of this state-of-the-art firefighting amphibian aircraft.   (Canadair Limited)

If you are interested in aircraft and rocketry, you may want to work in the aerospace industry, Fig. 1-10. Large numbers of highly skilled workers are employed in this segment of the economy.

Good SHEET METAL WORKERS are always in demand by the aircraft, building, heating, and air-conditioning industries.

These jobs are only a few of the many available in the metalworking industry that require the competence of the skilled worker. Regardless of what you choose to do, to be successful and advance in your job, you will need to complete high school. While there study mathematics, science, and vocational subjects.

TECHNICIANS

Advances in many areas of industry and science have brought about a demand for persons to do complex technical work. The men and women who do these jobs are called TECHNICIANS, Fig. 1-11. Technicians usually assist engineers in research, development, and design, Fig. 1-12. They construct and test experimental devices and machines, compile statistics, make cost estimates and prepare technical reports.

Fig. 1-11. This technician in training is learning the operation of CNC machine tools.
(William Schotta, Millersville University)

Fig. 1-12. This technician checks a device that speeds vehicle door alignment and installation on an auto assembly line. (Honda)

Where can you get training to become a technician? Start by taking all of the mathematics, science, computer, and vocational classes that you can fit into your high school schedule. English is also highly desirable. Most technicians must be able to understand and write technical reports.

Upon graduation from high school, you can get formal training from many sources. These include community colleges, vocational/technical centers, and technical institutes. In addition, many colleges offer two year technical programs. The programs stress mathematics, science, English, engineering, drafting, manufacturing processes, and computer science. Students also learn to handle tools, machines, and materials related to their area of interest.

THE PROFESSIONS

If you are planning to attend college, you may be interested in one of the professional occupations in metalworking.

There are many excellent jobs in TEACHING. This is a challenging career for those who like to work with people.

ENGINEERING, Fig. 1-13, is one of the largest professional occupations. The minimum requirement for becoming an engineer is a bachelor's degree in some type of engineering. However, some men and women have entered the profession after experience as group leaders and engineering technicians.

There are many types of engineers. The INDUSTRIAL ENGINEER finds the safest and most efficient way to use machines, materials, and personnel, Fig. 1-14.

The design and development of new machines and ideas is the responsibility of the MECHANICAL ENGINEER.

A TOOL AND MANUFACTURING ENGINEER devises the methods for manufacturing and assembling a product, Fig. 1-15.

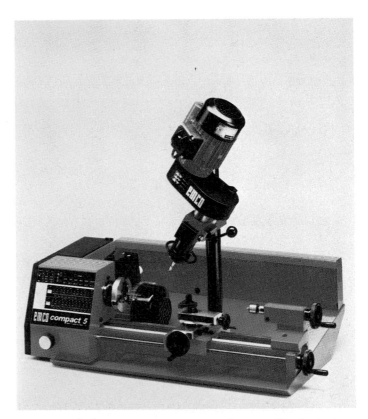

Fig. 1-13. Many engineering areas worked together to design, develop, and manufacture this combination lathe/milling machine used in many school metals technology labs. (Emco Maier Corp.)

Fig. 1-14. The industrial engineer planned the safest and most efficient way for this automated foundry to operate. It casts auto engine parts. (Saturn)

AERONAUTICAL ENGINEERS design and plan the manufacture of aerospace products. See Fig. 1-16.

METALLURGICAL ENGINEERS work with metals. They are concerned with the processing of metals, their conversion into commercial products, and the development of new metal alloys.

LEADERSHIP IN METALWORKING TECHNOLOGY

In your study of this unit, you have learned that metalworking technology requires a highly-skilled and educated work force. Strong and dynamic (energetic and aggressive) leadership is also needed.

LEADERSHIP is the ability to be a leader. A LEADER is a person who is in charge or command. The quality of leadership usually determines whether an organization will be a success or a failure.

WHAT MAKES A GOOD LEADER?

No one is quite sure what makes people work hard for one leader but not for another. This desire to work hard for a reason is MOTIVATION. It

Fig. 1-15. When a design is approved for production, tool and manufacturing engineers must devise the methods for manufacturing and assembling it. (Pontiac)

Fig. 1-16. The aeronautical engineer designs and plans the manufacture and testing of aircraft like this swept-forward-wing craft. (Grumman Aerospace Corp.)

is seldom based on standard rewards such as increased salaries and improved working conditions.

There are no tests to determine whether a person will be a strong leader. However, studies have shown that strong leaders have the following traits in common.

VISION. This is the talent for knowing what must be done. Short and long range goals with well defined plans are developed. A schedule is followed to achieve these goals. The status quo (existing conditions or situations) is always being updated and improved. Persons with vision enjoy the challenge of leadership and can inspire others.

COMMUNICATION. It is important to communicate in such a way that others will wish to assist and cooperate. A good leader has the ability to create situations where the energies and ideas of others can be used to reach desired goals or results.

PERSISTENCE. Persistence means the ability to stay on course, no matter what obstacles are met. However, a leader must still be flexible and willing to make changes when new ideas are presented. Persistent leaders will work long hours to achieve goals.

ORGANIZATIONAL ABILITIES. A good leader knows how to organize and direct the activities of a group. He or she learns from mistakes. The lesson learned is to improve efficiency and performance. It is important to be tactful (knowing the right thing to do or say). But it is also important to be autocratic (demand cooperation of others within the group) when necessary.

RESPONSIBILITY. Responsible persons accept responsibility for their actions. Also, they quickly give credit and recognition to others when deserved.

DELEGATES AUTHORITY. A good leader must not be afraid to assign tasks that may give others in the group opportunities for leadership. Helping others to develop leadership skills is an important quality.

GETTING LEADERSHIP EXPERIENCE

How can you get experience in leadership? The vocational education program and other clubs all need leaders. These activities offer good leadership training. See Fig. 1-17.

As a leader you will be expected to set a good example. Leadership also means that you get all

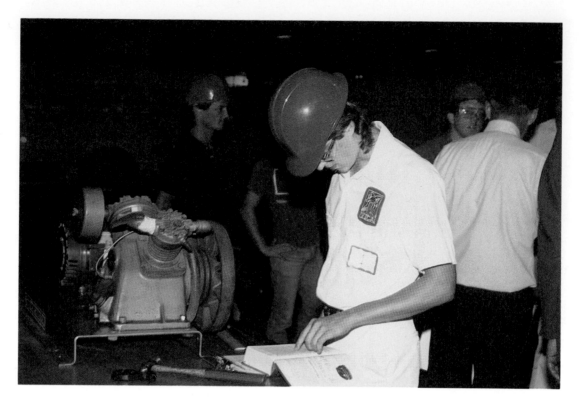

Fig. 1-17. Student groups like the Technology Student Association (TSA) and SkillsUSA-VICA offer opportunities to build leadership skills. This student is competing in the annual SkillsUSA Championships. (Howard Bud Smith)

members of the group involved in activities. This requires tact and encouragement.

A good leader does not have answers for every problem the group meets. Other members should give possible solutions. It will, however, be your job to decide which (if any) to use.

Leadership is not always easy. Decisions must be made that some of the group will not like. People, even friends, may have to be reproached or excused for poor behavior or attitude.

Do you have the traits (qualities) of a leader? Do you think you could develop them? Can you handle the unpleasant tasks that are part of leadership? Then, by all means, join a group where it will be possible for you to become a leader.

OCCUPATION AND LEADERSHIP INFORMATION

You can get more information on jobs in metalworking technology from many sources. The guidance office and your vocational education teacher are good starting points.

Most types of metalworking jobs are described in the OCCUPATIONAL OUTLOOK HANDBOOK. This is published by the United States Department of Labor. It can be found in libraries and guidance offices.

A local community college is also a good source. They will be glad to give information on the technical programs they offer.

Another outstanding source of information is your local State Employment Service. They can inform you of local job opportunities in metalworking.

The TECHNOLOGY STUDENT ASSOCIATION (TSA) will be happy to provide information on how to start a TSA chapter in your school. Included will be material on the leadership training program that is part of the club.

WHAT AN EMPLOYER EXPECTS OF YOU

Have you thought about what a company that employs you will look for in you? They often look for a return on the salary they pay you.

They will expect a fair day's work for a fair day's pay. They will count on you to do your assigned work to the best of your ability.

For your part, you should develop responsibility for the tools and equipment you use. These tools are very costly. Many employers have more than $100,000 invested in tools and equipment for each worker.

You must also remember that high school graduation is not the end of your training and/or education. If you want to keep and advance in a good job, you must keep up with technology advances. This may be done thorugh in-plant classes and/or advanced studies at community colleges or technical schools.

GETTING A JOB

Getting your first full-time job will be a very important task.

You must first decide what type of work you would like to do. Your guidance office and the local State Employment Service give tests that help find work areas where you may succeed. You can get more help by answering the following questions:

1. What have I accomplished with some degree of success?
2. What have I done that others have commended me for doing well?
3. What are the things I really like doing?
4. What are the things I DO NOT like to do?
5. What jobs have I held? Why did I leave them?

You might find that you have several areas of interest. List them. Start getting information on each of these areas. This can be done by reading, by talking with persons doing this kind of work, and/or by visiting industry.

The final step is to plan your school program. Prepare yourself for entry into the job or for advanced schooling, if it is required.

Now that you are ready to get a job, how would you go about it?

You must remember that some jobs are always available. Workers get promoted, retire, change jobs, quit, die, and are fired. Technological advances create new jobs. YOU MUST TRACK THEM DOWN.

Zero in on getting the job. MAKE YOUR JOB APPLICATION IN PERSON. Be specific about the type of job you are after. Do not just ask for a job or ask the employment office, "What job openings do you have?"

Prepare a JOB INFORMATION SHEET in advance, Fig. 1-18. This speeds up the task of filling out job applications. With this sheet, the responses you give will be the same on all applications. There will be little chance to supply confusing responses.

Finally, know where to look for a job. Review the "want ads" of local newspapers each day. Talk with friends and relatives. They may be aware of job openings before they become official or are advertised.

New office and factory buildings usually mean new job openings. You may also prepare a list of employers you would like to work for in your community. Visit their employment offices.

Remember, in most cases the job will not come to you. YOU must go to it.

JOB INFORMATION SHEET

This form has been designed to help you with the general information that will be required on most employment applications. Fill it out as completely as you can, making sure that the information is correct.

Name _____ Soc. Sec. #_____
        last   first   middle

Address _____
        Street   City   State   Zip

Telephone _____ How long living at present address? _____

Previous Address _____How long did you live there?_____

Age _____ Date of Birth _____

In case of emergency notify (usually nearest relative):

Name _____

Address_____ Phone _____

RECORD OF EDUCATION

School	No. Yrs.	Name of School	Course	Graduate
Elementary				Yes __ No __
High School				Yes __ No __
College				Yes __ No __
Other Type				Yes __ No __

EMPLOYMENT RECORD

	Dates Mo. Yr.	Type of work performed
Present or last previous employer		
Address	Reason for Leaving	
Rate of Pay		
Previous Employer		
Address	Reason for Leaving	
Rate of Pay		

THREE PERSONAL REFERENCES   (DO NOT GIVE RELATIVES)

Name	Address	Phone

Fig. 1-18. Prepare a job information sheet in advance to speed up the task of filling out job applications.

TEST YOUR KNOWLEDGE, Unit 1

Please do not write in the text. Place your answers on a separate sheet of paper.

1. Describe the meaning of technology.
2. Metalworking jobs fall into one of four classifications. Name them.
3. If you were a semiskilled worker in metalworking, what four jobs would you probably do?
4. There are disadvantages that semiskilled workers must face. Choose the disadvantages from the list below.
   a. Doing the same job over and over for long periods of time.
   b. Requires only a brief on-the-job training period.
   c. First to lose jobs when business slows down.
   d. All of the above.
   e. None of the above.
5. The formal apprentice program usually lasts _____ to _____ years.

MATCHING QUESTIONS: Match each of the following terms with their correct definitions.

a. Machinist.
b. Tool and die maker.
c. Instrument maker.
d. Set up specialist.
e. Technician.
f. Industrial engineer.
g. Mechanical engineer.
h. Aeronautical engineer.

6. __ Works between the shop and the engineering department
7. __ Finds the safest and most efficient way to use machines, materials, and personnel.
8. __ Designs and plans the manufacture of aerospace products.
9. __ Makes the tools and dies needed to cut, shape and form metal.
10. __ Designs and develops new machines and ideas.
11. __ Makes the parts for the instruments used by industry, medicine, science, and government.
12. __ Adjusts machines and tools so semiskilled workers can operate them.
13. __ Must be able to operate all kinds of machine tools.
14. The _____ engineer is concerned with the processing of metals and their conversion into commercial products.
15. What does the term leadership mean?
16. The person who is in charge or command is called a _____.
17. The quality of leadership often determines whether an organization will be a _____.

TECHNICAL TERMS TO LEARN

apprentice
careers
communications
creativity
design
enhance
industry
inspector
job application
leadership
manufacturing
metalworking
occupation

persistence
research
resources
semiskilled workers
skilled workers
skills
state-of-the-art
technician
technique
technology
vocational-technical center

ACTIVITIES

1. Metals play a very important part in our modern world. How many jobs can YOU name that make use of metal in some shape or form? Think of all those jobs that use metal directly (use metal to make things) or indirectly (only in machinery and tools).
2. Contact the TECHNOLOGY STUDENT ASSOCIATION (1914 Association Drive, Reston, VA 22091). Ask for information on how to start a TSA chapter in your school. Discuss the idea with your teacher before sending for the material.

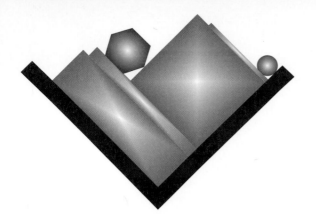

Unit 2
Metals We Use

In this unit you will learn about the many types of metals in use today. You will discover how metals are classified and which metals are used in school laboratories. You will be able to discuss the qualities of various metals we use and how these metals are shaped, measured, and purchased.

The metals we use have many properties. Some of them, like aluminum, magnesium, and titanium, are strong and light enough to be used to manufacture aircraft, Fig. 2-1.

Rocket and jet engines are subjected to great amounts of heat. The metals used on these engines must be able to withstand this heat. The fuel (uranium) that powers nuclear vessels, Fig. 2-2, is a metal. Only a few pounds are needed to power a ship around the world.

Fig. 2-1. The metals used to construct aircraft must be light and strong. (Airbus)

Fig. 2-2. A metal fuel (uranium) powers this nuclear aircraft carrier. Only a few pounds
are needed to propel it around the world. (U.S. Navy)

The racing car, Fig. 2-3, makes use of many kinds of metals. Each metal is selected for its special qualities: strength, lightness, and ability to dissipate heat.

Fig. 2-3. Many different kinds of metal must be used in high performance racing cars. (Scott Gauthier)

Hundreds of metals and alloys are used by industry. The small internal combustion engine, Fig. 2-4, that powers a model plane, car, or boat costs only a few dollars. However, its construction makes use of more than a dozen different metals.

How many other products can you name that need metals with special qualities?

HOW METALS ARE CLASSIFIED

The metals you will use in your shop work are the same as those used by industry. They are available in a large number of shapes and sizes, Fig. 2-5.

For identification purposes, metals fall into several categories.

BASE METALS are pure metals like gold, copper, lead, and tin. They contain no other metals.

ALLOYS are combinations of several metals. They are fused (blended) together while in a molten state. For example, brass is an alloy of copper and zinc.

21

Fig 2-4. Metals mined in all parts of the world (iron, aluminum, copper, brass, chrome, lead, tin, zinc, platinum, etc.) are needed to manufacture the various parts of this miniature engine. (Cox Hobbies, Inc.)

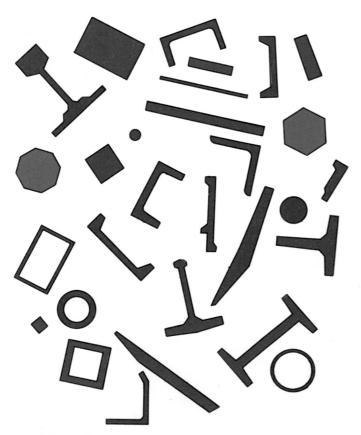

Fig. 2-5. A few of the hundreds of different metal shapes available to industry.

Metals are further grouped into the following classes.

FERROUS METALS are alloys that contain iron as a major part of their composition. Steel is a ferrous metal.

NONFERROUS METALS contain no iron, except in very small amounts as impurities. Metals like aluminum, brass, and tin are nonferrous metals.

METALS USED IN SCHOOL LABORATORIES

The ferrous metals include carbon steels, tin plate, and galvanized sheet. Carbon steels are classified according to the amount of carbon they contain. The carbon is measured in PERCENTAGE or in POINTS (100 points equal 1 percent).

1. LOW CARBON STEELS do not contain enough carbon to be hardened (less than 0.30 percent or 30 points). They are relatively soft and are often called MILD STEEL. As they are easy to machine, weld, and form, they have

Fig. 2-6. This trivet is made from mild steel.

Fig. 2-8. Tools such as punches are made from tool steel.

many applications in bench metal work, Fig. 2-6. Mild steels are available as rods, bars, strips, and sheets.

2. MEDIUM CARBON STEELS contain 0.30 to 0.60 percent carbon (30 to 60 points). They are good for projects that need machining, Fig. 2-7.

3. HIGH CARBON STEELS are sometimes called TOOL STEELS. They contain 0.60 to 1.00 percent carbon. These steels are used to make tools because they can be heat-treated (this is the process of controlling the heating and cooling of metal to bring about certain desirable characteristics, such as hardness and toughness), Fig. 2-8.

Hot finished steel has a characteristic black coating (oxide). When cold finished, the steel has a surface that is smooth with no trace of the black scale, Fig. 2-9. TIN PLATE (tin cans are made from it) is a mild steel sheet to which a tin coating has been applied. GALVANIZED SHEET is a mild steel sheet on which a coating of zinc has been deposited.

Fig. 2-7. Medium carbon steel machines easily. It was used to make this small lathe.

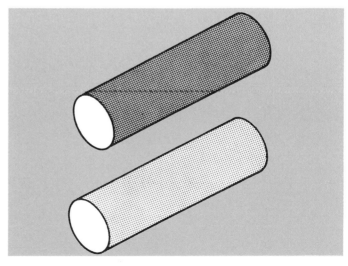

Fig. 2-9. Hot finished steel (top) has a characteristic black coating, while cold finished steel (bottom) has a smooth surface, with no trace of the black scale.

A recent addition is a mild steel sheet with a vinyl coating. It can be cut, shaped, formed, and joined like other steel sheets. Various colors and textured finishes are available.

Many nonferrous metals are also used in the school laboratory.

ALUMINUM

Aluminum is a term used to identify an entire family of metals (there are over 100 different aluminum alloys). They range from almost pure aluminum which is very soft, to an aluminum alloy tough enough to be used as armor plate, Fig. 2-10.

Aluminum is similar to silver in color, but has a bluish tint. It is lighter weight than steel. Aluminum melts at about 1200°F (649°C), and can be worked using regular metalworking techniques.

Aluminum can be purchased as foil, sheet, rod, bar, plate, wire, or tubing.

BRASS

Brass is an alloy of copper and zinc. The color varies from a reddish bronze to a yellowish gold. Color depends upon the percentage of zinc the brass contains. Brass is easy to shape, cut, etch, solder, electroplate, and chemically color. Brass is used a great deal in art metal work, Fig. 2-11.

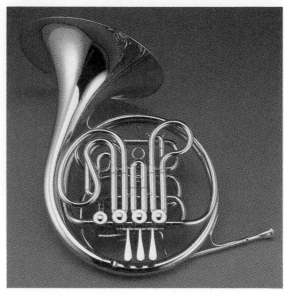

Fig. 2-11. Brass is widely used for making musical instruments. such as this French horn. (G. Leblanc Corp.)

Brass is available in soft, half-hard, three-quarter, and hard tempers.

COPPER

Copper is an easily-worked metal. It is reddish brown in color and melts at 1981°F (1083°C). The metal takes a brilliant polish but, like brass, it must be sprayed with clear lacquer to keep the finish. Copper is worked like brass.

Fig. 2-10. Aluminum provides the strength and light weight needed to construct aircraft like this forest-fire-fighting water bomber. (Canadair Limited)

24

Fig. 2-12. A great deal of copper is used in the windings of electric motors like these industrial models. (Baldor)

Fig. 2-14. Modern pewter contains no lead. It can be used to serve food and drink.

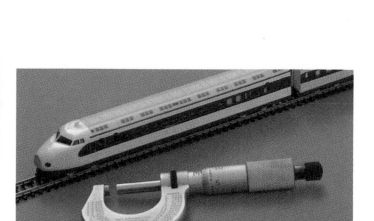

Fig. 2-13. Copper makes it possible to produce the small powerful motor that powers this miniature train.

and zinc. It works much like brass but is a bit more brittle.

STERLING SILVER

Silver combined with a small amount of copper (7 1/2 percent) is known as sterling silver. When polished, it is shiny, silvery-white in color. Sterling silver has outstanding working characteristics. It can be easily shaped and formed, and it hard solders well, Fig. 2-15.

Much copper is used in electric wiring and in electric motors. See Figs. 2-12 and 2-13.

PEWTER

Modern pewter, or Britannia metal, is an alloy of tin (91 percent), copper (1 1/2 percent) and antimony (7 1/2 percent). When polished, it has a fine silvery sheen. Modern pewter DOES NOT contain lead and can be used to serve food and drink, Fig. 2-14. However, old pewter may contain lead and should not be used to serve food or drink. Pewter is easy to work, but a good deal of skill is needed to join it properly.

NICKEL SILVER

Nickel silver, or German silver, is used as a substitute for silver in inexpensive jewelry. It is a copper base alloy with varying quantities of nickel

Fig. 2-15. Sterling silver has outstanding working characteristics. It was used to make this bowl and vase.

SHAPES OF METALS WE USE

Metals are produced in many shapes and sizes. A few of the standard metal shapes are illustrated in Fig. 2-16.

Fig. 2-17 presents several sheet metals commonly found in the school laboratory. The chart shows how they are measured and purchased.

Have you ever wondered how metal is shaped? Many shapes start as ingots which are heated and passed through rolling mills. There they are shaped into semifinished products called slags, blooms, or billets, Fig. 2-18. These pass on through other rolling mills, where rolls of different types and sizes turn them into plates, sheets, rods, and bars. The rolling process not only shapes the metal but also makes it tougher and stronger.

	SHAPES	LENGTH	HOW MEASURED	*HOW PURCHASED	OTHER
	Sheet less than 1/4 in. thick	to 144 in.	Thickness x width widths to 72 in.	Weight, foot, or piece	Available in coils of much longer lengths
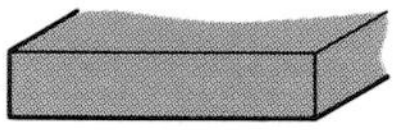	Plate more than 1/4 in. thick	to 20 ft.	Thickness x width	Weight, foot, or piece	
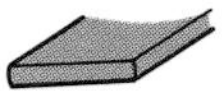	Band	to 20 ft.	Thickness x width	Weight or piece	Mild steel with oxide coating
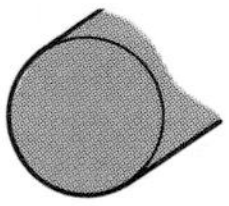	Rod	12 to 20 ft.	Diameter	Weight, foot, or piece	Hot-rolled steel to 20 ft. length; cold-finished steel to 12 ft. length; steel drill rod 36 in.
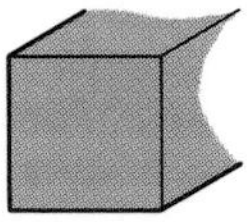	Square	12 to 20 ft.	Width	Weight, foot, or piece	
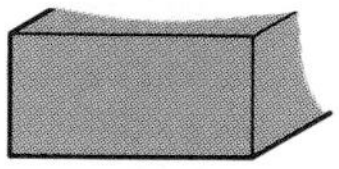	Flats	Hot rolled 20-22 ft. Cold finished	Thickness x width	Weight, foot, or piece	
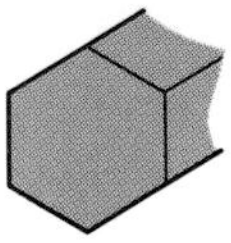	Hexagon	12 to 20 ft.	Distance across flats	Weight, foot, or piece	
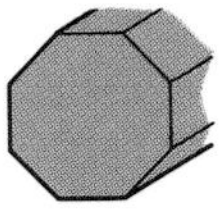	Octagon	12 to 20 ft.	Distance across flats	Weight, foot, or piece	
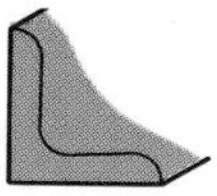	Angle	Lengths to 40 ft.	Leg length x leg length x thickness of legs	Weight, foot, or piece	
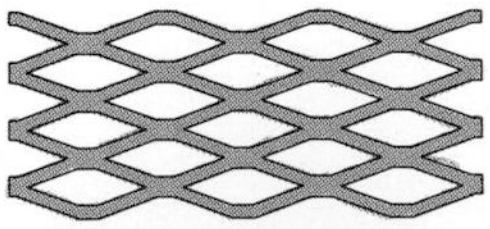	Expanded sheet	to 96 in.	Gauge number (U.S. Standard)	36 x 96 in. and size of openings	Metal is pierced and expanded (stretched) to diamond shape; also available rolled to thickness after it has been expanded
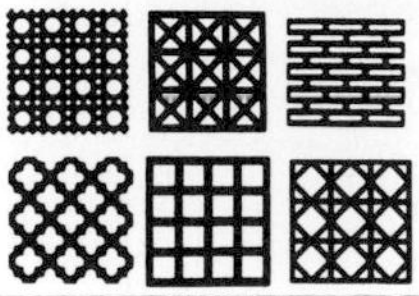	Perforated Sheet	to 96 in.	Gauge number (U.S. Standard)	30 x 36 in. 36 x 48 in. 36 x 96 in.	Design is cut in sheet; many designs available.

*Charge made for cutting to other than standard lengths.

Fig. 2-16. Metals we use . . . available shapes, how they are measured and purchased, and some of their characteristics.

MATERIAL (Sheet less than 1/4 in. thick)	HOW MEASURED	HOW PURCHASED	CHARACTERISTICS
Copper	Gauge number (Brown & Sharp & Amer. Std.)	24 x 96 in. sheet or 12 or 18 in. by lineal feet on roll	Pure metal
Brass	Gauge number (B & S and Amer. Std.)	24 x 76 in. sheet or 12 or 18 in. by lineal feet on roll	Alloy of copper and zinc
Aluminum	Decimal	24 x 72 in. sheet or 12 or 18 in. by lineal feet on roll	Available as commercially pure metal or alloyed for strength, hardness, and ductility
Galvanized Steel	Gauge number (U.S. Std.)	24 x 96 in. sheet	Mild steel sheet with zinc plating, also available with zinc coating that is part of sheet
Black Annealed Steel Sheet	Gauge number (U.S. Std.)	24 x 96 in. sheet	Mild steel with oxide coating- hot rolled
Cold Rolled Steel Sheet	Gauge number (U.S. Std.)	24 x 96 in. sheet	Oxide removed and cold rolled to final thickness
Tin Plate	Gauge number (U.S. Std.)	20 x 28 in. sheet 56 or 112 to pkg.	Mild steel with tin coating
Nickel Silver	Gauge number (Brown & Sharp)	6 or 12 in. wide by lineal sheet	Copper 50%, zinc 30%, nickel 20%

Fig. 2-17. Metals we use . . . how they are measured and purchased, and some of their characteristics.

Sheet stock is made by passing a metal slab between rolls, Fig. 2-19, until the desired thickness is obtained. As the slab becomes thinner, its length increases. The slab can be made wider by passing it crosswise through the rolls.

Bar, rod, and other forms are also shaped by rolling. However, the process differs from the rolling of sheet in that the rolls are grooved to produce the specific shape desired. The rolling process

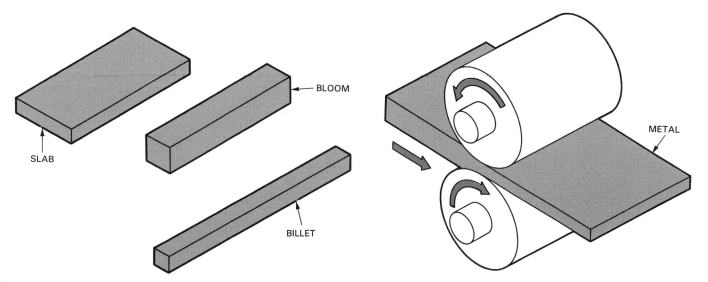

Fig. 2-18. Most metal shapes start as slabs, blooms or billets.

Fig. 2-19. How sheet metal and plate are rolled to thickness.

starts with a billet which is gradually worked into the required shape. See Fig. 2-20.

Wire is made by a drawing process, Fig. 2-21. In wire drawing, the end of the rod is pointed and pulled through dies. These dies reduce the diameter of the rod until the required wire diameter is reached.

Seamless tubing is produced as shown in Fig. 2-22. A cylinder is cupped at one end and drawn through well-lubricated dies that reduce its diameter and increase its length.

Seamless tubing can also be made by the extrusion process, Fig. 2-23. Pressure is employed to force the metal through a die of the required shape and size. The extrusion process is also employed to produce other complex metal shapes, Fig. 2-24.

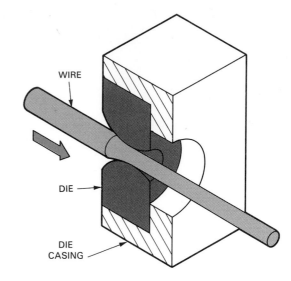

Fig. 2-21. Wire is made by pulling a rod through dies to reduce its diameter to the required size.

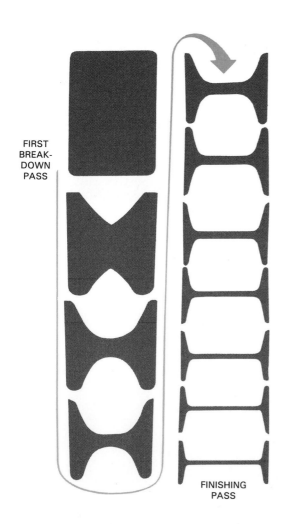

Fig. 2-20. When rolling structural shapes, the process starts with a billet that is gradually worked into the required shape.

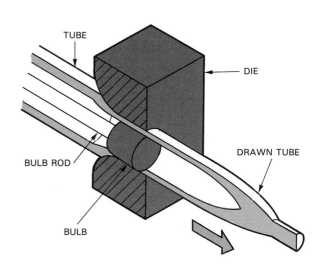

Fig. 2-22. One technique used to draw seamless tubing.

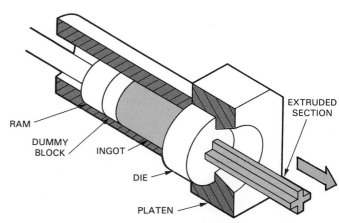

Fig. 2-23. In the extrusion process, metal is forced through a die of the required shape and size.

28

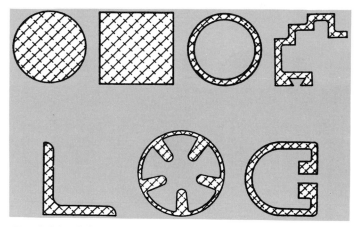

Fig. 2-24. A few of the extruded shapes available. Almost any shape can be produced by this process.

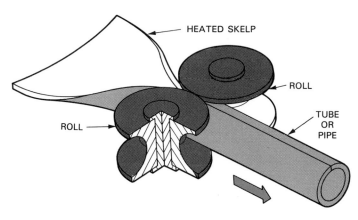

Fig. 2-25. How some pipe and tubing is made from flat metal sheet.

Butt welded pipe is made from a heated strip called skelp. Skelp is formed and welded together as shown in Fig. 2-25.

Metal is shaped by other techniques too. Many of the methods used will be described in this book.

MAJOR PROCESSES IN THE STEEL INDUSTRY

It is hard to imagine a person who does not use at least one of the many items made of steel during the course of a day. People depend on steel for such things as buildings, bridges, safety pins, and paper clips.

Although steel is familiar in our daily lives, it is also found at the frontiers of science. Few people understand the human effort required to produce it. The following flow charts, courtesy of American Iron and Steel Institute, show some of the things that must be done to produce steel.

The United States is the world's largest producer of raw steel. Markets in this country support many companies operating steelmaking or finishing facilities in 38 states. More than half of these companies make raw steel and finish it themselves. Others buy semifinished steel from which they make bars, wire, and wire products, hot and cold rolled sheets, plates, pipe and tubing. A wide variety of other products are also made. However, the public seldom sees these because they are processed further by the customers of the steel industry.

The modern steel industry is concerned with producing huge amounts of steel. It is also concerned with being a partner with individuals, communities, and governments working for ecological (relationship between living things and surrounding conditions) improvement. Each of the processes which follow has air and water quality control equipment designed into it.

FLOWLINE OF STEELMAKING

From iron ore, limestone, and coal in the earth's crust to space-age steels — this fundamental flowline shows only major steps in an intricate progression of processes with their many options.

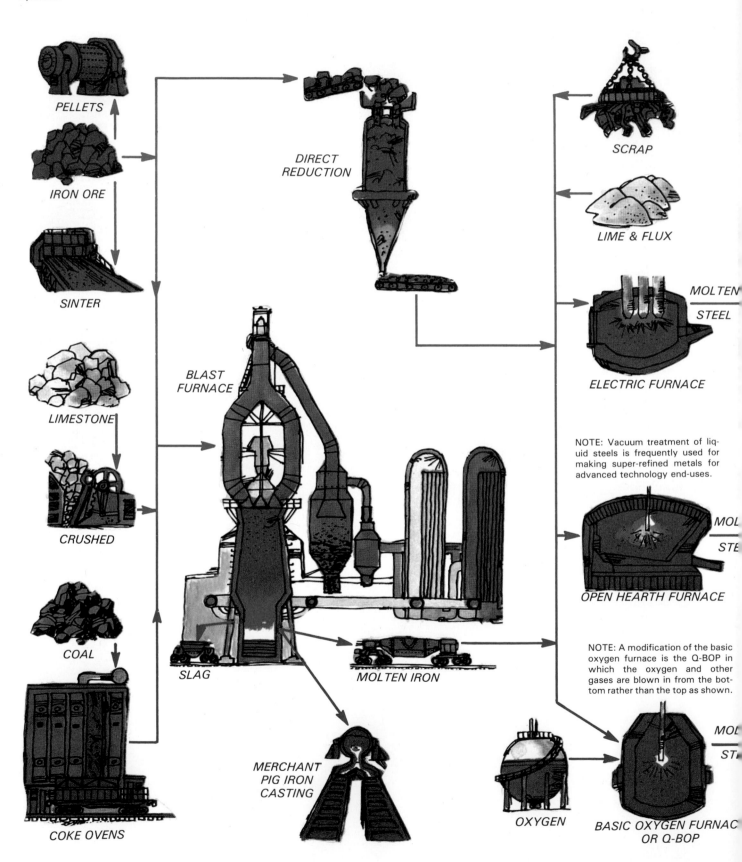

PELLETS

IRON ORE

SINTER

LIMESTONE

CRUSHED

COAL

COKE OVENS

BLAST FURNACE

SLAG

MERCHANT PIG IRON CASTING

DIRECT REDUCTION

MOLTEN IRON

OXYGEN

SCRAP

LIME & FLUX

ELECTRIC FURNACE

MOLTEN STEEL

NOTE: Vacuum treatment of liquid steels is frequently used for making super-refined metals for advanced technology end-uses.

OPEN HEARTH FURNACE

MOL STE

NOTE: A modification of the basic oxygen furnace is the Q-BOP in which the oxygen and other gases are blown in from the bottom rather than the top as shown.

BASIC OXYGEN FURNAC OR Q-BOP

MOL ST

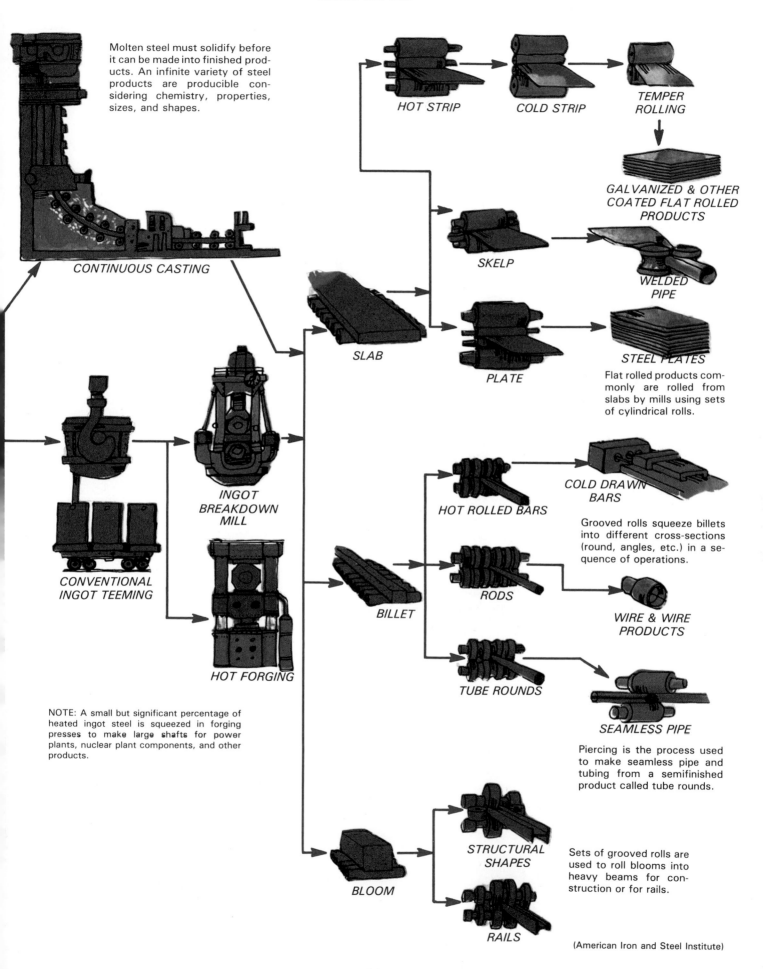

Molten steel must solidify before it can be made into finished products. An infinite variety of steel products are producible considering chemistry, properties, sizes, and shapes.

CONTINUOUS CASTING

HOT STRIP

COLD STRIP

TEMPER ROLLING

GALVANIZED & OTHER COATED FLAT ROLLED PRODUCTS

SKELP

WELDED PIPE

SLAB

PLATE

STEEL PLATES

Flat rolled products commonly are rolled from slabs by mills using sets of cylindrical rolls.

INGOT BREAKDOWN MILL

CONVENTIONAL INGOT TEEMING

HOT FORGING

HOT ROLLED BARS

COLD DRAWN BARS

Grooved rolls squeeze billets into different cross-sections (round, angles, etc.) in a sequence of operations.

BILLET

RODS

WIRE & WIRE PRODUCTS

TUBE ROUNDS

SEAMLESS PIPE

Piercing is the process used to make seamless pipe and tubing from a semifinished product called tube rounds.

NOTE: A small but significant percentage of heated ingot steel is squeezed in forging presses to make large shafts for power plants, nuclear plant components, and other products.

BLOOM

STRUCTURAL SHAPES

Sets of grooved rolls are used to roll blooms into heavy beams for construction or for rails.

RAILS

(American Iron and Steel Institute)

BLAST FURNACE IRONMAKING

A blast furnace is a cylindrical steel vessel, often as tall as a ten-story building, lined inside with a heat-resistant brick. Once it is fired up, the furnace runs continuously until this lining is worn out and must be replaced. Coke, iron ore, and limestone are charged into the top of the furnace. They proceed slowly down through the furnace being heated as they are exposed to a blast of super-hot air that blows upward from the bottom of the furnace. The blast of air burns the coke, releasing heat and gas which remove oxygen from the ore. The limestone acts as a cleansing agent.

Freed from its impurities, the molten iron collects in the bottom where it is drawn off every three-to-five hours as a white-hot stream of liquid iron. From there, the molten iron will be further refined in a steelmaking furnace.

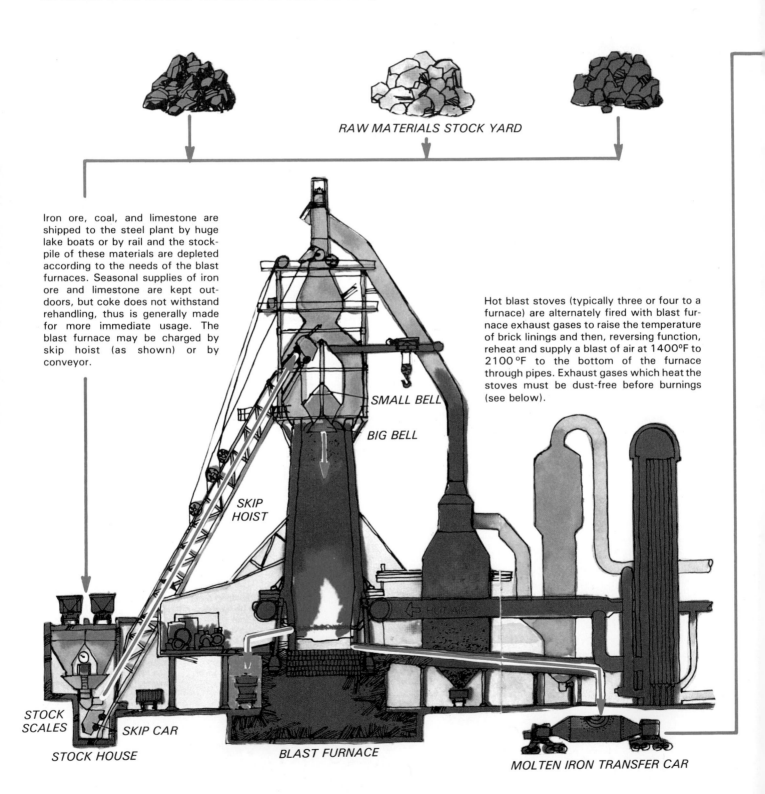

RAW MATERIALS STOCK YARD

Iron ore, coal, and limestone are shipped to the steel plant by huge lake boats or by rail and the stock-pile of these materials are depleted according to the needs of the blast furnaces. Seasonal supplies of iron ore and limestone are kept out-doors, but coke does not withstand rehandling, thus is generally made for more immediate usage. The blast furnace may be charged by skip hoist (as shown) or by conveyor.

Hot blast stoves (typically three or four to a furnace) are alternately fired with blast furnace exhaust gases to raise the temperature of brick linings and then, reversing function, reheat and supply a blast of air at 1400°F to 2100°F to the bottom of the furnace through pipes. Exhaust gases which heat the stoves must be dust-free before burnings (see below).

SMALL BELL

BIG BELL

SKIP HOIST

HOT AIR

STOCK SCALES

SKIP CAR

STOCK HOUSE

BLAST FURNACE

MOLTEN IRON TRANSFER CAR

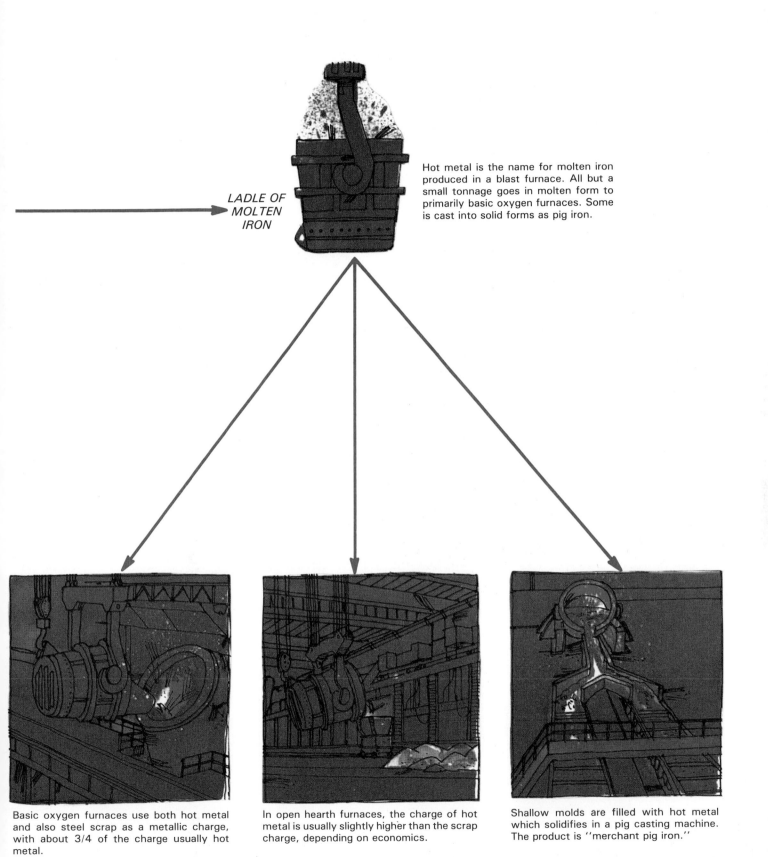

LADLE OF MOLTEN IRON

Hot metal is the name for molten iron produced in a blast furnace. All but a small tonnage goes in molten form to primarily basic oxygen furnaces. Some is cast into solid forms as pig iron.

Basic oxygen furnaces use both hot metal and also steel scrap as a metallic charge, with about 3/4 of the charge usually hot metal.

In open hearth furnaces, the charge of hot metal is usually slightly higher than the scrap charge, depending on economics.

Shallow molds are filled with hot metal which solidifies in a pig casting machine. The product is ''merchant pig iron.''

(American Iron and Steel Institute)

BASIC OXYGEN STEELMAKING

In the United States more steel is produced by basic oxygen furnaces (BOF's) than by any other means. They consume large amounts of oxygen to support the combustion of unwanted elements and so eliminate them. No other gases or fuels are used. BOF's make steel very quickly compared with the other major methods now in use — for example, 300 tons in 45 minutes as against several hours in open hearth and electric furnaces. Most grades of steel can be made in BOF's.

The actual furnaces are a small part of the facility, as this schematic drawing shows. Gas cleaning devices and materials handling equipment occupy most of the space.

GAS CLEANING EQUIPMENT

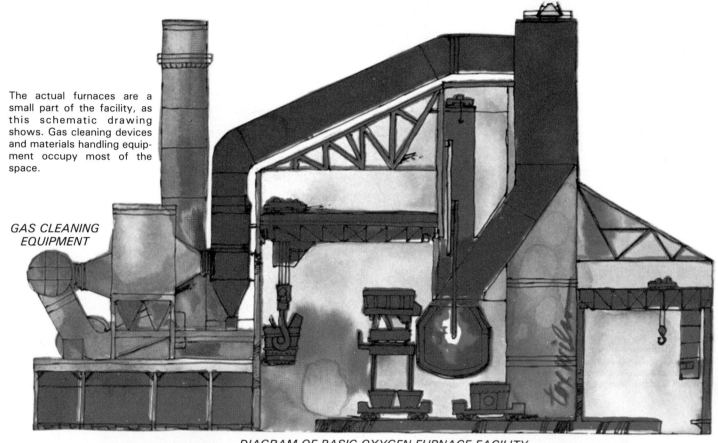

DIAGRAM OF BASIC OXYGEN FURNACE FACILITY

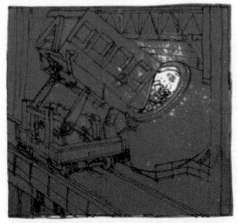

The first step for making a heat of steel in a BOF is to tilt the furnace and charge it with scrap. The furnaces are mounted on trunnions and can be rotated through a full circle.

Hot metal from the blast furnace accounts for up to 80 percent of the metallic charge and is poured from a ladle into the top of the tilted furnace.

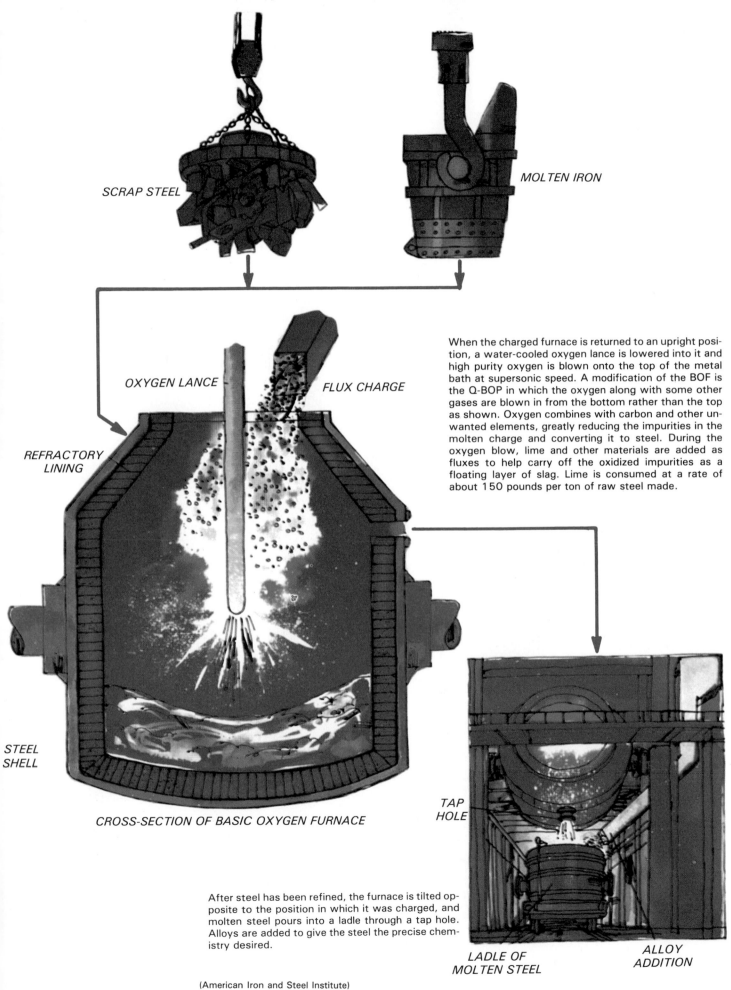

SCRAP STEEL

MOLTEN IRON

OXYGEN LANCE

FLUX CHARGE

When the charged furnace is returned to an upright position, a water-cooled oxygen lance is lowered into it and high purity oxygen is blown onto the top of the metal bath at supersonic speed. A modification of the BOF is the Q-BOP in which the oxygen along with some other gases are blown in from the bottom rather than the top as shown. Oxygen combines with carbon and other unwanted elements, greatly reducing the impurities in the molten charge and converting it to steel. During the oxygen blow, lime and other materials are added as fluxes to help carry off the oxidized impurities as a floating layer of slag. Lime is consumed at a rate of about 150 pounds per ton of raw steel made.

REFRACTORY
LINING

STEEL
SHELL

CROSS-SECTION OF BASIC OXYGEN FURNACE

TAP
HOLE

After steel has been refined, the furnace is tilted opposite to the position in which it was charged, and molten steel pours into a ladle through a tap hole. Alloys are added to give the steel the precise chemistry desired.

LADLE OF
MOLTEN STEEL

ALLOY
ADDITION

(American Iron and Steel Institute)

35

OPEN HEARTH STEELMAKING

Open hearth furnaces are so named because the limestone, scrap steel and molten iron charged into the shallow steelmaking area (the hearth) are exposed (open) to the sweep of flames. A furnace that will produce a fairly typical 350 tons of steel in five to eight hours may be about 90 feet long and 30 feet wide.

The cutaway drawing below shows several steps simultaneously that would normally occur in sequence. First the long-armed charging machine picks up boxes of limestone and steel scrap, thrusts them through the furnace doors and dumps the contents. The flame of burning fuel oil, tar, or gases partially melts the solid charge, after which molten iron (lower right) is poured into the furnace. High-temperature reactions cause several unwanted elements to combine with the limestone to form a slag.

When tests of samples show the steel to be of specified chemistry, the tap hole is opened by an explosive charge and the steel runs into a ladle. The slag, which is lighter than steel, floats on the metal and overflows into a slag thimble during pouring. Alloy additions are made to the steel in the ladle.

FURNACE
ROOF

OXYGEN
LANCE
(see upper right)

CHARGING MACHINE

FLAME

FUEL
PORT

AIR
PORT

CHARGING
BOXES

TAP HOLE

SPOUT

STEEL
LADLE

AIR
PORT

SLAG
THIMBLE

In recent years practically all open hearth furnaces have been converted to the use of oxygen. The gas is fed into the open hearth through the roof by means of retractable lances. The use of gaseous oxygen in the open hearth increases flame temperature, and thereby speeds the melting process.

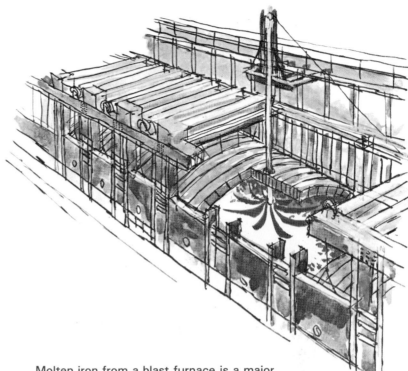

Molten iron from a blast furnace is a major raw material for the open hearth furnaces. A massive "funnel" is wheeled to an open hearth door and the contents of a ladle or iron are poured through it into the furnace hearth. The principal addition of molten metal is made after the original scrap charge has begun to melt.

CONTROL PANELS

BRICK CHECKER CHAMBERS

Brick checker chambers are located on both ends of the furnace. The bricks are arranged to leave a great number of passages through which the hot waste gases from the furnace pass and heat the brickwork prior to going through the cleaner and stack. Later on, the flow is reversed and the air for combustion passes through the heated bricks and is itself heated on its way to the hearth.

(American Iron and Steel Institute)

ELECTRIC FURNACE STEELMAKING

Traditionally used to produce alloy, stainless, tool, and specialty steels, electric furnaces have been developed in size and capability to become high-tonnage makers of carbon steel, too.

The heat within the furnace is precisely controlled as the electric current arcs from one electrode (of the three inserted through the furnace roof) to the metallic charge and back to another electrode.

ELECTRIC FURNACE FACILITY

ELECTRODES

At left, this cutaway drawing shows an electric furnace with its carbon electrodes attached to support arms and electrical cables and extending into the furnace. Molten steel and the rocker mounting on which the furnace may be tilted is also shown.

CHARGING BUCKET

At right, the roof of a furnace is pivoted aside so that a charging bucket of scrap may be lowered into position for bottom-dumping. Alternatively, direct reduced pellets may be fed continuously during the meltdown.

Metals We Use

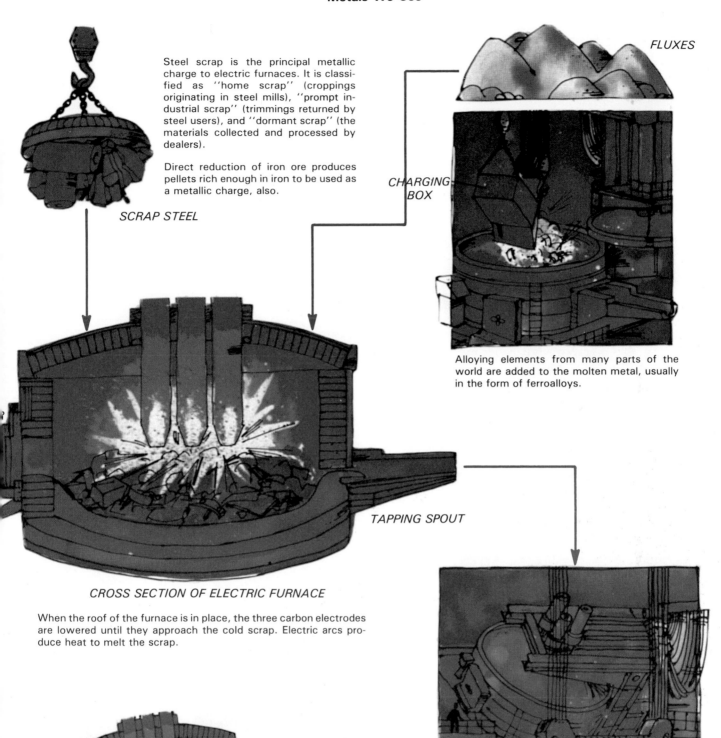

Steel scrap is the principal metallic charge to electric furnaces. It is classified as "home scrap" (croppings originating in steel mills), "prompt industrial scrap" (trimmings returned by steel users), and "dormant scrap" (the materials collected and processed by dealers).

Direct reduction of iron ore produces pellets rich enough in iron to be used as a metallic charge, also.

SCRAP STEEL

FLUXES

CHARGING BOX

Alloying elements from many parts of the world are added to the molten metal, usually in the form of ferroalloys.

TAPPING SPOUT

CROSS SECTION OF ELECTRIC FURNACE

When the roof of the furnace is in place, the three carbon electrodes are lowered until they approach the cold scrap. Electric arcs produce heat to melt the scrap.

SLAG

SLAG THIMBLE

Limestone and flux are charged after the scrap becomes molten. Impurities in the steel rise into a floating layer of slag, some or most of which can be poured off.

(American Iron and Steel Institute)

LADLE OF MOLTEN STEEL

When the chemical composition of the steel meets specifications, the furnace is tilted backward. Molten steel pours out the spout into a ladle. The slag follows the steel and serves as an insulating layer.

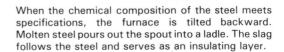

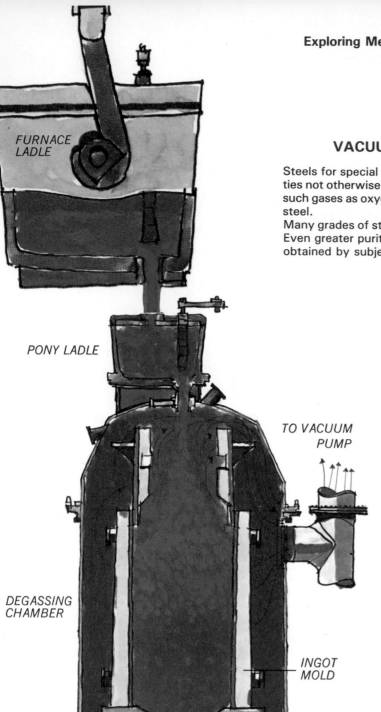

FURNACE
LADLE

PONY LADLE

TO VACUUM
PUMP

DEGASSING
CHAMBER

INGOT
MOLD

VACUUM PROCESS OF STEELMAKING

Steels for special applications are often processed in a vacuum to give them properties not otherwise obtainable. The primary purpose of vacuum processing is to remove such gases as oxygen, nitrogen, and hydrogen from molten metal to make higher-purity steel.

Many grades of steel are degassed by processes similar to those shown on this page. Even greater purity and uniformity of steel chemistry than available by degassing is obtained by subjecting the metal to vacuum melting processes.

The Vacuum Degasses

In vacuum stream degassing (left), a ladle of molten steel from a conventional furnace is taken to a vacuum chamber. An ingot mold is shown within the chamber. Larger chambers designed to contain ladles are also used. The conventionally melted steel goes into a pony ladle and from there into the chamber. The stream of steel is broken up into droplets when it is exposed to vacuum within the chamber. During the droplet phase, undesirable gases escape from the steel and are drawn off before the metal solidifies in the mold.

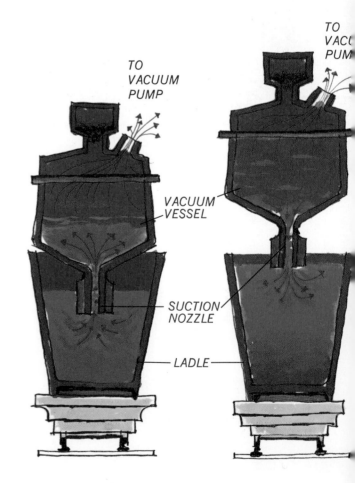

TO
VACUUM
PUMP

TO
VAC
PUM

VACUUM
VESSEL

SUCTION
NOZZLE

LADLE

Ladle degassing facilities (right) of several kinds are in current use. In the left-hand facility, molten steel is forced by atmospheric pressure into the heated vacuum chamber. Gases are removed in this pressure chamber, which is then raised so that the molten steel returns by gravity into the ladle. Since not all of the steel enters the vacuum chamber at one time, this process is repeated until essentially all the steel in the ladle has been processed.

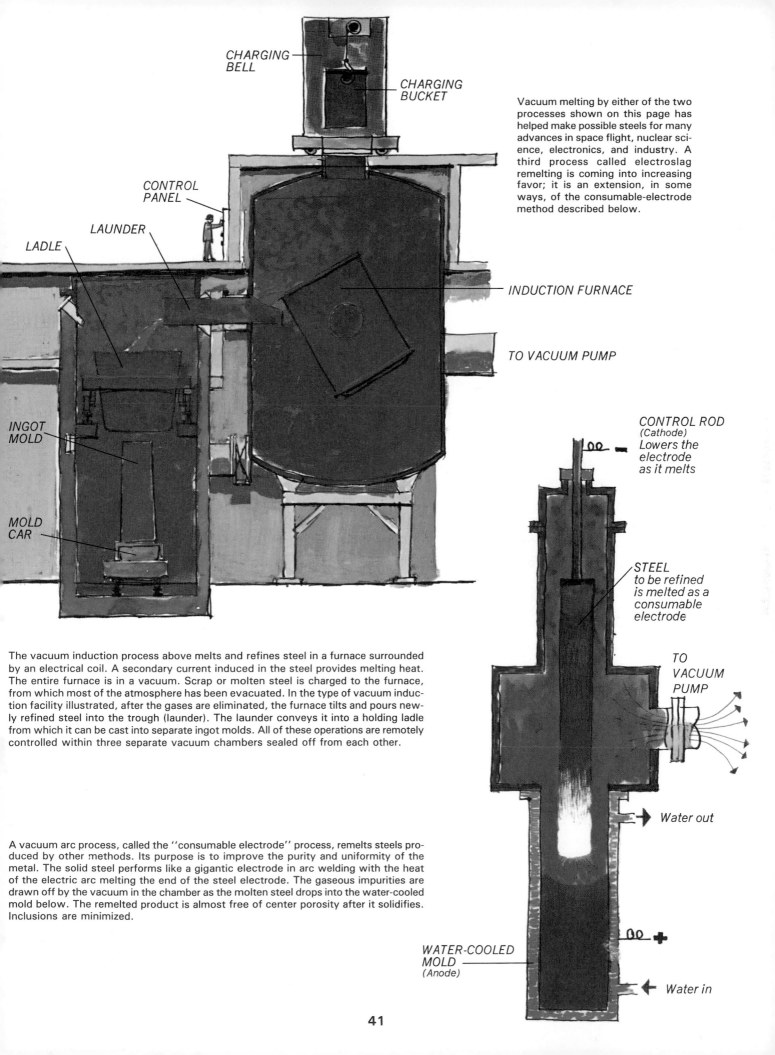

CHARGING BELL

CHARGING BUCKET

CONTROL PANEL

LAUNDER

LADLE

INGOT MOLD

MOLD CAR

INDUCTION FURNACE

TO VACUUM PUMP

CONTROL ROD
(Cathode)
Lowers the electrode as it melts

STEEL
to be refined is melted as a consumable electrode

TO VACUUM PUMP

Water out

WATER-COOLED MOLD
(Anode)

Water in

Vacuum melting by either of the two processes shown on this page has helped make possible steels for many advances in space flight, nuclear science, electronics, and industry. A third process called electroslag remelting is coming into increasing favor; it is an extension, in some ways, of the consumable-electrode method described below.

The vacuum induction process above melts and refines steel in a furnace surrounded by an electrical coil. A secondary current induced in the steel provides melting heat. The entire furnace is in a vacuum. Scrap or molten steel is charged to the furnace, from which most of the atmosphere has been evacuated. In the type of vacuum induction facility illustrated, after the gases are eliminated, the furnace tilts and pours newly refined steel into the trough (launder). The launder conveys it into a holding ladle from which it can be cast into separate ingot molds. All of these operations are remotely controlled within three separate vacuum chambers sealed off from each other.

A vacuum arc process, called the "consumable electrode" process, remelts steels produced by other methods. Its purpose is to improve the purity and uniformity of the metal. The solid steel performs like a gigantic electrode in arc welding with the heat of the electric arc melting the end of the steel electrode. The gaseous impurities are drawn off by the vacuum in the chamber as the molten steel drops into the water-cooled mold below. The remelted product is almost free of center porosity after it solidifies. Inclusions are minimized.

THE FIRST SOLID FORMS OF STEEL

Molten steel from basic oxygen, electric, and open hearth furnaces flows into ladles and then follows one or the other of two major routes to the rolling mills that form most of the industry's finished products. Both processes shown on these pages produce solid semifinished products—one, ingots (below); the other, cast slabs (far right).

SOAKING PIT　　　　　*BUGGY*

Stripped ingots are taken to furnaces called soaking pits. There they are "soaked" in heat until they reach uniform temperature throughout. Then the reheated ingots are lifted out of the pit and carried to the roughing mill on a buggy.

LADLE

STRIPPER CRANE

INGOT MOLDS

The traditional method of handling molten steel is to position the ladle via an overhead crane above a line of ingot molds (left). Then the operator opens a stopper rod within the ladle, and a stream of steal flows through a hole in the bottom of the ladle to "teem" or fill the cast iron molds which rest on special ingot railroad cars.

Molten steel in an ingot mold cools and solidifies from the outside towards the center. When the steel is solid enough, a stripper crane lifts away the mold while a plunger holds the steel ingot down on the ingot car.

ROUGHING MILL

Roughing mills are the first stage in shaping the hot steel ingot into semifinished steel—usually blooms, billets, or slabs. Some roughing mills are the first in a series of continuous mills, feeding sequences of finishing rolls.

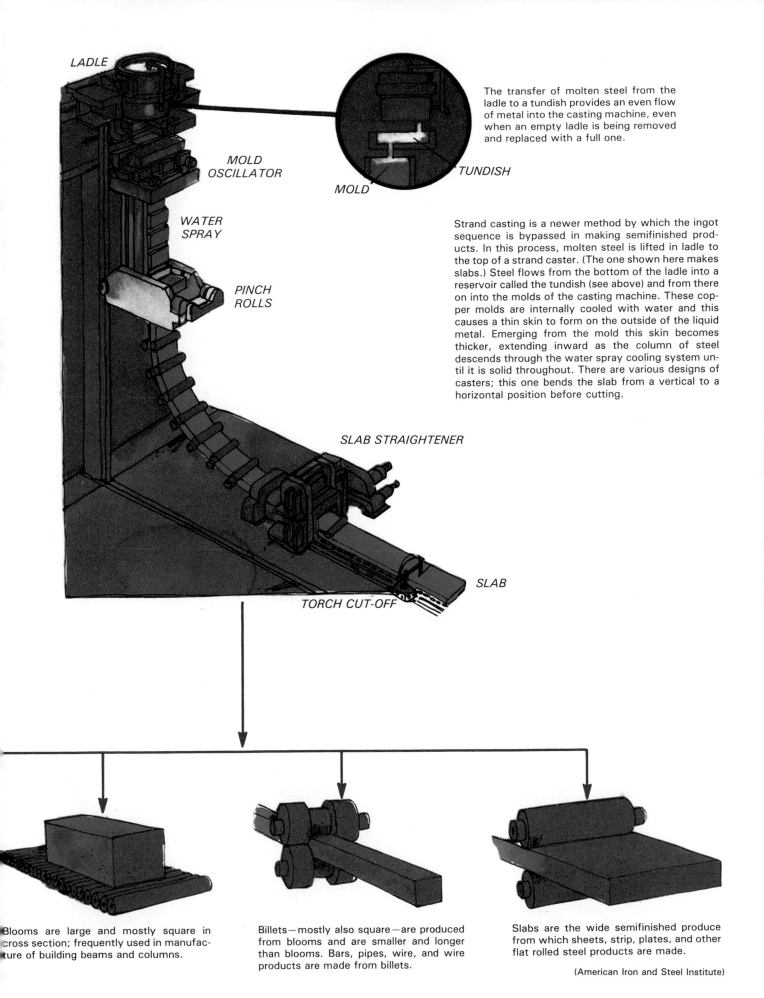

LADLE

MOLD
OSCILLATOR

MOLD

TUNDISH

The transfer of molten steel from the ladle to a tundish provides an even flow of metal into the casting machine, even when an empty ladle is being removed and replaced with a full one.

WATER
SPRAY

PINCH
ROLLS

Strand casting is a newer method by which the ingot sequence is bypassed in making semifinished products. In this process, molten steel is lifted in ladle to the top of a strand caster. (The one shown here makes slabs.) Steel flows from the bottom of the ladle into a reservoir called the tundish (see above) and from there on into the molds of the casting machine. These copper molds are internally cooled with water and this causes a thin skin to form on the outside of the liquid metal. Emerging from the mold this skin becomes thicker, extending inward as the column of steel descends through the water spray cooling system until it is solid throughout. There are various designs of casters; this one bends the slab from a vertical to a horizontal position before cutting.

SLAB STRAIGHTENER

SLAB

TORCH CUT-OFF

Blooms are large and mostly square in cross section; frequently used in manufacture of building beams and columns.

Billets—mostly also square—are produced from blooms and are smaller and longer than blooms. Bars, pipes, wire, and wire products are made from billets.

Slabs are the wide semifinished produce from which sheets, strip, plates, and other flat rolled steel products are made.

(American Iron and Steel Institute)

43

TEST YOUR KNOWLEDGE, Unit 2

Please do not write in the text. Place your answers on a separate sheet of paper.

1. Give two reasons why aluminum, magnesium, and titanium are used to manufacture airplanes.
2. Base metals are _____ metals.
3. Alloys are _____ of several metals.
4. Ferrous metals are those metals which contain _____ as the major element in their composition.
   a. Copper.
   b. Iron.
   c. Aluminum.
   d. Zinc.
5. What is the difference between ferrous and nonferrous metals?
6. How are carbon steels classified?
7. _____ _____ steel has a black coating. _____ _____ steel has a smooth surface.
8. Answer each of the following questions and also make drawings of the processes involved.
   a. How is metal in sheet or plate form made?
   b. Wire is made by _____ process.
   c. Seamless tubing is manufactured by _____ _____.

TECHNICAL TERMS TO LEARN

aluminum	nickel silver
alloy	nonferrous
base metal	oxide
brass	pewter
carbon steel	steel
cold finished steel	steerling silver
copper	tin
ferrous	tin plate
galvanized sheet	titanium
hot finished steel	tool steel
iron	uranium
magnesium	zinc
metals	

ACTIVITIES

1. How many applications can you name that need metals with special qualities? Classify them as base metals or alloys.
2. Secure samples of as many metals as you can. Mount them on hardboard and label each of the samples.

Steel mills are usually located on waterways to allow efficient delivery of iron ore and other materials by ship or barge.   (Jack Klasey)

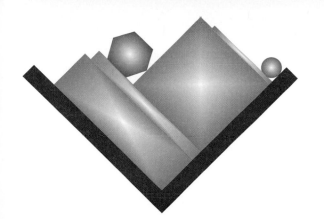

Unit 3
Planning Your Project

This unit introduces the steps to follow when planning a project. You will learn the purpose of a working drawing, a Bill of Material, and a Plan of Procedure. You will be able to use these steps in planning your own project.

The experienced artisans plan their jobs carefully before starting. You should plan your projects with the same concern. Planning will save time, effort, and material.

If plans are not available for the project you want to make, your first task will be to prepare working drawings. These drawings may be sketched or made with drafting instruments, Fig. 3-1. The plans should be drawn full size if possible.

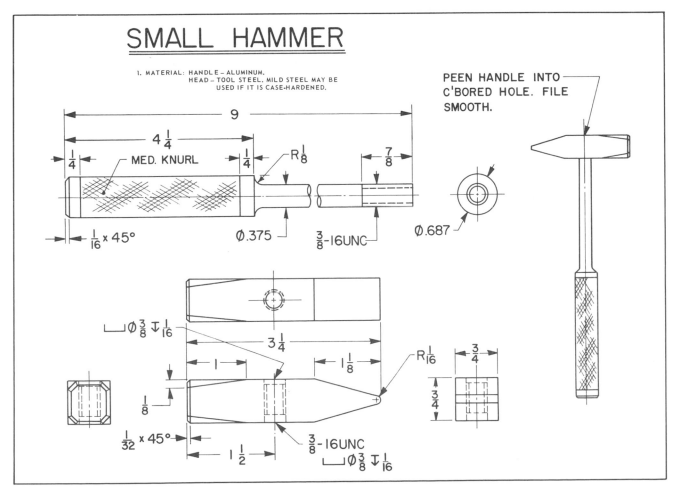

Fig. 3-1. Working drawing of the project that will be described in this unit.

While preparing the working drawings, you will need to decide on the methods of construction and fabrication you will use on your project.

The final steps in planning the project should be the preparation of a BILL OF MATERIAL and a PLAN OF PROCEDURE. A Bill of Material is a list of the materials needed to construct the project. A Plan of Procedure is a list of operations, in the order they will be done. These operations are those you will follow in the construction of your project. You may want to combine both of them on the same sheet, Fig. 3-2.

PLANNING A PROJECT

Using the project plan sheet from Fig. 3-2, the following operations are necessary to manufacture the hammer shown in Fig. 3-1. The completed plan sheet is shown in Fig. 3-3. See Figs. 3-4 and 3-5 for a pictorial outline for making the hammer. The completed hammer is shown in Fig. 3-6.

PROJECT PLAN SHEET

Name _____ Period _____

Name of Project _____

Date Started _____ Date Completed _____

Source of Idea or Project _____

BILL OF MATERIAL

Part Name	No. of Pieces	Material	Size (T x W x L)	Unit Cost	Total Cost
				Total Cost	

PLAN OF PROCEDURE

List the operations to be performed in their sequential order.
Indicate the tool(s) and equipment needed to accomplish the job.

No.	Operation	Tools and Equipment

Fig. 3-2. A typical project plan sheet includes a Bill of Material and a Plan of Procedure.

PROJECT PLAN SHEET

Name __RICHARD J. WALKER__   Period __3 MON, WED, & FRI.__

Name of Project __MACHINIST'S HAMMER__

Date Started __SEPTEMBER 15__   Date Completed __DECEMBER 10__

Source of Idea or Project __EXPLORING METALWORKING__

BILL OF MATERIAL

Part Name	No. of Pieces	Material	Size (T x W x L)	Unit Cost	Total Cost
HAMMER HEAD	1	C. F. STEEL	¾ x ¾ x 3¼		
HANDLE	1	ALUMINUM	¾ DIA. x 9¼		
				Total Cost	

PLAN OF PROCEDURE

List the operations to be performed in their sequential order.
Indicate the tool(s) and equipment needed to accomplish the job.

No.	Operation	Tools and Equipment
	HAMMER HEAD	
1.	CUT STOCK TO LENGTH AND REMOVE BURRS.	RULE, SCRIBE, HACK SAW AND FILE
2.	FILE HEAD SQUARE.	FILE AND SQUARE
3.	LAYOUT AS PER PLANS.	SCRIBE, RULE AND SQUARE
4.	FILE CHAMFERS AND BEVELS ON HEAD.	FILE
5.	CUT WEDGE END.	HACK SAW
6.	FILE WEDGE END TO SIZE.	FILE AND SQUARE
7.	DRILL ⅜ DIA. HOLE ¼ DEEP.	DRILL PRESS, VISE, PARALLELS, CENTER FINDER, CENTER DRILL AND ⅜ DIA. DRILL
8.	DRILL "F" DRILL THROUGH.	SAME SET UP BUT FINISH DRILLING WITH "F" DRILL
9.	TURN HEAD OVER IN VISE AND C'BORE ⅜" DIA. BY ⅛ DEEP.	SAME SET UP. ⅜ DIA. DRILL
10.	TAP HOLE ⅜-16 NC.	⅜-16 NC TAP, TAP HANDLE AND SQUARE
11.	CLEAN AND POLISH.	ABRASIVE CLOTH AND OIL
12.	CASE HARDEN.	FURNACE AND CASE HARDENING COMPOUND
13.	FINAL POLISH.	ABRASIVE CLOTH AND OIL
	HANDLE	
1.	CUT STOCK TO LENGTH.	RULE, SCRIBE AND HACK SAW
2.	FACE ENDS AND CENTER DRILL.	LATHE, 3-JAW UNIVERSAL CHUCK, JACOBS CHUCK, CENTER DRILL AND R.H. TOOL HOLDER
3.	TURN A SECTION 5 IN. LONG BY ¹¹⁄₁₆ IN. DIAMETER.	L.H. TOOL HOLDER, CALIPER AND RULE OR MICROMETER
4.	MARK OFF SECTION TO BE KNURLED AND KNURL.	HERMAPHRODITE CALIPER, RULE AND KNURLING TOOL
5.	MOVE HANDLE CLOSE INTO CHUCK AND MACHINE CHAMFER.	R.H. TOOL HOLDER AND RULE
6.	REVERSE WORK IN CHUCK AND TURN SHANK SECTION TO .375 IN. DIAMETER.	L.H. TOOL HOLDER, RULE AND MICROMETER
7.	THREAD SHANK ⅜-16 NC.	⅜-16 NC DIE AND DIE STOCK
8.	ASSEMBLE.	
9.	PEEN HANDLE. FILE SMOOTH.	BALL PEEN HAMMER AND SINGLE CUT FILE

Fig. 3-3. Completed project plan sheet for the hammer.

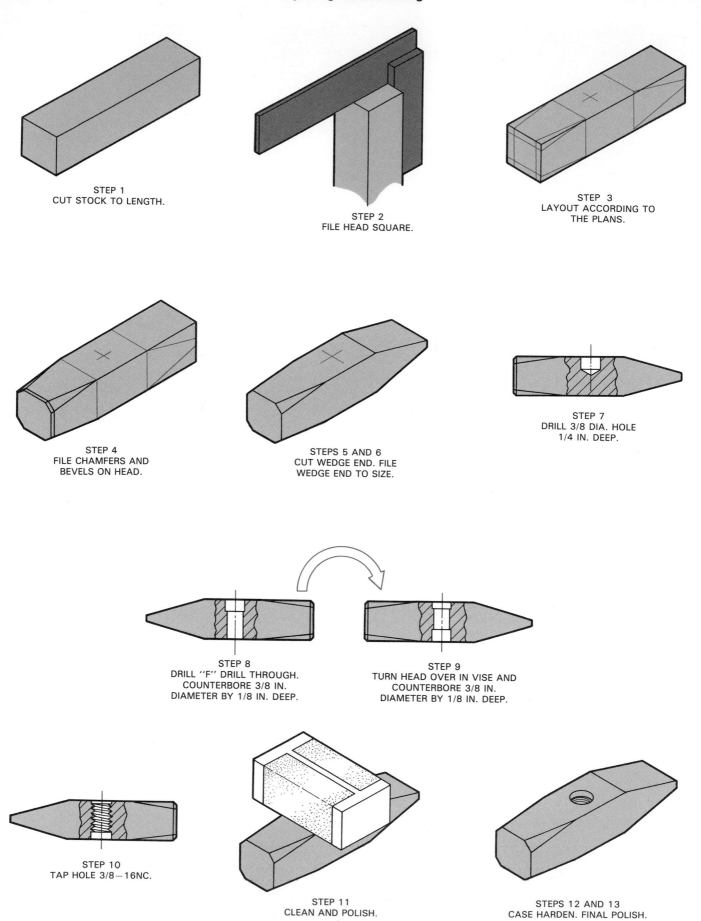

STEP 1
CUT STOCK TO LENGTH.

STEP 2
FILE HEAD SQUARE.

STEP 3
LAYOUT ACCORDING TO
THE PLANS.

STEP 4
FILE CHAMFERS AND
BEVELS ON HEAD.

STEPS 5 AND 6
CUT WEDGE END. FILE
WEDGE END TO SIZE.

STEP 7
DRILL 3/8 DIA. HOLE
1/4 IN. DEEP.

STEP 8
DRILL "F" DRILL THROUGH.
COUNTERBORE 3/8 IN.
DIAMETER BY 1/8 IN. DEEP.

STEP 9
TURN HEAD OVER IN VISE AND
COUNTERBORE 3/8 IN.
DIAMETER BY 1/8 IN. DEEP.

STEP 10
TAP HOLE 3/8 — 16NC.

STEP 11
CLEAN AND POLISH.

STEPS 12 AND 13
CASE HARDEN. FINAL POLISH.

Fig. 3-4. Steps in making the hammer head.

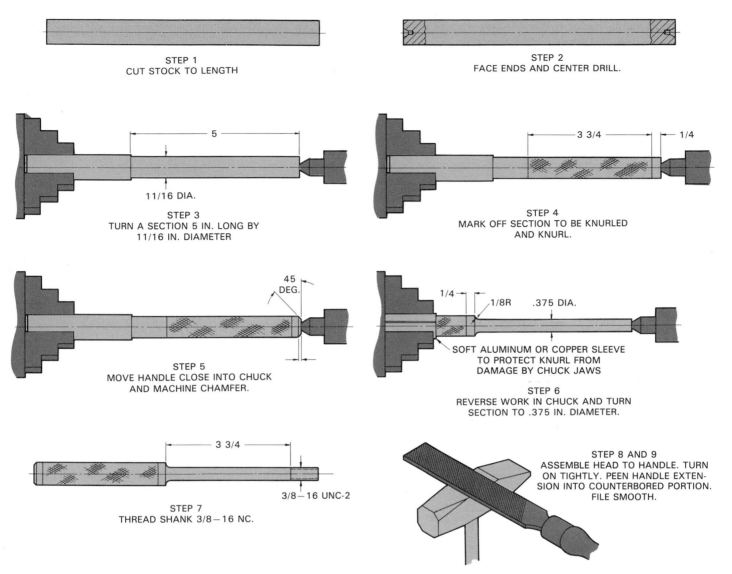

STEP 1
CUT STOCK TO LENGTH

STEP 2
FACE ENDS AND CENTER DRILL.

5

11/16 DIA.

STEP 3
TURN A SECTION 5 IN. LONG BY
11/16 IN. DIAMETER

3 3/4   1/4

STEP 4
MARK OFF SECTION TO BE KNURLED
AND KNURL.

45
DEG.

STEP 5
MOVE HANDLE CLOSE INTO CHUCK
AND MACHINE CHAMFER.

1/4   1/8R   .375 DIA.

SOFT ALUMINUM OR COPPER SLEEVE
TO PROTECT KNURL FROM
DAMAGE BY CHUCK JAWS

STEP 6
REVERSE WORK IN CHUCK AND TURN
SECTION TO .375 IN. DIAMETER.

3 3/4

3/8 — 16 UNC-2

STEP 7
THREAD SHANK 3/8 — 16 NC.

STEP 8 AND 9
ASSEMBLE HEAD TO HANDLE. TURN
ON TIGHTLY. PEEN HANDLE EXTEN-
SION INTO COUNTERBORED PORTION.
FILE SMOOTH.

Fig. 3-5. Steps in making the hammer handle.

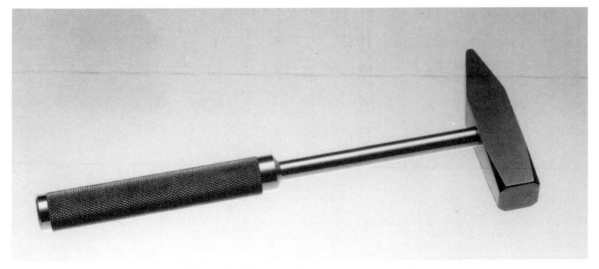

Fig. 3-6. The completed hammer project.

TEST YOUR KNOWLEDGE, Unit 3

Please do not write in the text. Place your answers on a separate sheet of paper.

1. Planning your work will save _____, _____, and _____.
2. When plans are not available, first prepare _____ _____.
3. A _____ _____ _____ is a list of items needed to construct a project.
4. What is a Plan of Procedure?
5. The following are steps recommended for planning a project. Put them in their correct order.
   a. Prepare a Plan of Procedure.
   b. Follow plans.
   c. Make working drawings.
   d. Prepare a Bill of Materials.

TECHNICAL TERMS TO LEARN

artisan	full size
Bill of Material	manufacture
construction	Plan of Procedure
fabricate	project
fabrication	working drawing

ACTIVITIES

1. Design a plan of procedure for use in your metals lab. Design it in such a way that it moves step-by-step through the schedule you follow during a typical class.
2. Secure samples of project plan sheets used by othe labs in your school. Review plan sheets done in other vocational departments in other schools. How do these plan sheets differ from those that you have done? Do the other plan sheets list steps that you may have forgotten?
3. Ask a local industry for a drawing of a simple assembly. If possible, also obtain a sample of this assembly. Based on the drawing and the sample, compile a sample project plan sheet that will allow the production of this piece.

The pilots of the U.S. Navy's BLUE ANGELS carefully plan their precision maneuvers before each flight. The more carefully *you* plan your work, the less chance there will be of *you* making a mistake. (U.S. Navy)

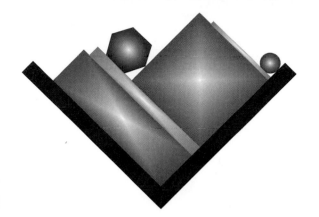

Unit 4
Designing a Project

After studying this unit, you will be able to identify the design guidelines to follow for making a good design. You will also be able to identify the elements of good design. And you will be able to explain how industry designs a project.

Good design is hard to define. Like most things, good design means different things to different people. What looks like good design to you may not appear so to a classmate. If good design cannot be defined and is different things to different people, then how do we know what is good design?

In many ways, good design may be thought of as a plan for a direct solution to a problem. Good design uses certain guidelines which point the way toward a well-designed product.

DESIGN GUIDELINES

Good design is characterized by the following guidelines.

FUNCTION. How well does the design fit the purpose for which it was planned? Does it fulfill a need?

ORGANIZATION. Do the individual parts of the design create interest when they are brought together? Are the proportions in balance? Do the parts seem to belong together? Does the basic shape look "chopped up," rather than blended together? Is the product pleasing in appearance?

QUALITY. Quality is also part of good design. A quality product is made well by people who know what they are doing. Quality must be built into the product. It cannot be added after the product is manufactured.

ELEMENTS OF GOOD DESIGN

Certain basic elements and principles are common to good designs.

LINES. Lines are used to define and give shape to an object, Fig. 4-1.

Fig. 4-1. Lines define and give shape to an object. Note how the straight lines show the unusual shape of these Lockheed F-117 aircraft.
(Lockheed Advanced Development Company)

Fig. 4-2. The shape of an object is often determined by its intended use. This arrangement of computer-controlled machining centers is known as the "Doughnut" system, because the units are arranged in a circle. Rather than the operator going to the machines, the machines come to the operator. This optimizes productivity. (Kurt Manufacturing Company)

Fig. 4-3. Form is the three-dimensional shape of the object. These candle holders illustrate how different shapes can be used.

SHAPE. The shape of an object may be detemined by the way it will be used. Refer to Fig. 4-2.

FORM. Form is the three-dimensional shape of the object. It may be round, square, or some other geometric shape, Fig. 4-3.

PROPORTION. Proportion is the balance between parts. Each part should be made the size that is best suited for its purpose, Fig. 4-4.

BALANCE. An object has balance when its parts appear to be of equal weight. The parts should be neither top-heavy nor lopsided. When the parts on each side of the centerline are alike in shape and size this is called SYMMETRICAL BALANCE, Fig. 4-5. INFORMAL BALANCE presents a design in such a manner that the balance cannot be measured. However, there is a feeling that the design is balanced, Fig. 4-6.

UNITY. A design with unity brings the various parts together as a whole. Each part of the object seems to have a relationship to another part.

Fig. 4-4. This camera is an example of a well-proportioned product. Its size and weight make it easy to handle and use.

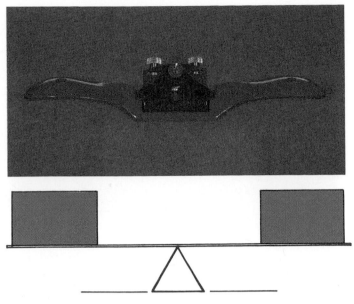

Fig. 4-5. An example of symmetrical balance.

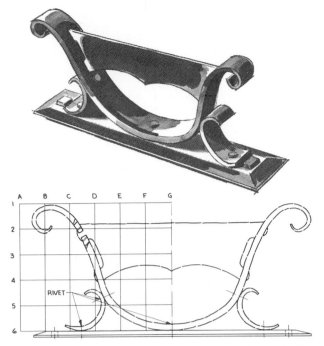

Fig. 4-7. The curves of this foot scraper show rhythm.

EMPHASIS. This is where the design is given a point of interest.

RHYTHM. Rhythm is achieved by the repetition of lines, curves, forms, colors, and textures

Fig. 4-6. Example of informal balance. The front end of the truck is long to counteract or balance the weight of the big mixer located in the rear.

within the design, Fig. 4-7. It gives an object a pleasing appearance.

TEXTURE. Texture is the condition of the surface of a material. Texture can be added by cutting, pressing, perforating, rolling, or expanding.

COLOR. All metals have a color of their own. Colors may also be added using chemicals, paints, lacquers, or other finishing materials. Selection of color is important.

HOW INDUSTRY DESIGNS A PRODUCT

How does industry go about developing a new product? Let us look at the automotive industry as an example.

New automobile designs are first developed as a series of sketches. These sketches are based on specifications made by management, Fig. 4-8. Specifications are usually based on market research.

A full size outline of the vehicle is prepared. This outline shows the placement of mechanical parts, Fig. 4-9. Line drawings and illustrations, while useful for early studies, do not provide the three-dimensions needed to test and prove design ideas. Therefore, a clay model is made, Fig. 4-10. After

Fig. 4-8. New automobile designs start as a series of sketches in the company's design studio. (General Motors Design Studio)

Fig. 4-10. Clay model being formed. Clay is used because changes can be made quickly to check new design concepts. (General Motors Design Studio)

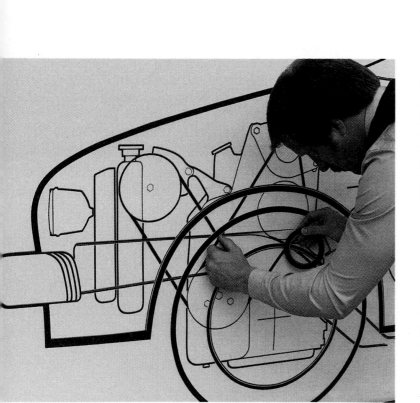

Fig. 4-9. A full-scale outline of the vehicle design is laid out. This will help in planning seating arrangement, placement of mechanical parts, and the general "look" of the car. Often, "tape-drawings," are used so that shapes and lines can be quickly changed as the design changes. (General Motors Design Studio)

further work, if the design is approved, a fiberglass model is constructed, Fig. 4-11.

Production planning uses precise, full-size wood die models to prepare permanent dies (to stamp out body panels), fabricate fixtures (devices to hold body panels while they are being welded together) and to check fixtures (measuring tools needed to maintain quality control). See Fig. 4-12. Finally the new model is made available to the public.

DESIGNING YOUR PROJECT

When designing your project, follow the pattern used by professional designers. Think of the project as a DESIGN PROBLEM. As such, there are often many ways to solve a particular design problem, Fig. 4-13. Never be afraid to experiment with new ideas that are different from the "usual" way of doing things.

SOLVING A DESIGN PROBLEM

1. STATE THE PROBLEM. What is the purpose for which the project is to be used?
2. THINK THROUGH THE PROBLEM. What must the project do? How can it be done? What

Fig. 4-11. Fiberglass model of an "idea" car. Only a few idea cars ever go into production but they are used for design studies. Many proposed changes are used on vehicles already being produced.   (General Motors Corp.)

Fig. 4-12. Wood models are used to check seating and equipment placement. Similar models are used to check part fit and molding and trim design accuracy.   (General Motors Corp.)

are the limitations, if any, that must be considered? How have others solved the problem?

3. DEVELOP YOUR IDEAS. Start with a simple design problem, such as a combination hat and coat hanger. The easiest solution might be to drive a nail into the door or wall. However, it would not be very neat and may damage the hat or coat.

After studying how other designers have done the job, make sketches of your ideas. See Fig. 4-14. Develop the ideas and discuss them with your teacher and fellow students. Determine the material(s) that will be most suitable.

4. PREPARE WORKING DRAWINGS. When you feel you have developed a good design, prepare working drawings, Fig. 4-15.

5. CONSTRUCT THE PROJECT. Make the project the best you are able. Do a job you will be proud to show others. See Fig. 4-16.

A

B

C

Fig. 4-13. There are many ways to solve a design problem. Shown are several ways aerospace designers have solved the problem of vertical take-off and landing (in addition to the familiar helicopter). A—Tilt-wing aircraft. Rotors fitted to the wing tilt vertically for take-off, as shown at left. When proper altitude is reached, right, they tilt forward for high-speed horizontal flight. (Bell Helicopter Textron/Boeing Helicopters) B—The jet engine outlets are tilted downward to lift the craft from the ground. The outlets are slowly pivoted as the craft climbs and goes into horizontal flight. The sequence is reversed for landing. The aircraft can hover like a helicopter, yet is able to attain near supersonic forward flight. (McDonnell-Douglas) C—The rotor/wing on this craft rotates for vertical takeoff. When sufficient altitude is reached, the rotor/wing is locked as shown and functions as a typical wing. This craft is also capable of high-speed forward flight.   (Sikorsky Helicopter, Div. of United Technologies)

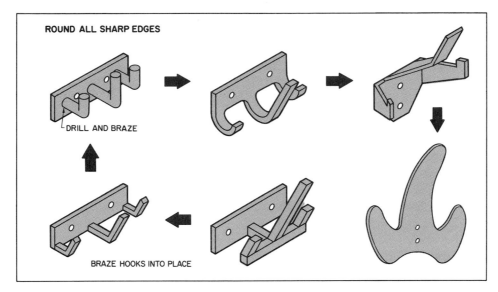

Fig. 4-14. PROBLEM: Design a simple hat and coat hanger. The first step is to study how others have solved the problem. Then, make sketches of your ideas.

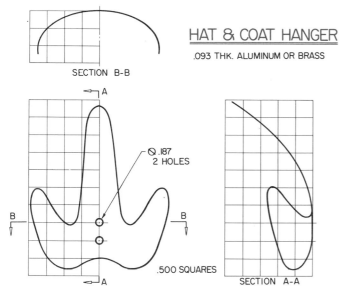

Fig. 4-15. Working drawing of hat and coat hanger.

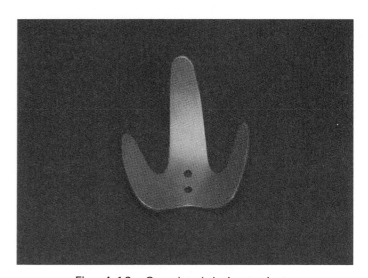

Fig. 4-16. Completed design project.

It takes time to acquire the skill to develop well-designed projects. YOU LEARN BY DOING. Keep a notebook of your ideas. Include photos of the projects you have designed and constructed. By reviewing your notebook it will be easy to see how your design skills improve.

TEST YOUR KNOWLEDGE, Unit 4

Please do not write in the text. Place your answers on a separate sheet of paper.

1. Use of _____ will help you on the way to a well-designed project.
2. Name three characteristics of good design.
3. Name four of the elements of good design and briefly explain each.
4. Design _____ are based on specifications made by management.
5. Which model is constructed first, the clay model or the fiberglass model?
6. List, in order, the steps to follow when solving a design problem.

ACTIVITIES

1. Develop and make plans for a wall rack to display model cars. Your design should hold at least three cars.
2. Prepare sketches of your ideas for a coffee table that can be made from wrought metal.
3. What are your ideas for book ends or a book rack? Prepare sketches of your ideas.
4. Design a model plane rack that is made from metal turned on the lathe.
5. Bird feeders that keep out squirrels are difficult to design. Prepare your ideas in sketch form. When you are satisfied that it will work, make a model of it.
6. Design a trivet (device used to prevent hot dishes and pots from scorching the table tops) that can be made from band iron. You may want to use ceramic tile on the top.
7. Prepare sketches for a post lamp made from copper. Include in your design the electrical fixtures.

TECHNICAL TERMS TO LEARN

balance	guidelines
characteristics	informal balance
color	organization
define	proportion
design	quality
elements	rhythm
emphasis	shape
form	solution
function	symmetrical

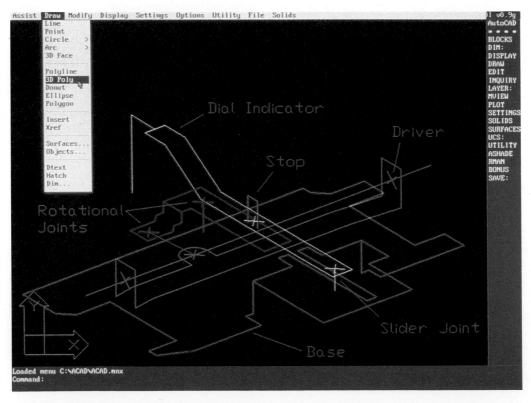

Computer-Aided Design and Drafting (CADD) programs are widely used in industry today to help designers develop new products and improve existing ones.   (Autodesk, Inc.)

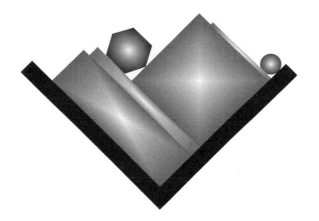

Unit 5
Metalworking Safety

After studying this unit, you will be able to use safe work habits in the metals lab. You will also be able to explain general safety rules and safety rules for hand tools.

The nature of metalworking makes safety necessary. You must exercise extreme care when working in the metals lab, Fig. 5-1. It can be a dangerous place for the careless.

Most accidents can be avoided. Develop safe work habits. ALWAYS BE ALERT. You are not a "sissy" if you wear goggles and follow safety rules. You are using good judgment. DON'T TAKE CHANCES.

Remember, IT HURTS WHEN YOU GET HURT. Become familiar with the safety rules shown and stressed by your teacher and become familiar with those included in this text. OBSERVE SAFETY AT ALL TIMES.

Fig. 5-1. Here is some common safety equipment used in the metals lab. When would these pieces of equipment be used?

GENERAL SAFETY RULES

Astronauts dress for the hazards of space flight, Fig. 5-2. The padding football players wear is designed to protect them from bumps and jolts on the playing field. When YOU work in the lab, dress for the job, Fig. 5-3.

The following are safety rules and guidelines to help you maintain safety in the lab.
1. Remove necklaces, scarves, or other dangling items that might be caught in machinery.
2. Roll up your sleeves, Fig. 5-4.
3. Wear approved safety goggles. Wear special goggles or use a shield when welding. Visitors must also obey this rule.
4. Remove wristwatches, rings, and other jewelry that might get caught in moving machinery. Also tie up or contain long hair with a cap or bandana.

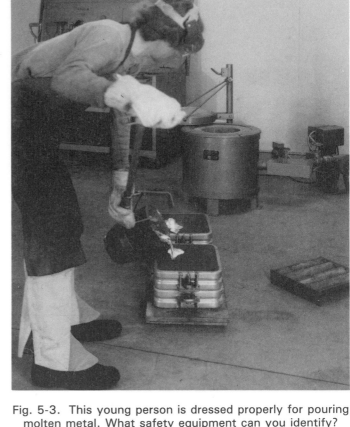

Fig. 5-3. This young person is dressed properly for pouring molten metal. What safety equipment can you identify? (McEnglevan Mfg. Co.)

Fig. 5-2. Always dress safely when working in the lab, Wear an apron or shop coat and remove rings, watch, or necklaces. Above all, WEAR APPROVED SAFETY GLASSES. (Millersville University)

Fig. 5-4. Roll your sleeves to or above your elbows. There will be no danger of them getting caught in moving parts of the machine.

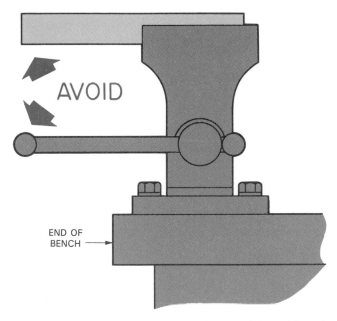

Fig. 5-6. Avoid this type of setup when working with a vise on the end of a bench.

Fig. 5-5. Wear special clothing when the job requires it. Note that this welder is wearing a face shield, leather jacket, and gloves.

Fig. 5-7. Be sure all guards and safety devices are installed on a machine before working with it. This machine should not be used until the guards are closed and locked in place.

5. Wear protective clothing when working in the foundry, forge, and welding areas. See Fig. 5-5.
6. An apron or shop coat will protect your school clothing.
7. The metals lab is no place for practical jokes or "horseplay." Tricks and pranks are dangerous to you and your fellow students.
8. Vise jaws should be left open slightly. The handle will then be in a vertical position when not in use. When working with a vise, avoid the setup shown in Fig. 5-6. Position the work and vise handle to the INSIDE of the bench.
9. Before starting any work, make certain all machines are fitted with guards and are in good working condition, Fig. 5-7.
10. Do not operate portable electric power tools in areas where thinners and solvents are used. A serious fire or explosion might result.
11. Always walk in the lab. Running is dangerous. You might stumble or collide with a fellow student who is operating a machine or pouring molten metal.
12. Get immediate medical attention for cuts, burns, bruises, and injuries no matter how minor. Report any accident to your teacher.
13. It is recommended that you NOT wear canvas shoes, sandals, or open footwear in the metals lab.

14. Place scrap metal and shavings in proper containers. Do not throw them on the floor.
15. Oil-soaked and greasy cleaning rags should be provided and placed in safety containers provided for them. DO NOT STORE THEM IN YOUR LOCKER.
16. Use care when handling long sections of metal. Watch for others who may be in your path. When removing or replacing long pieces from a vertical storage rack, be alert for people and objects that may be in your way, Fig. 5-8.
17. See your instructor before using any fluids in poorly-marked containers. Do not determine what is in a container by smelling the fumes.
18. Clean machines with brushes provided for that purpose. NEVER USE YOUR FINGERS OR HANDS TO REMOVE METAL CHIPS OR SHAVINGS FROM YOUR MACHINE.

SAFETY WHEN USING HAND TOOLS

Painful accidents can result from using hand tools incorrectly or using tools not in good repair.
1. Avoid using a tool until you have received instruction in its proper use.

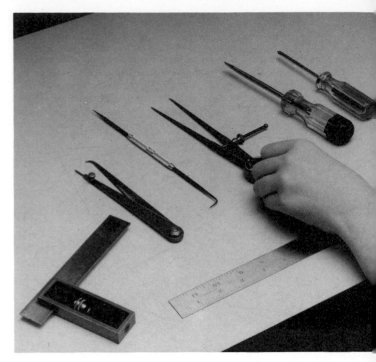

Fig. 5-9. Place sharp or pointed tools on the bench so there is no danger of being injured when you reach for them.

2. Keep tools sharp. Dull tools will not work properly and are dangerous.
3. Check each tool before use. Report defects such as loose or split handles on hammers. Grind mushroomed heads from chisels and punches.
4. A file without a handle should NEVER be used.
5. Use extreme care when carrying pointed hand tools or those with sharp edges. They should not be carried in your pockets.
6. When using dividers, scribes, etc., place them on the bench with the points and sharp edges directed AWAY from you, Fig. 5-9.
7. Tools should not be substituted for one another. For example, avoid using a wrench for a hammer, a screwdriver for a chisel, or pliers for wrenches.

There is very little chance of your being injured in the metals lab if you remember the ABCs of safety . . . ALWAYS BE CAREFUL.

TEST YOUR KNOWLEDGE, Unit 5

Please do not write in the text. Place your answers on a separate sheet of paper.

Fig. 5-8. This safety poster stresses the importance of remaining alert at all times.

1. What hazard is avoided by rolling your sleeves up to or above your elbows?

2. Leave vise jaws _____ (open, closed) when not in use so the handle will be in a vertical position.
3. Guards need not be in place when doing a "quick" job on a machine. True or False?
4. What is the danger in identifying liquids by smelling?
5. What are two causes of hand tool accidents?
6. The ABCs of safety are _____ _____ _____.
7. The same type of eye protection can be worn in all areas of the lab. True or False?

TECHNICAL TERMS TO LEARN

ABCs of safety	goggles
dangerous	protective clothing
eye protection	safety

ACTIVITIES

1. Design safety posters using one or more of the following themes.
   IT HURTS WHEN YOU GET HURT
   ALWAYS BE ALERT
   WEAR YOUR GOGGLES
   ALWAYS BE CAREFUL
   Or use your own idea. Then sketch your ideas on scrap paper. Draw your final design(s) in color on 8 1/2 in. by 11 in. poster board.
2. Design a bulletin board display on eye safety.
3. Using a tool catalog, calculate the cost of the safety equipment shown in Fig. 5-1.
4. Form a group to inspect the safety equipment in your lab. Prepare a report on the condition of this equipment and how much it will cost to replace damaged items. The report should also contain recommendations on how to improve the existing safety program.

No matter what you are doing in the metals lab, think *first* about safety. Wear proper protective clothing and equipment and follow all safety rules.

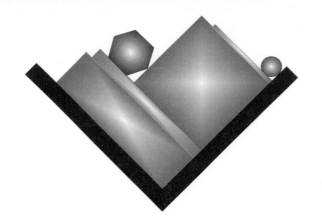

Unit 6
Measurement and Layout Tools

After studying this unit, you will be able to identify common measuring and layout tools. You will also be able to use the measuring tools to correctly measure various objects. And you will be able to pick the correct layout tool for a certain job.

To do metalworking with accuracy, you must be able to correctly read and use basic measuring and layout tools, Fig. 6-1.

RULE

The STEEL RULE is the simplest of the measuring tools. Various styles and lengths are used in school labs, Fig. 6-2.

The rule usually has fractional divisions on all four edges. These divisions are in 1/8, 1/16, 1/32, and 1/64 in., Fig. 6-3. Lines representing the divisions are called GRADUATIONS.

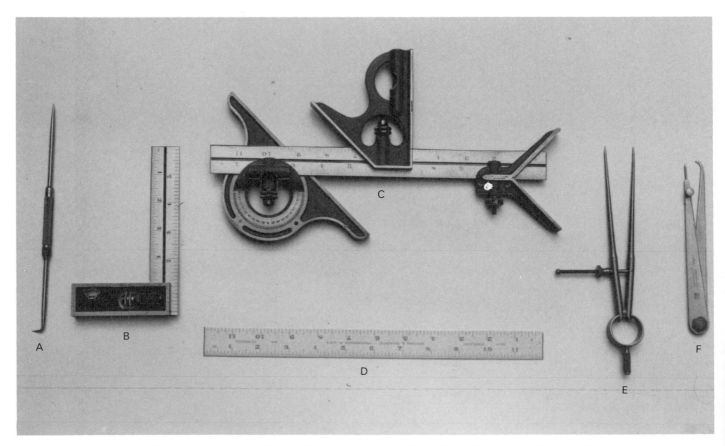

Fig. 6-1. Basic measuring and layout tools: A—Scriber. B—Adjustable square. C—Combination set. D—12 in. rule. E—Divider. F—Hermaphrodite caliper.

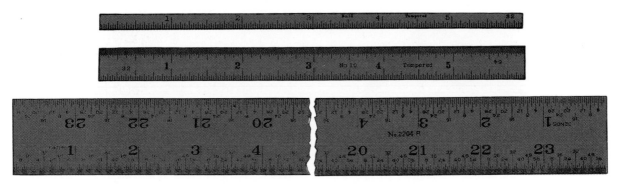

Fig. 6-2. A few of the many types of rules available to the metals student.

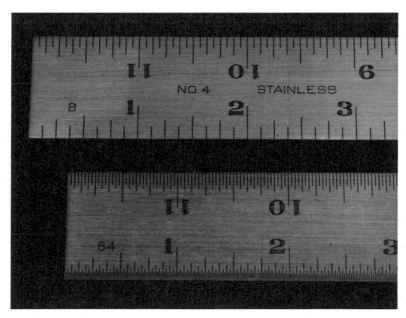

Fig. 6-3. The fractional divisions of the standard No. 4 steel rule.

The best way to learn to read a rule is to:

1. Practice making measurements with the 1/8 and 1/16 in. graduations until you are thoroughly familiar with them.
2. The same is then done with the 1/32 and 1/64 in. graduations.
3. Practice until the measurements can be made accurately and quickly.

How many measurements can you read correctly on the problem in Fig. 6-4?

Always reduce fractional measurements to their lowest terms. For example, a measurement of 6/8 reduces to 3/4, 6/16 reduces to 3/8.

Fig. 6-4. How many of the divisions can you read correctly?

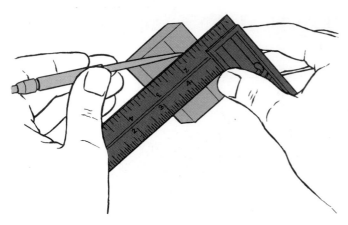

Fig. 6-5. In addition to being used to check the squareness of stock, the machinist's square can also be used for layout work.

SQUARES

The accuracy of a 90 deg. angle can be checked with a SQUARE, Fig. 6-5. The tool is also used for layout work and simple machine setups. Squares are manufactured in a variety of sizes and types.

COMBINATION SET

Many different measuring and layout operations can be performed with the COMBINATION SET, Fig. 6-6. The tool is composed of four parts: BLADE (rule), SQUARE HEAD, CENTER HEAD, and BEVEL PROTRACTOR. The blade fits all three heads.

The SQUARE HEAD, when fitted with the blade, will serve as a try square and miter square, Fig. 6-7. The tool can be used as a depth gauge by projecting the blade the desired distance beyond the edge of the unit. Simple leveling operations can be performed utilizing the SPIRIT LEVEL built into one edge.

The center of round stock can be located with the blade fitted in the CENTER HEAD, Fig. 6-8.

Various angles can be checked and/or laid out with the blade fitted in the PROTRACTOR HEAD. Refer to Fig. 6-9.

MICROMETER

The MICROMETER or "MIKE," is a precision measuring tool, Fig. 6-10. Study the major parts of the tool. A micrometer can be used to take measurements about one-thirtieth the thickness of the paper on which this is printed. The "mike" is made in many size and types, Fig. 6-11.

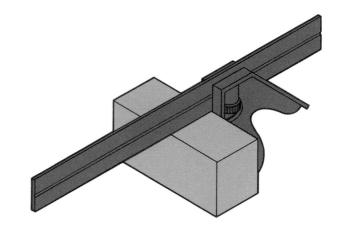

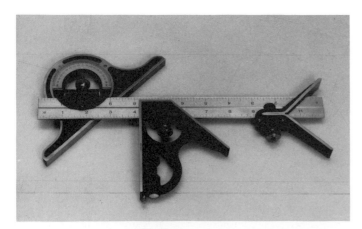

Fig. 6-6. The combination set. Left. Protractor head. Center. Square head. Right. Center head. These tools surround a 12 in. (300 mm) steel rule.

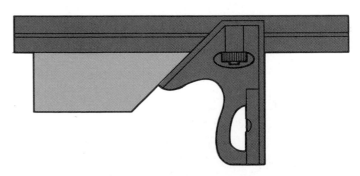

Fig. 6-7. In addition to being used for layout work, the square head of the combination set can be used to check squareness and accuracy of 45 deg. angle machined surfaces.

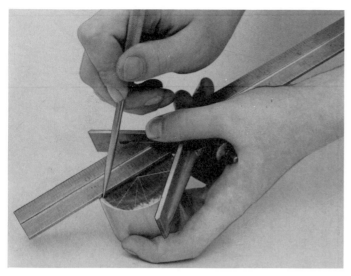

Fig. 6-8. Using center head to locate the center of round stock.

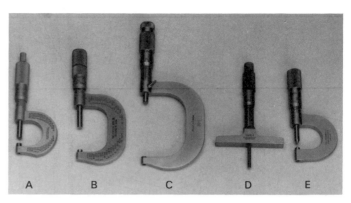

Fig. 6-11. A few of the many sizes and types of micrometers available: A—1 in. micrometer. B—2 in. micrometer. C—3 in. micrometer. D—Depth micrometer. E—Thread micrometer. Metric micrometers are similar in appearance but are sized in 25.0 mm increments.

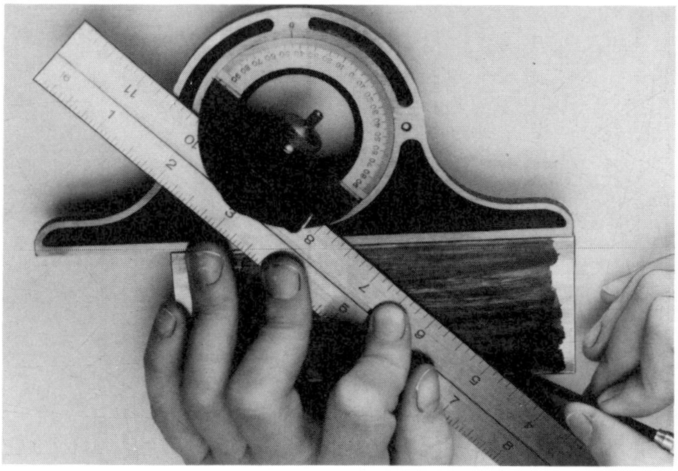

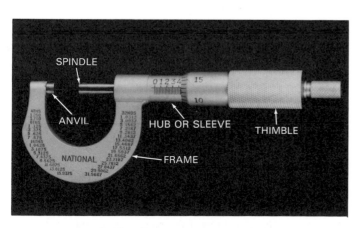

Fig. 6-9. Angular lines can be laid out using the protractor head of a combination set.

HOW TO READ AN INCH BASE MICROMETER

A micrometer has 40 precision threads per inch on its SPINDLE. When the THIMBLE (which is attached to the spindle) is rotated one complete turn, the spindle will move 1/40 in. (0.025 in.).

The line on the SLEEVE is divided into 40 equal parts per inch, Fig. 6-12. This corresponds to the number of threads on the spindle. Every fourth division is numbered 1, 2, 3, etc., representing 0.100 in., 0.200 in., 0.300 in., etc.

The THIMBLE has 25 equally spaced graduations. Each represents 1/1000 in. (0.001 in.). On some micrometers, each graduation is numbered. On others, every fifth graduation is numbered.

Fig. 6-10. The micrometer or "mike."

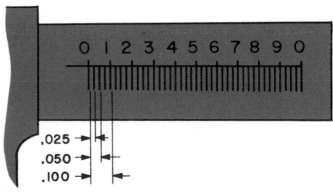

Fig. 6-12. The micrometer hub or sleeve is divided into 40 equal parts per inch. Each division is equal to 0.025 in. Every fourth division is numbered 1, 2, 3, etc., representing 0.100 in., 0.200 in., 0.300 in., etc.

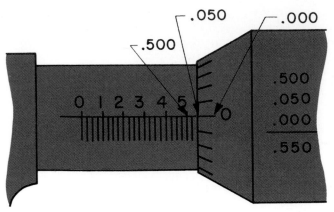

Fig. 6-13. The micrometer is read by recording the highest figure visible on the hub; 5 = 0.500 in. To this is added the number of vertical lines visible between the number and the thimble edge; 2 = 2 × 0.25 or 0.050 in. To this total add the number of thousandths indicated by the line on the thimble. This line coincides with the horizontal line on the hub.

To read the micrometer, do the following. Also, see Fig. 6-13.

1. Record the highest figure visible on the SLEEVE, such as 1 = 0.100 and 2 = 0.200.
2. To the above, add the number of vertical lines that can be seen between the number and the THIMBLE edge, 1 = 0.025, 2 = 0.050, and 3 = 0.075.
3. Finally, add the total number of thousandths indicated by the line on the THIMBLE that corresponds with the horizontal line on the SLEEVE.

Practice this sequence by reading the measurements seen in Figs. 6-14 and 6-15.

USING METRIC UNITS

Using metric units (meter, millimeter, etc.) as a basis for measurement is known as METRICA-

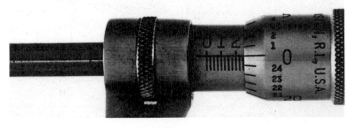

Fig. 6-14. Can you read this micrometer measurement?

Fig. 6-15. A section of sheet metal was measured with this micrometer. How thick is the metal?

TION. In metalworking, the millimeter (mm) is almost always used for measurement and layout work.

Over a period of years, the United States has slowly begun using the metric system along with the English (or standard) system for measurement. Because of this, it is to your advantage to learn to use metric-based measuring tools.

The best way to learn to use the metric system is to NOT compare metric measurements to English (fractional and decimal).

METRIC MEASURING TOOLS

The rule and micrometer are available graduated in metric units. The metric rule is compared with fractional and decimal rules in Fig. 6-16.

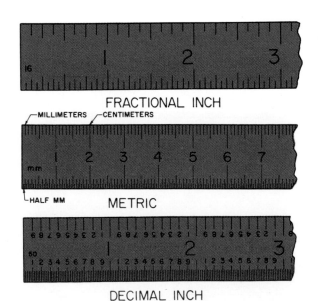

FRACTIONAL INCH

METRIC

DECIMAL INCH

Fig. 6-16. A comparison of the metric rule with fractional and decimal rules.

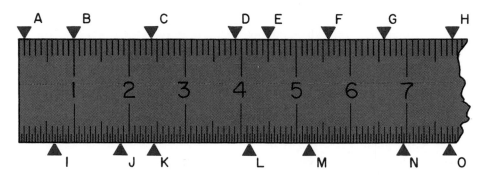

Fig. 6-17. How many of the dimensions marked can you read correctly?

How many of the dimensions marked on the metric rule, shown in Fig. 6-17, can you read? For example, measurement B would read 10.0 mm. Measurement J would be read 18.5 mm.

The metric micrometer, Fig. 6-18, is used in the same manner as the English type, except the graduations are in metric measures. The readings are obtained as follows:

1. Since the pitch of the spindle screw in a metric micrometer is 0.5 mm, one complete revolution of the thimble advances the spindle towards or away from the anvil exactly 0.5 mm.
2. The lengthwise line on the sleeve is graduated in millimeters from 0 to 25 mm. Each millimeter is subdivided in 0.5 mm. Therefore, two revolutions of the thimble are required to advance the spindle towards or away from the anvil a distance equal to 1.0 mm.
3. The beveled edge of the thimble is graduated into 50 divisions, with every fifth line numbered. Since a complete revolution of the thimble moves the spindle 0.5 mm, each graduation on the thimble is equal to 1/50 of 0.5 mm or 0.01 mm. Two graduations equal 0.02 mm, three equals 0.03 mm, and so on.

4. To read a metric micrometer, see Fig. 6-19. Add the total reading in millimeters (visible on the sleeve) to the reading in hundredths of a millimeter (indicated by the graduation on the thimble). The hundredths measurement lines up with the lengthwise line on the sleeve.

METRIC-DIMENSIONED DRAWINGS

Since 1960, the metric system has been referred to throughout the world as "Systeme International d'Unites" or the International Systems of Units. The abbreviation SI is universally used to indicate this system.

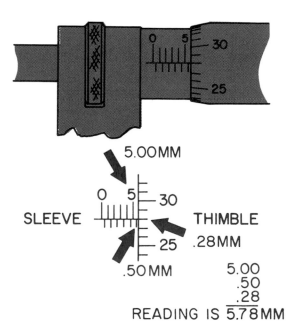

Fig. 6-19. Reading a metric micrometer. Find the total reading in millimeters shown on the sleeve. It is 5.5 mm. Add 5.5 to the reading in hundredths of a millimeter shown by the graduation on the thimble. In this case, it is .28 mm. This coincides with the lengthwise line on the sleeve.

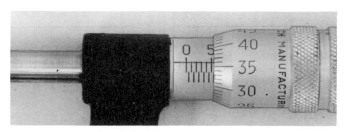

Fig. 6-18. Metric-based micrometer.

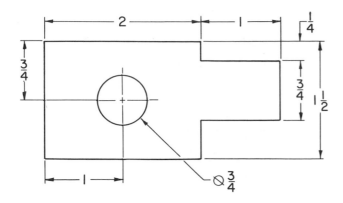

Fig. 6-20. A drawing dimensioned with the standard inch dimensioning system.

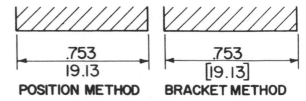

Fig. 6-22. Current method of indicating inches and millimeters on a dual-dimensioned drawing.

You are familiar with drawings that use the inch dimensioning system, Fig. 6-20. DUAL DIMENSIONED drawings, Fig. 6-21, were the first step in the changeover to metric dimensioning. If the part was made in the United States, the inch dimension appeared as shown in Fig. 6-22. In metric countries, the millimeter dimension appear in the top position.

If the part is designed and dimensioned in the inch system, there is seldom any reason to convert inch dimensions to millimeters. However, if the product is designed using metric-sized material, metric dimensions must be used, Fig. 6-23. A comparison of inch sizes with the preferred metric-sized material is shown in Fig. 6-24.

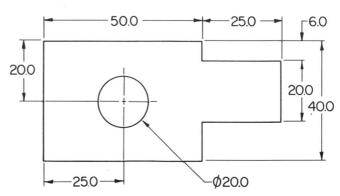

Fig. 6-23. This drawing shows a part similar to that in Fig. 6-21. This one, however, is designed for metric-sized materials. Note how the metric dimensions differ from those shown on the dual-dimensioned drawing, where metric values were exact conversions from inch dimensions.

HELPER MEASURING TOOLS

A helper measuring tool must be used with a steel rule or micrometer. A helper measuring tool cannot be used without one of these two tools as a guide.

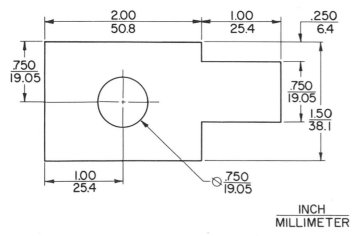

Fig. 6-21. A drawing based on dual dimensioning. Note how the location of the inch/millimeter dimensions are indicated. While many dual-dimensioned drawings are still in use, they are seldom drawn this way anymore.

INCH	mm
1/8	3.0
1/4	6.0
3/8	10.0
1/2	12.0
5/8	16.0
3/4	20.0
1	25.0
1 1/4	30.0
1 1/2	40.0
1 3/4	45.0
2	50.0

Fig. 6-24. This chart shows a comparison of inch sizes to metric sizes.

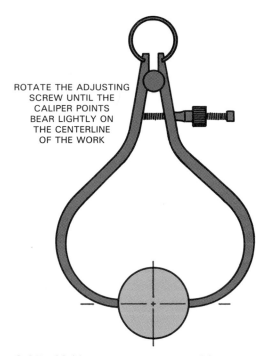

ROTATE THE ADJUSTING
SCREW UNTIL THE
CALIPER POINTS
BEAR LIGHTLY ON
THE CENTERLINE
OF THE WORK

Fig. 6-25. Making a measurement with an outside caliper.

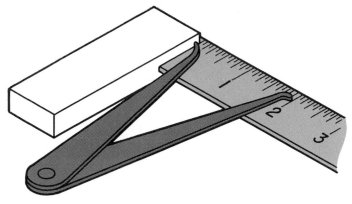

Fig. 6-27. Making a measurement with an inside caliper.

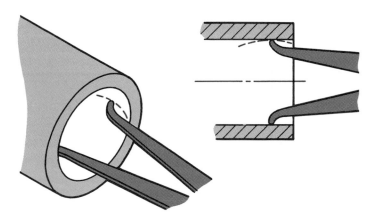

Fig. 6-28. Suggested way to read a measurement taken with an inside caliper.

OUTSIDE CALIPER

External measurements can be made with an OUTSIDE CALIPER, Fig. 6-25, when a 1/64 in. (0.4 mm) tolerance is permitted. The 1/64 in. (0.4 mm) tolerance means that the part can be 1/64 in. (0.4 mm) larger or 1/64 in. (0.4 mm) smaller than the size shown on the plans and still be used.

Measurement of round stock is made by setting the caliper legs lightly on the centerline of the stock. Hold the caliper to the rule and read the size, Fig. 6-26.

Never force a caliper over the work. This will "spring" the caliper legs and give an inaccurate measurement.

INSIDE CALIPER

Interior dimensions are gauged with the INSIDE CALIPER, Fig. 6-27. Measurement is made by inserting the caliper legs into the opening. The legs are opened until they "drag" slightly when moved in or out, or from side to side. To read hole size, hold the caliper to the rule as shown in Fig. 6-28.

LAYOUT TOOLS

A LAYOUT is a series of reference points and lines that show the shape the material is to be cut or machined, Fig. 6-29. The layout may also include the location of openings and holes that must be made in the work.

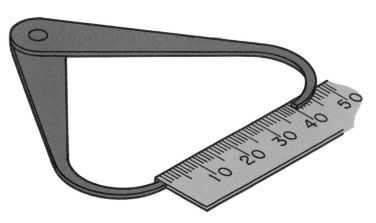

Fig. 6-26. Reading the measurement taken with an outside caliper.

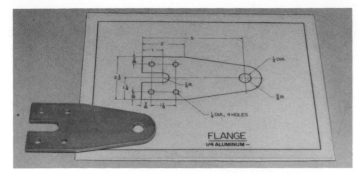

Fig. 6-29. A typical layout drawing. The part made using the drawing as a guide is also shown.

Accuracy depends, to a large extent, on your skill in properly using and caring for the layout tools.

MAKING LINES ON METAL

It is not easy to see layout lines on shiny metal. Therefore, a LAYOUT DYE, Fig. 6-30, is used. This coating is applied to the metal to provide contrast between the metal and the layout lines.

Chalk can be used on hot rolled steel as a layout background.

Fig. 6-30. Layout dye is used to make scribed lines stand out on metal surfaces.

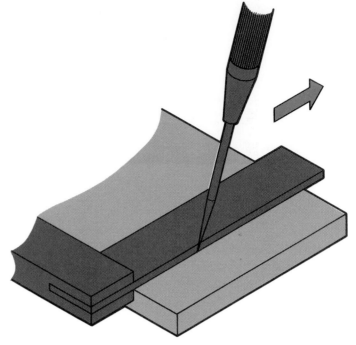

Fig. 6-31. Make sure the scriber is held firmly against the straightedge before making the layout line.

SCRIBER

Accurate layouts require scribing fine lines on metal surfaces. The lines may be made with a SCRIBER, Fig. 6-31. The tool point is made of hardened steel. It should be kept needle-sharp by frequent honing on a fine oilstone. There are many styles of scribers.

A pencil should never be used to make layout lines. The pencil line is too wide and rubs off easily.

CAUTION: NEVER CARRY OPEN SCRIBERS IN YOUR POCKET. PLACE THEM ON THE BENCH WITH THE POINT FACING AWAY FROM YOU.

DIVIDERS

Circles and arcs are drawn on metal with the DIVIDER, Fig. 6-32. The tool may also be employed to measure for equal distances.

To set a divider to the desired dimension, place one point on the inch/10.0 mm mark of a steel rule. Then open the divider until the other leg is set to the required size, Fig. 6-33.

72

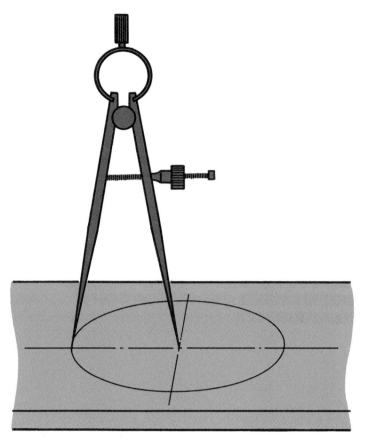

Fig. 6-32. Circles and arcs are drawn on metal with a divider.

CAUTION: NEVER CARRY DIVIDERS IN YOUR POCKET. PLACE THEM ON THE BENCH WITH THE POINTS FACING AWAY FROM YOU.

HERMAPHRODITE CALIPER

A HERMAPHRODITE CALIPER, Fig. 6-34, is a cross between an inside caliper and a divider.

A hermaphrodite caliper is used to lay out lines parallel to an edge. It is also used to locate the approximate center of irregularly shaped stock.

CAUTION: NEVER CARRY A HERMAPHRODITE CALIPER IN YOUR POCKET. ALSO PLACE THEM ON THE BENCH WITH THE POINT FACING AWAY FROM YOU.

PUNCHES

Two types of PUNCHES are commonly employed in making layouts. They are the following:

1. PRICK PUNCH, Fig. 6-35, top. A hardened, pointed steel rod is used to "spot" the point where centerlines intersect on a layout. The sharp point (60 deg.) makes it easy to locate points.

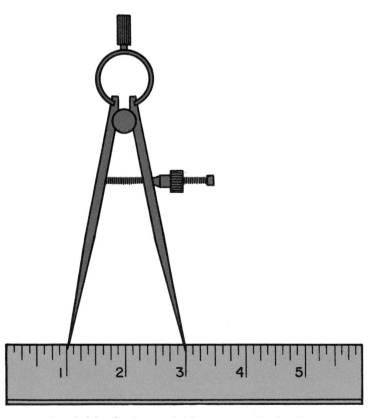

Fig. 6-33. Setting a divider to a required radius.

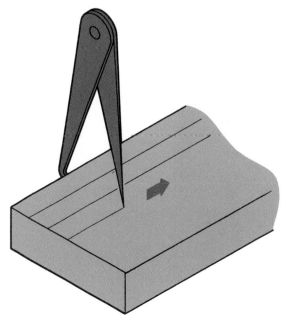

Fig. 6-34. Scribing a line parallel to the edge of the work with a hermaphrodite caliper.

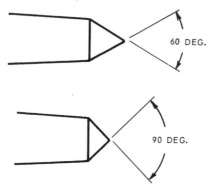

Fig. 6-35. Top. Prick punch. Bottom. Center punch.

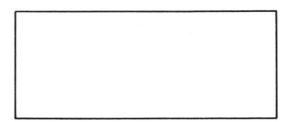

CUT METAL TO APPROXIMATE SIZE

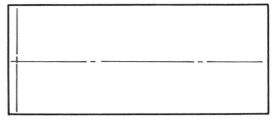

LOCATE CIRCLE AND ARC CENTERLINES

2. CENTER PUNCH, Fig. 6-35, bottom. This is similar to a prick punch, but has a more blunt point (90 deg.). It is used to enlarge the prick punch mark after it has been checked and found to be on center.

HOW TO MAKE A LAYOUT

Each layout job will have its own problems. This will require careful planning before starting. The following sequence is recommended. Also refer to Figs. 6-36 and 6-37.

1. Carefully study the plans.
2. Cut the stock to size and remove all burrs and sharp edges.
3. Clean the metal and apply layout dye.

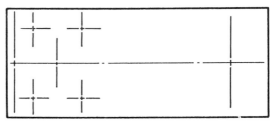

LOCATE AND SCRIBE ANGULAR LINES

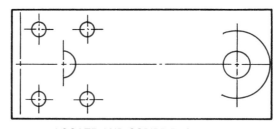

LOCATE AND SCRIBE BASE LINES

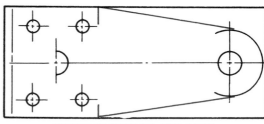

SCRIBE IN CIRCLES AND ARCS

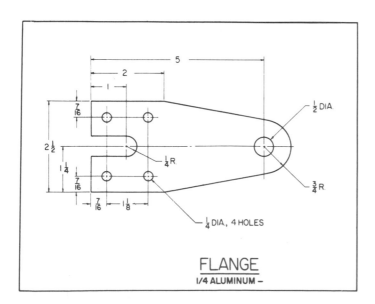

FLANGE
1/4 ALUMINUM –

Fig. 6-36. This job requires layout in its manufacture.

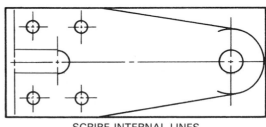

SCRIBE INTERNAL LINES

Fig. 6-37. Steps in laying out the job shown in Fig. 6-36.

4. Locate and scribe BASE or REFERENCE LINES. All measurements are made from these lines. If the material has one true edge, it can serve as one of the base or reference lines.
5. Locate all center and arc centerlines.
6. Locate and scribe angular lines.
7. Scribe in all other internal openings.

LAYOUT SAFETY

1. NEVER carry scribers, dividers, or other pointed tools in your pocket.
2. Cover sharp pointed tools with a cork when they are not being used.
3. When using sharp pointed tools, lay them on the bench with the pointed ends facing away from you. There will then be no danger of reaching into the points when you pick up the tool.
4. Wear goggles when repointing layout tools.
5. Remove burrs and sharp edges from the stock before starting to lay it out.
6. Secure prompt medical attention for any cut, scratch, or bruise, no matter how minor it may appear.

TEST YOUR KNOWLEDGE, Unit 6

Please do not write in the text. Place your answers on a separate sheet of paper.

1. The _____ _____ is the simplest and most widely used measuring tool.
2. The accuracy of 90 deg. angles can be checked with a _____.
3. The combination set is composed of four parts. Name them and briefly describe the use of each part.
4. Define metrication.
5. What is a dual dimensioned drawing?
6. A _____ is a series of reference points and lines that show the shape the material is to be cut or machined.
7. Straight layout lines are drawn using a _____.
8. To provide contrast between metal surfaces and layout lines, _____ _____ should be applied to the surface being laid out.
9. Make the readings on the micrometer drawings shown.

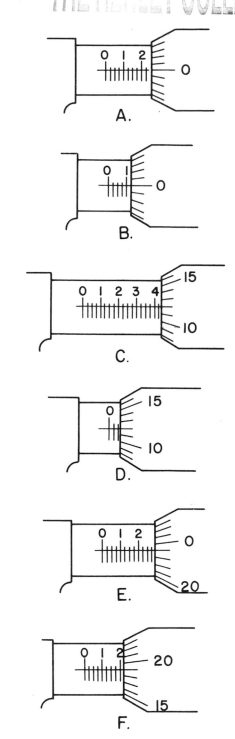

10. Determine the measurements on the metric rule section shown. Give your answers in millimeters.

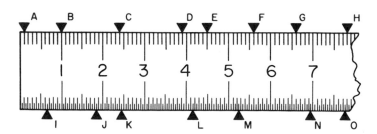

11. Make the readings on the metric micrometer drawings shown.

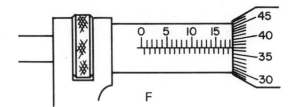

F

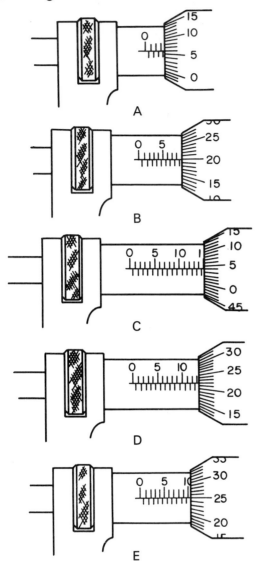

A

B

C

D

E

TECHNICAL TERMS TO LEARN

accuracy
arc
base line
center head
center punch
circle
combination set
divider
dual-dimensioned
fractional
graduation
hermaphrodite caliper
inside caliper
layout

layout tools
measure
metric
metrication
micrometer
millimeter
outside caliper
protractor
punch
reference line
rule
scriber
square

ACTIVITIES

1. Secure a metric rule and micrometer from the science department at your school. Instruct the class in their use.
2. Make a drawing of a project to be made in the school shop. Use metric dimensioning.
3. Discuss the metric system with a representative of a nearby company. Find out whether the company would find the system an advantage or a disadvantage.

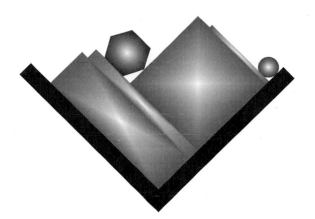

Unit 7
Basic Metalworking Tools and Equipment

After studying this unit, you will be able to identify basic metalworking tools and equipment. You will be able to demonstrate the safe and proper use of these tools. You will also be able to safely use those machines that cut metal.

The proper and safe use of hand tools and machines is basic to all areas of metalworking.

VISE

Metal is normally held in a VISE while it is being worked, Fig. 7-1. There are many sizes and types of vises.

CAPS made of soft metal, Fig. 7-2, should be fitted over the vise jaws. These will protect the work from being damaged or marred by the jaw serrations (teeth).

HAMMERS

The BALL PEEN HAMMER, Fig. 7-3, is usually used in metalworking. Its size is usually determined by the weight of the head. For example, a ball peen hammer with a 4 oz. head is known as a 4 oz. hammer.

Commonly, the lightest hammer that will do the job easily and safely should be used.

SCREWDRIVERS

Select the screwdriver that fits the screw being driven, Fig. 7-4. Two types of screwdrivers that are commonly used in the metals lab are the

Fig. 7-1. Vise used in metalworking.

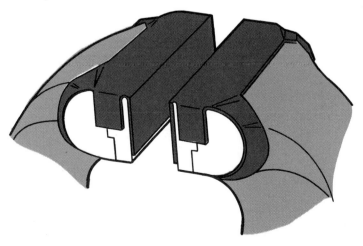

Fig. 7-2. Caps protect material from being damaged by the serrations on the vise jaws. The caps are made of soft metal.

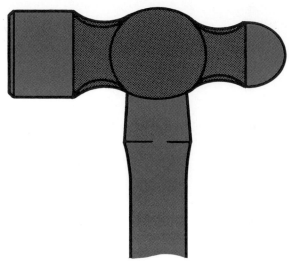

Fig. 7-3. Ball peen hammer.

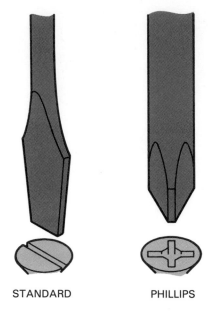

STANDARD   PHILLIPS

Fig. 7-5. Two types of widely used screwdrivers.

STANDARD TYPE for slotted heads, and the PHILLIPS TYPE for X-shaped recessed heads, Fig. 7-5.

PLIERS

Many holding, bending, and cutting jobs can be done with PLIERS, Fig. 7-6. Pliers are made in many sizes and styles to handle a variety of jobs. PLIERS SHOULD NEVER BE USED AS A SUBSTITUTE FOR A HAMMER OR A WRENCH.

Pliers commonly found in the metals lab include those listed in the next several paragraphs.

COMBINATION PLIERS (also known as SLIP-JOINT PLIERS) have many uses. The slip joint makes it possible to adjust the jaws and grip both large and small work. The jaws have serrations or teeth. On some slip-joint pliers, a short cutting edge for cutting wire is located near the hinge.

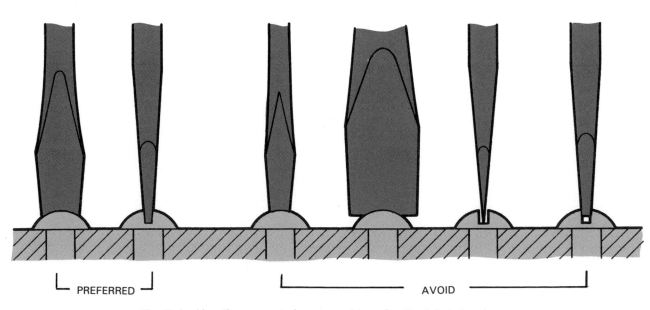

PREFERRED   AVOID

Fig. 7-4. Use the correct size screwdriver for the job to be done.

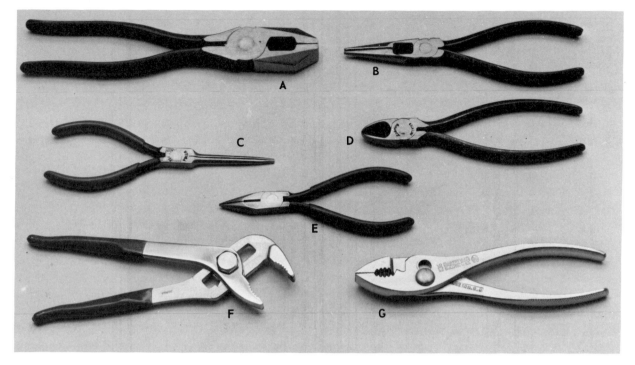

Fig. 7-6. A few of the many types and sizes of pliers available. A—Lineman's side cutting pliers. B—Long nose cutting pliers. C—Needle nose pliers. D—Diagonal cutting pliers. E—Short reach needle nose pliers. F—Groove joint pliers. G—Slip-joint or combination pliers.

The cutting edges of DIAGONAL PLIERS are at an angle. This angle makes cutting flush with the work surface possible.

Heavier wire and pins can be cut with SIDE-CUTTING PLIERS.

Wire and light metal can be bent and formed with ROUND NOSE PLIERS. The smooth jaws will not mar the work.

LONG or NEEDLE NOSE PLIERS are useful when space is limited, or where small work is to be held.

When using pliers keep your fingers clear of the cutting edges. When doing electrical work be sure to insulate the handles with rubber tape, or use specially manufactured rubber grips.

WRENCHES

Many types of wrenches are available to the metalworker. Each is designed for a specific use.

Many wrenches are adjustable to fit different sizes of nuts and bolts. However, the wrench shown in Fig. 7-7, is usually known as an AD-JUSTABLE WRENCH. It is made in a range of sizes. Use the smallest wrench that will fit the nut or bolt being worked.

The PIPE WRENCH, Fig. 7-8, is also adjustable. The movable jaw has a small amount of

Fig. 7-7. Adjustable wrench.   (Diamond Tool Co.)

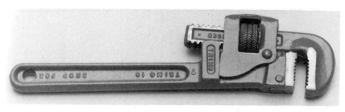

Fig. 7-8. Pipe wrench.

79

play built in it so that it can take a "bite" on round stock. A pipe wrench will leave marks on the stock. This wrench should not be used on nuts and bolts unless the corners have been so damaged a regular wrench cannot be used.

The size of the space between the jaws of the OPEN-END WRENCH, Fig. 7-9, determines its size. Open-end wrenches are made in many different sizes and styles.

The BOX-WRENCH, Fig. 7-10, completely surrounds the bolt head or nut. It is preferred over other wrenches because it will not slip. These wrenches are available in sizes to fit standard nuts and bolts.

The COMBINATION OPEN AND BOX-WRENCH, Fig. 7-11, has one open end and one box end.

SOCKET WRENCHES, Fig. 7-12, are like box-wrenches because they surround the bolt head or nut. However, they are made as detachable tools that fit many different types of handles. A typical socket wrench set contains various handles and a wide range of socket sizes and styles.

PULL, NEVER PUSH, on any wrench. Pushing is considered dangerous. If the nut loosens suddenly, you may strike your knuckles on the work (known as "knuckle dusting"). The movable jaw of a wrench should ALWAYS face the direction the fastener is being turned, Fig. 7-13.

FILES AND FILING

FILES are frequently used to remove extra metal.

FILE CLASSIFICATION

Files are classified by their shape, Fig. 7-14, by the cut of their teeth, Fig. 7-15, and by the coarseness of the teeth. Types of coarseness include rough, coarse, bastard, second-cut, smooth, and dead smooth.

Fig. 7-9. Open-end wrench.

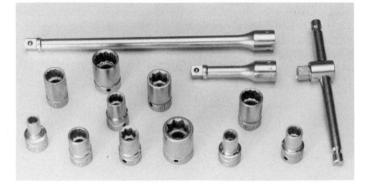

Fig. 7-12. Socket wrench set with extensions.

Fig. 7-10. Box wrench.

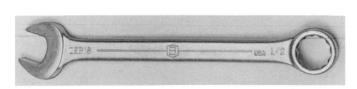

Fig. 7-11. Combination open and box wrench.

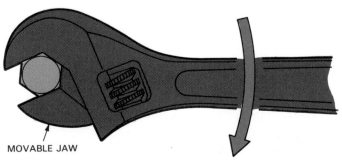

MOVABLE JAW

Fig. 7-13. The movable jaw of the wrench should ALWAYS face the direction the fastener is being turned.

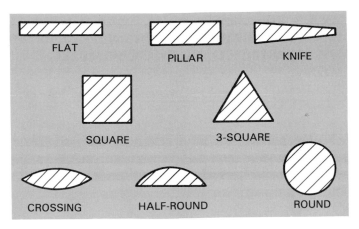

Fig. 7-14. File shapes.

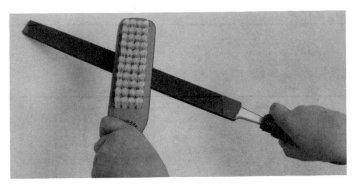

Fig. 7-16. Clean a file with a file card.

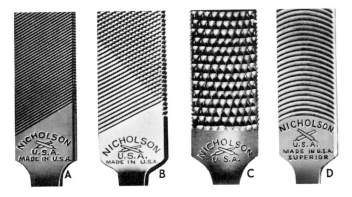

Fig. 7-15. These files are classified by the cut of their teeth.
A—Single-cut. B—Double-cut. C—Rasp. D—Curved-tooth.

Fig. 7-17. The proper way to hold the file for straight or cross
filing.

Fig. 7-18. When done properly, draw filing improves the surface finish.

FILE CARE

Your file should be cleaned frequently using a FILE CARD, Fig. 7-16. This will prevent PINNING. Pinning is a problem in which small slivers of metal clog the file and cause scratches on the work. DO NOT try to clean a file by striking it against the bench top or vise.

FILE SELECTION

The nature of the work will determine the size, shape, and cut of the file that should be used.

A SINGLE-CUT file is generally used to produce a smooth surface finish. DOUBLE-CUT files remove metal rapidly but produce a rougher surface finish. RASPS are used on wood and some plastics. Flat surfaces of steel and aluminum are worked with a CURVED-TOOTH file.

USING THE FILE

Most filing is done while the work is held in a vise. Mount the work at about elbow height for general filing. STRAIGHT or CROSS FILING is done by pushing the file lengthwise, straight ahead, or at a slight angle across the work, Fig. 7-17. DRAW FILING, Fig. 7-18, usually produces a finer finish than straight filing.

To avoid injury, NEVER USE A FILE WITHOUT A HANDLE. Avoid running your fingers over a newly filed surface. You might cut yourself on the sharp burr formed by the file.

CUTTING METAL BY HAND

Some metal cutting can be done easily and safely by hand using a chisel or hacksaw.

CHISELS

CHISELS are used to cut or shear metal. The four types shown in Fig. 7-19 are often used. These are usually referred to as COLD CHISELS.

NEVER USE A CHISEL WITH A MUSHROOMED HEAD, Fig. 7-20. Correct this dangerous condition by grinding.

Shearing is done with the stock held in a vise, Fig. 7-21. Flat stock should be cut on a soft steel backing plate. NEVER cut flat stock on the vise slide or on top of the anvil.

HACKSAWS

Most hacksaws can be adjusted to fit various blade lengths, Fig. 7-22. Some are made so the blade can be installed in either a vertical or horizontal position, Fig. 7-23.

A fine-tooth blade should be used to cut thin stock. Heavier work is cut with a coarse-tooth blade. When selecting a blade, keep in mind the THREE-TOOTH RULE. That rule says that at least three teeth of the blade should be in contact with the work at all times.

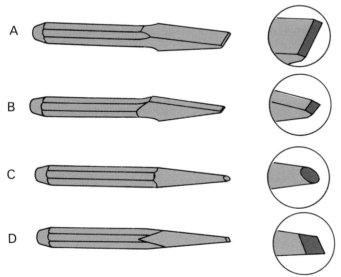

Fig. 7-19. COLD CHISELS. A—Flat chisel is used for general cutting. B—Cape chisel has a narrower cutting edge and is used to cut grooves. C—Round nose chisel is used to cut round grooves and radii. D—Diamond point chisel is used to square corners.

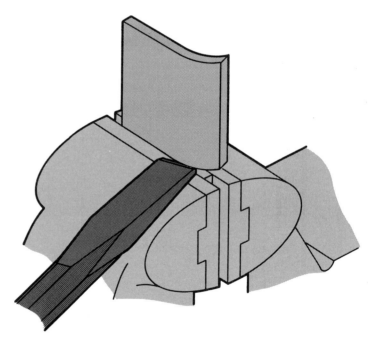

Fig. 7-21. Shearing work held in a vise with a chisel.

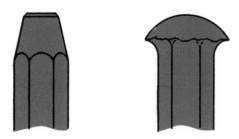

Fig. 7-20. Never use a chisel with a mushroomed head.

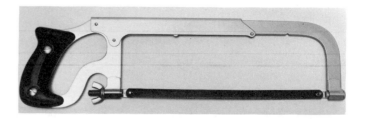

Fig. 7-22. The adjustable hacksaw.

Install blades with the teeth pointing AWAY from the handle, Fig. 7-24. Tighten until the blade "pings" when snapped with your finger.

HOLDING WORK FOR SAWING

Clamp the work so the cut to be made is close to the vise, Fig. 7-25. In this way, you can avoid "chatter" (vibration that dulls the teeth). Mount work so the cut is started on a flat side rather than on a corner or edge, Fig. 7-26. Start the cut using the thumb of your free hand to guide the blade. Methods for holding different metal shapes for cutting are shown in Fig. 7-27.

CUTTING METAL

Hold the saw firmly (but comfortably) by the handle and the front of the frame. Apply pressure on the cutting (forward) stroke. Lift the saw slightly on the return stroke. Make about 40 strokes per minute using the full length of the blade.

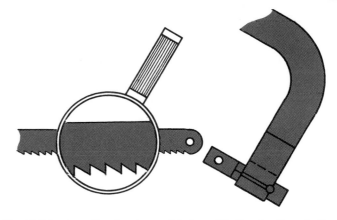

Fig. 7-24. Install a hacksaw blade with the teeth pointing away from the handle.

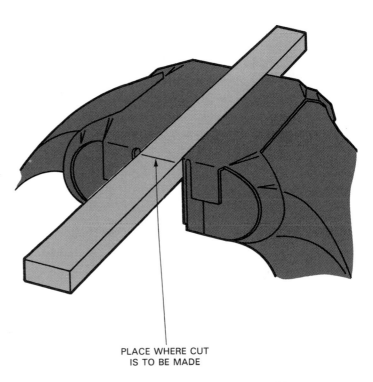

PLACE WHERE CUT
IS TO BE MADE

Fig. 7-25. Make cut as close to vise as possible to prevent chatter. Chatter can ruin the saw blade.

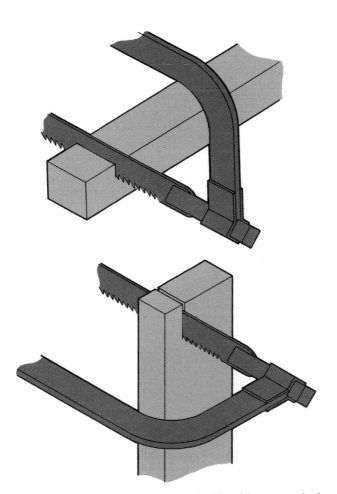

Fig. 7-23. This blade can be installed in either a vertical or horizontal position.

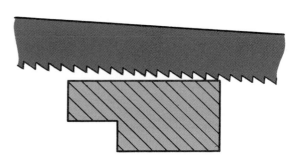

Fig. 7-26. Start the cut on a flat edge of the work rather than on a sharp corner.

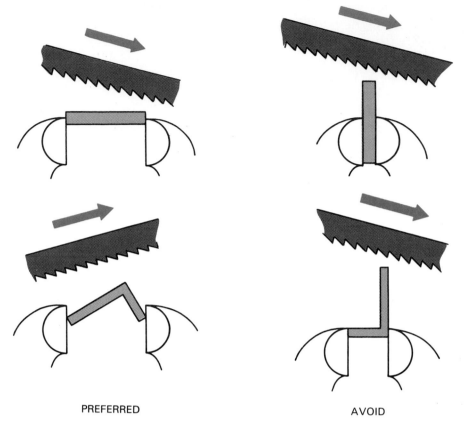

PREFERRED   AVOID

Fig. 7-27. Recommended ways to hold work of various shapes for sawing.

Saw slower when the blade has cut almost through the metal. Support the work so it will not drop when the cut is completed.

There may be times when a blade breaks or becomes dull. The blade must then be changed. After changing the blade, if possible, position the work so a new cut can be made. If a new cut is not made, the new blade will be pinched in the old cut, dulling quickly.

DO NOT TEST SAW BLADE SHARPNESS BY RUNNING YOUR FINGER ACROSS THE TEETH!

CUTTING METAL WITH MACHINES

Large metal sections can be cut quickly with a POWER HACKSAW. See Fig. 7-28.

As with hand sawing, the THREE-TOOTH RULE also applies to power sawing. In general, large pieces and soft material require saws with coarse teeth. Small or thin work, or hard materials, require a fine tooth blade.

Fig. 7-28. Top. Reciprocating type power hacksaw. Bottom. Band type power hacksaw.

84

Fig. 7-29. Sabre saw. (Makita)

Mount the blade so it cuts on the power stroke. This is the back stroke on most reciprocating (back and forth) power saws.

Be sure the work is mounted solidly before starting the cut.

KEEP YOUR HANDS CLEAR OF THE MOVING BLADE. STOP THE MACHINE BEFORE MAKING ADJUSTMENTS. PULL THE PLUG WHEN CHANGING BLADES OR MAKING ADJUSTMENTS.

Thin material can be cut with a SABRE SAW, Fig. 7-29. The type and thickness of the metal determines the type of blade to use. If a cutting chart for a saw is available, use this as a guide.

DRILLING METAL

The DRILL PRESS, Fig. 7-30, is mostly used to cut round holes in metal. It operates by rotating a cutting tool (the drill) against material with

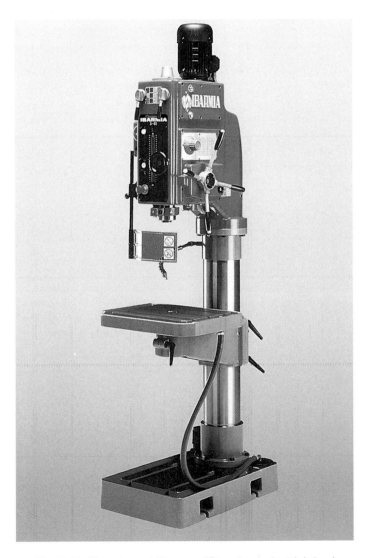

Fig. 7-30. Floor type drill press. (Clausing Industrial, Inc.)

STRAIGHT SHANK

Fig. 7-31. The portable electric power drill has a built-in power source. The batteries are rechargeable; length of service with each charging will depend upon drill size and the material being drilled. ALWAYS WEAR SAFETY GLASSES WHEN USING POWER TOOLS.

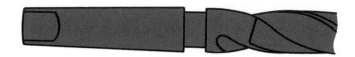

TAPER SHANK

Fig. 7-34. Types of drill shanks.

Fig. 7-32. Hand drill.

Drills are made of HIGH SPEED STEEL (HSS) or CARBON STEEL. High speed drills are costly when purchased. However, if used properly, they will cut faster and last longer than carbon steel drills. Proper use of drills includes following recommended cutting speeds. These are the speeds at which a certain drill should rotate. The speeds for several metals are given in Fig. 7-33.

enough pressure to cause the drill to cut into the material. The PORTABLE ELECTRIC DRILL, Fig. 7-31, and the HAND DRILL, Fig. 7-32, are also used to drill holes in all kinds of material.

STRAIGHT SHANK DRILLS, Fig. 7-34, top, must be held in a chuck. TAPER SHANK DRILLS, Fig. 7-34, bottom, mount directly in the drill press spindle.

CUTTING SPEEDS FOR HIGH SPEED DRILLS					
DRILL DIAMETER	ALUMINUM	BRASS	CAST IRON	MILD STEEL	TOOL STEEL
1/16	4500	4500	4500	4000	3500
1/8	2000	3000	1800	1800	1500
3/16	1800	2900	1500	1400	1200
1/4	1700	2300	1200	1100	1000
5/16	1500	1900	950	850	725
3/8	1200	1500	750	700	600
7/16	1100	1300	650	600	525
1/2	1000	1150	575	525	375
9/16	850	1000	500	525	350
5/8	750	900	450	425	300
3/4	600	750	375	350	250
7/8	550	650	325	300	225
1.000	450	575	280	265	185

PLEASE NOTE:

These cutting speeds are recommended. It may be necessary because of the characteristics of the material to increase or decrease the drill speed to do a satisfactory cutting job. Use a cutting fluid on all metals EXCEPT cast iron.

Fig. 7-33. Suggested cutting speeds for drilling various types of metal.

DRILL SIZES

Drill sizes are expressed by one of the following drill series:

NUMBER DRILLS—No. 80 to No. 1 (0.0135 in. to 0.2280 in. diameter).

LETTER DRILLS—A to Z (0.234 in. to 0.413 in. diameter).

FRACTIONAL DRILLS—1/64 in. to 3 1/2 in. diameter.

METRIC DRILLS—3.0 mm to 76.0 mm diameter.

Number drills and letter drills are often needed for drilling holes that are to be tapped (threaded) or reamed.

The drill size, except on very small drills, is stamped on the drill shank. Should the size wear away, drill diameter can be checked using a DRILL GAUGE, Fig. 7-35, or a micrometer.

DRILLING PROCESS

The following steps should be followed when drilling a hole.

1. Lay out work to be drilled as shown on the layout plans.
2. Mount the work solidly on the drill press. If possible, use a vise, Fig. 7-36, or clamp the work to the drill press table. Be careful not to drill into the vise or table. SAFETY NOTE: When drilling, do not hold short pieces or thin stock by hand. A "merry-go-round" may result and cause a painful injury.
3. Center the work using a "wiggler" (center finder) or a center. See Fig. 7-37.
4. Select the proper size drill. Check it for size and sharpness. Mount it in the chuck. SAFETY NOTE: Be sure to remove the chuck key from the chuck.
5. Set the drill press to the correct speed. In general, use slow speeds for large drills and hard materials. Use faster speeds for small drills and soft materials. Check a DRILL SPEED

Fig. 7-36. Work mounted for drilling.

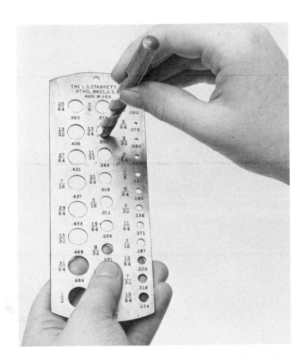

Fig. 7-35. Drill size can be checked quickly with a drill gauge.

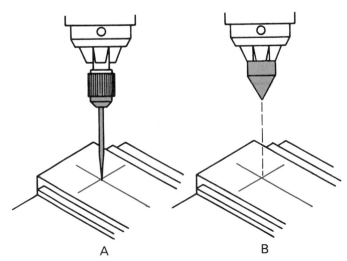

A   B

Fig. 7-37. Two techniques for locating the hole center when drilling. A—Wiggler or center finder. B—Small center.

CHART to determine the correct speed for the drill and material being used, Fig. 7-33.

6. Turn on the machine and bring the drill into contact with the work. Apply a few drops of cutting fluid (machine or lard oil) to the drill point from time to time to improve cutting action. Reduce pressure as the drill starts through the work. SAFETY NOTE: When using a drill press, be sure long hair is contained, sleeves are rolled up, hanging jewelry is removed, and safety goggles are worn.

7. Turn off the machine. Clean away chips. Unclamp the work and remove any burrs. SAFETY NOTE: Remove chips with a brush—NOT WITH YOUR HANDS.

8. Clean the drill press and drill. Return tools to their proper place.

When drilling holes larger than 1/2 in. (12 mm) in diameter, it is often best to first drill a smaller PILOT HOLE, Fig. 7-38.

Holes that will hold flat head fasteners should be COUNTERSUNK. This is done with a COUNTERSINK, Fig. 7-39. Countersink the hole deep enough so that the head of the fastener is flush with the work surface. See Fig. 7-40.

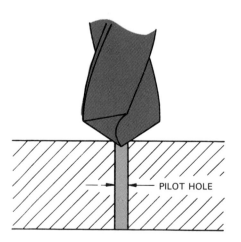

Fig. 7-38. Drilling large diameter holes is made easier by using a pilot hole.

Fig. 7-39. Countersink.

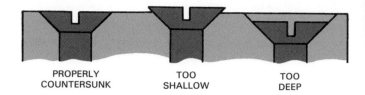

Fig. 7-40. Countersink deeply enough so that the screw head is flush with the work surface.

HAND THREADING

Spiral grooves found on nuts, bolts, and screws are called THREADS. The operation that cuts this groove in material is called THREADING.

Threads are cut by hand with either taps or dies, Fig. 7-41. INTERNAL THREADS (like those inside a nut) are cut with a TAP. EXTERNAL THREADS (like those on a bolt) are cut with a DIE.

THREAD SERIES

There are two widely used series of threads. The FINE THREAD SERIES, Fig. 7-42, left, has more threads per inch of length for a given thread diameter than the COARSE THREAD SERIES, Fig. 7-42, right. Both are part of the AMERICAN NATIONAL THREAD SYSTEM. The FINE THREAD SERIES is indicated by NF and the COARSE THREAD SERIES is indicated by NC.

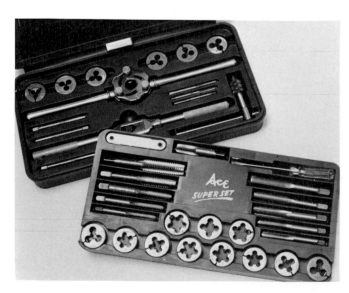

Fig. 7-41. Tap and die sizes.

Fig. 7-42. Left. National Fine Thread (NF). Right. National Coarse Thread (NC).

In the late 1940s, Canada, Great Britain and the United States adopted a thread form (shape) that is very similar to the American National system. This new system is known as the UNIFIED SYSTEM. When used, the UNIFIED FINE THREAD SERIES is identified as UNF and the UNIFIED NATIONAL COARSE SERIES as UNC.

METRIC THREADS are now being used in some American-made products. Metric threads are made to standards set by the INTERNATIONAL ORGANIZATION FOR STANDARDIZATION (ISO). ISO is composed of 80 nations. The groups establish standards that define and allow measurements of length, volume, weight, and other values. Metric thread size is identified differently than in the Unified Thread System. See Fig. 7-43.

Metric threads have the same basic shape as UNF and UNC threads. While they may appear to be similar in diameter, ISO and UNF/UNC threads cannot be used in place of each other, however. See Fig. 7-44.

Because metric and UNF/UNC threads are similar in appearance, mismatching often occurs. To control this problem, metric fasteners can be identified as shown in Fig. 7-45.

CUTTING INTERNAL THREADS

A hole must be drilled before an internal thread can be cut. The diameter of the hole must be smaller than the tap size. The hole is made with a TAP DRILL. A tap drill is a number, letter,

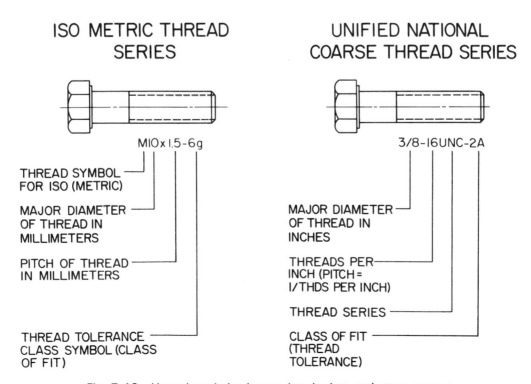

Fig. 7-43. How thread size is noted and what each term means.

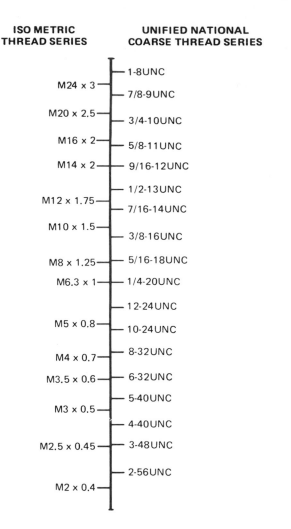

ISO METRIC THREAD SERIES	UNIFIED NATIONAL COARSE THREAD SERIES
	1-8UNC
M24 x 3	7/8-9UNC
M20 x 2.5	3/4-10UNC
M16 x 2	5/8-11UNC
M14 x 2	9/16-12UNC
	1/2-13UNC
M12 x 1.75	7/16-14UNC
M10 x 1.5	3/8-16UNC
M8 x 1.25	5/16-18UNC
M6.3 x 1	1/4-20UNC
	12-24UNC
M5 x 0.8	10-24UNC
M4 x 0.7	8-32UNC
M3.5 x 0.6	6-32UNC
	5-40UNC
M3 x 0.5	4-40UNC
M2.5 x 0.45	3-48UNC
	2-56UNC
M2 x 0.4	

ISO and Unified National Thread Series ARE NOT INTERCHANGEABLE

Fig. 7-44. A comparison of ISO metric coarse pitch and Unified Coarse (UNC) inch based thread sizes. Even though several of them seem to be the same size, they are not switchable.

fractional, or metric drill. Using this drill, the diameter of the hole will be small enough to permit the tap to cut threads. A TAP DRILL CHART must be used to choose the proper drill size (tap drill) needed with a specified tap. You can find such a chart on page 260 of this text.

TAPS

Taps are made in sets of three, Fig. 7-46. In this set is a TAPER, PLUG, and BOTTOM TAP.

The TAPER TAP is so called because the point has a pronounced taper. This tool permits easy starting of holes. It is employed for tapping THROUGH HOLES. In through holes the threads are cut all of the way through the work.

The PLUG TAP is used to thread blind holes if the holes are drilled deeper than the threads are to be cut. Blind holes do not go all the way through the material.

The BOTTOM TAP is used when the thread must be cut to the bottom of a blind hole. Start the thread with a taper tap, cut with a plug tap, and complete with the bottom tap.

TAP HOLDERS

Taps are turned with a TAP HOLDER or TAP WRENCH. See Figs. 7-47 and 7-48. Tap size will determine which is to be used.

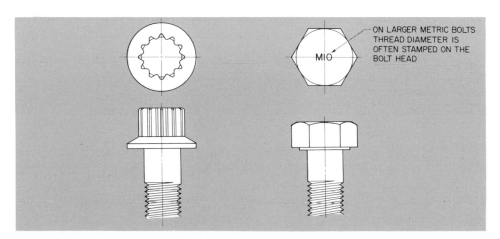

Fig. 7-45. Much study is being done to find an easy way to identify metric-based fasteners from inch-based fasteners. The 12-element spline head and imprinted hex head (thread size is stamped on the head) are two methods being considered.

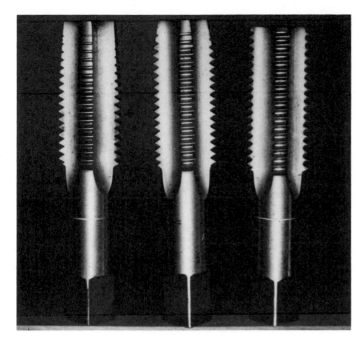

Fig. 7-46. Tap set. Left. Taper tap. Center. Plug tap. Right. Bottoming tap.

A T-HANDLE TAP WRENCH is used with small taps (under 1/4 in. or 6.5 mm). It allows a sensitive "feel" when tapping and reduces the danger of tap breakage.

More leverage is required with larger taps, so the HAND TAP WRENCH should be used.

NOTE: A conventional wrench is NEVER used to turn a tap when cutting threads.

TAPPING A HOLE

To keep tap breakage to a minimum, the following steps are recommended:

1. Drill the required size hole.
2. Start the tap square, Fig. 7-49. A drop of cutting oil will improve cutting ability.
3. Turn the tap into the work a partial turn. Back it off (counterclockwise) until you feel the chips break loose. DO NOT FORCE THE TAP. Continue this until the hole is tapped.
4. Use cutting oil for tap lubrication. If a blind hole is being threaded, back the tap out of the hole often to remove chips.
5. Use care. Broken taps are difficult to remove from work.

Fig. 7-47. T-handle tap wrench is used with small taps.

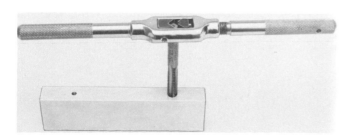

Fig. 7-48. Hand tap wrench.

Fig. 7-49. The tap must be started square.

SAFETY NOTE: Use a brush to remove chips made by a tap. Also, avoid running your finger over a newly tapped hole. A burr is often present and can cause a painful cut.

CUTTING EXTERNAL THREADS

External threads are cut with a DIE, Fig. 7-50. A DIE STOCK, Fig. 7-51, holds the die and serves as a wrench for turning it.

When cutting external threads, remember the following rules.

1. Stock diameter is the same size as the required threads. That is, 3/8-16UNC threads would be cut on 3/8 in. diameter rod.
2. Grind a small chamfer on the end of the stock, Fig. 7-52. This makes it easier to start the die.
3. Mount the work solidly in a vise.
4. Start the cut with the tapered end of the die. Again, see Fig. 7-52.
5. If an adjustable die is used, make trial cuts on scrap stock to see if the die is properly adjusted.

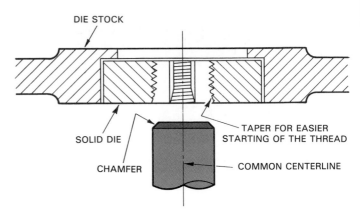

Fig. 7-52. A small chamfer ground or turned on the end of the stock will permit the die to be started easier.

6. Back off the die every turn or two to break the chips and to allow them to fall free.
7. Apply liberal amounts of cutting fluid. Be careful not to spill any on the floor.
   SAFETY NOTE: Do not remove chips from newly cut threads with your fingers. A brush will do it better and more safely.

REAMERS

Drilled holes may be enlarged to the desired size with a REAMER. A HAND REAMER, Fig. 7-53, is turned with a tap wrench. Like the machine reamer that is used with the drill press, it is available with STRAIGHT or SPIRAL FLUTES, Fig. 7-54.

EXPANSION REAMERS, Fig. 7-55, permit slight size adjustment.

Fig. 7-50. Two types of adjustable dies.

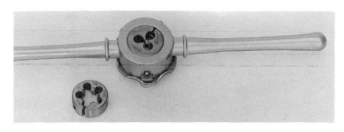

Fig. 7-51. Die stock.

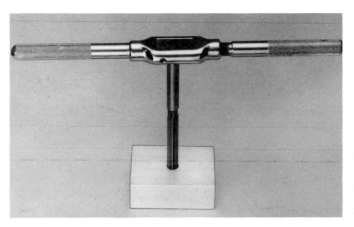

Fig. 7-53. Hand reamer mounted in tap wrench.

Fig. 7-54. Cutting portion of machine reamers.

Fig. 7-55. Adjustable reamer.

These allowances (how much smaller the hole must be drilled) are recommended for reaming:

a. To 1/4 in. diameter allow 0.010 in.

b. 1/4 to 1/2 in. diameter allow 0.015 in.

c. 1/2 to 1.0 in. diameter allow 0.020 in.

d. 1.0 to 1 1/2 in. diameter allow 0.025 in.

Use sharp reamers.

Fig. 7-56. Reaming table.

After drilling a hole to the correct size for reaming, the following steps are suggested to hand ream a hole. See Fig. 7-56 for recommended allowances.

1. Mount the work solidly in a vise.
2. Position the reamer in the hole. The end of the tool is tapered slightly so it will fit easily into the hole. If proper alignment is a problem, use a square.
3. Slowly turn the reamer CLOCKWISE until it is centered in the hole.
4. After the reamer is centered, keep turning the wrench clockwise with a firm, steady pressure

until the reamer is through the work. Lubricate the reamer with cutting or machine oil. Remove the tool from the hole by turning the wrench CLOCKWISE and raising the reamer at the same time.

CAUTION: Do not turn the reamer counterclockwise at any time. This will dull the tool.

TEST YOUR KNOWLEDGE, Unit 7

Please do not write in the text. Place your answers on a separate sheet of paper.

MATCHING QUESTIONS: Match the following sentences with the words listed below.

a. Hacksaw.	h. Die.
b. Double-cut.	i. Countersink.
c. Standard.	j. UNC.
d. Vise caps.	k. Tap.
e. File card.	l. Straight shank.
f. Single-cut.	m. UNF.
g. Phillips.	n. Reamer.

1. ____ The soft covers used to protect work held in a vise.
2. ____ The screwdriver used to drive slotted head screws.
3. ____ The screwdriver used to drive "X" slotted head screws.
4. ____ The file cut used to produce a smooth surface finish.
5. ____ The file cut that removes metal quickly but makes a rough surface finish.
6. ____ Used to clean files.
7. ____ Tool used to cut metal.
8. ____ Enlarges drilled holes.
9. ____ Drill held in a drill chuck.
10. ____ Cuts internal threads.
11. ____ Cuts external threads.
12. ____ Tool used to permit flat head screws to be mounted flush with the work surface.
13. ____ Used to indicate Unified National Coarse Thread Series.
14. ____ Used to indicate Unified National Fine Thread Series.

15. Name the four commonly used drill series.
16. A _____ tap is used to start the thread.
17. The _____ tap is used to cut threads to the bottom of the hole.

18. Identify the taps shown below.

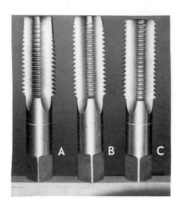

TECHNICAL TERMS TO LEARN

adjustable wrench
ball peen hammer
bottom tap
box wrench
burr
center finder/wiggler
chisel
clockwise
combination set
combination pliers

combination wrench
counterclockwise
cross filing
diagonal pliers
die
die stock
double-cut
draw filing
drill
drill press

file
file card
hacksaw
needle-nose pliers
open-end wrench
Phillips screwdriver
pilot hole
pipe wrench
pliers
plug tap
reamer
round nose pliers
sabre saw

screwdriver
side cutting pliers
single-cut
socket wrench
straight shank
tap
tap drill
taper shank
taper tap
tap wrench
threads
vise

ACTIVITIES

1. Example the tools in your metals lab. Repair those that are damaged.
2. Prepare a series of safety posters on the proper way to use tools.
3. Inventory the hand tools in your lab. Using tool catalogs, determine how much it would cost if they had to be replaced.
4. Design a new tool panel and storage facility for your lab.
5. Devise a method for quickly checking whether all tools were returned to their proper place at the end of a work period.

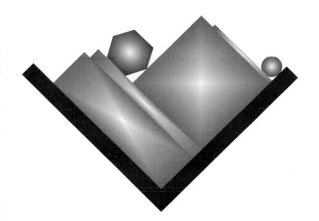

Unit 8
Wrought Metal

After studying this unit, you will be able to identify the hand tools and specially designed tools used for bending metals. In addition, you will demonstrate how to twist metal and shape curves in metal. And, finally, you will be able to create your own piece of wrought metalwork.

In Colonial America, wrought metalwork was hammered and worked into shape by BLACK-SMITHS, Fig. 8-1. Their products were not only useful, but also decorative, Fig. 8-2.

Today, wrought metalwork is also known as ornamental ironwork and bench metal.

The metal most commonly used is wrought iron. It is almost pure iron and contains very little carbon (carbon makes iron harder and tougher).

Fig. 8-1. In Colonial America, blacksmiths hammered and worked metal into useful shapes.
(Colonial Williamsburg)

Fig. 8-2. Reproductions of hand wrought work made by Colonial blacksmiths are not only useful, but also decorative. (Old Guilford Forge)

Wrought iron is easy to bend (either hot or cold) and weld, but is expensive. Standard hot rolled steel shapes are often used as substitutes.

BENDING WITH HAND TOOLS

Metal up to 1/4 in. in thickness can be bent cold. Heavier metal bends easier when heated.

Cut the metal to length. This can be done with a hacksaw or a ROD PARTER, Fig. 8-3.

In working with wrought iron, the length of the metal is reduced slightly with each bend. To make up for this, add one-half the metal thickness to the length of the piece for each bend.

Some projects require that several bends be made in the same piece of metal. The bending sequence must be planned carefully. If it is poorly planned, it may be difficult to make all the bends with accuracy, Fig. 8-4.

Many bends in wrought ironwork can be made using a heavy vise and a ball peen hammer. Make the necessary layout. Place the metal in the vise, with the extra material allowed for the bend, projecting above the jaws, Fig. 8-5. Start the bend by striking the metal near the vise, with the flat of the hammer, Fig. 8-6.

Right angle bends are squared by using one corner of the vise jaw as a form, Fig. 8-7. Strike the metal near the bend.

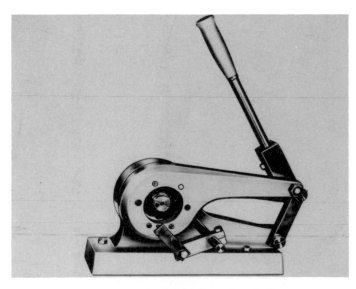

Fig. 8-3. Bar and rod stock can be easily sheared to length by using a ROD PARTER. (Di-Acro)

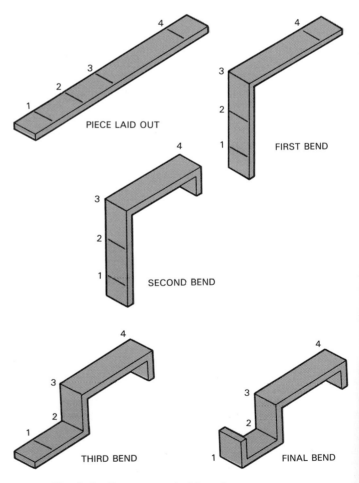

Fig. 8-4. Recommended bending sequence.

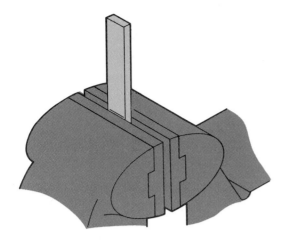

Fig. 8-5. Add one-half the thickness of the metal to the total length of the piece, for each bend. Place the section with this extra metal above the vise.

ACUTE ANGLES (angles less than 90 deg.) can also be made in the vise. However, this is done after the first bend has been made, Fig. 8-8.

OBTUSE ANGLES (angles more than 90 deg.) may be made using a MONKEY WRENCH (an adjustable wrench that resembles a pipe wrench, but has smooth jaws) as a bending tool, Fig. 8-9. It is also called a SPUD WRENCH.

Thin stock can be bent in a vise using two pieces of angle iron or wood to position the metal. Make the bend with a mallet, Fig. 8-10.

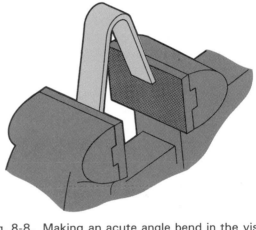

Fig. 8-8. Making an acute angle bend in the vise.

Fig. 8-6. Many bends can be made using a vise with a hammer.

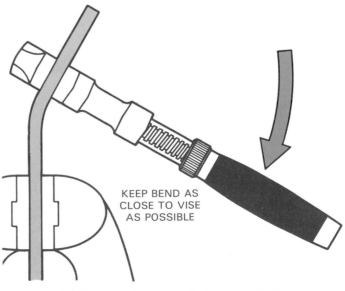

KEEP BEND AS CLOSE TO VISE AS POSSIBLE

Fig. 8-9. Making an obtuse angle bend with the monkey wrench.

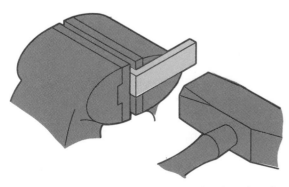

Fig. 8-7. Making a right angle bend using the vise.

Fig. 8-10. Thin stock can be bent by placing it between two pieces of angle iron and making the bend with a mallet.

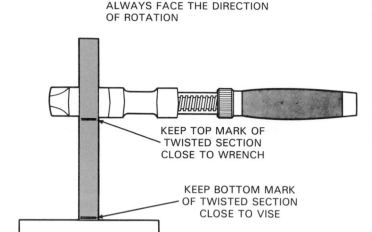

SAFETY NOTE:
THE MOVABLE JAW SHOULD ALWAYS FACE THE DIRECTION OF ROTATION

KEEP TOP MARK OF TWISTED SECTION CLOSE TO WRENCH

KEEP BOTTOM MARK OF TWISTED SECTION CLOSE TO VISE

VISE

Fig. 8-12. Making a twisted section. Place a monkey wrench at the top mark and make the required number of turns.

TWISTING METAL

Wrought metal may be twisted for increased strength and rigidity, for decorative purposes, or to alter the look of long flat sections. See Fig. 8-11.

Lay out the section to be twisted. Allow extra material because twisting shortens the metal slightly.

Short sections may be twisted in a vise. Place the metal in the vise. Place the bottom layout mark flush with the jaws. Place a monkey wrench at the top mark and make the required number of turns. See Fig. 8-12.

Long sections of metal often bend out of line when twisted. To avoid this problem, slide a section of snug fitting pipe over the portion to be twisted. When you have finished twisting, remove the pipe. Minor bends can be straightened with a mallet.

BENDING CIRCULAR SHAPES

Some wrought ironwork makes use of curved sections for decorative purposes, Fig. 8-13. The

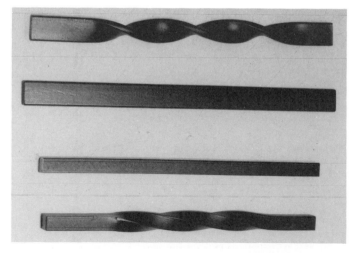

Fig. 8-11. Twisted sections have extra strength and rigidity. They are also more decorative than long, flat pieces.

Fig. 8-13. A wrought iron trivet that makes use of curved sections.

curves can be made over the anvil horn or with a BENDING JIG, Fig. 8-14. For best results, make a full size pattern of the proposed curve.

Stock length can be found by forming a piece of wire over the pattern. Straighten the wire and measure its length.

The SCROLL, Fig. 8-15, is frequently used. It is a curved section with a constantly expanding radius that looks like a loose clock spring.

The scroll is formed with the aid of a bending jig. One section is formed at a time, Fig. 8-16. Check the curve against the pattern during the forming operation to insure accuracy.

Often, a scroll is required on both ends of the piece, Fig. 8-17. One curve should blend smoothly into the other curve.

Scroll ends are often flared or decorated, Fig. 8-18. This is done before the scroll is formed.

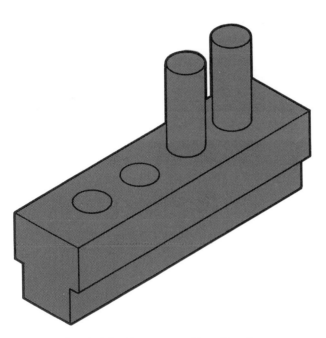

Fig. 8-14. One type of bending jig.

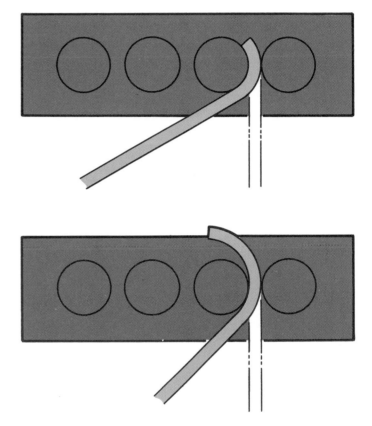

Fig. 8-16. Making a scroll on a bending jig.

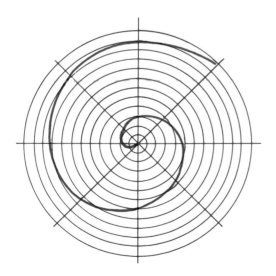

Fig. 8-15. Scrolls may be developed using this technique.

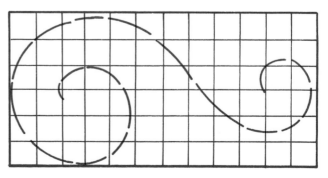

Fig. 8-17. A pattern for a double end scroll.

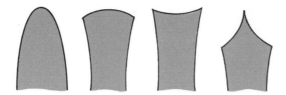

Fig. 8-18. Typical scroll ends.

Curves of a given radius can be formed using a vise and a piece of pipe or rod. The diameter of the pipe must equal the inside diameter of the required curve.

Clamp the metal in the vise as shown in Fig. 8-19. Pull the metal forward. As the curve takes shape, move the work farther in and around the pipe.

Curves can also be formed using a hammer and a rod clamped in the vise, Fig. 8-20.

BENDING WITH SPECIAL TOOLS

Bending can be done by using special machine. See Figs. 8-21, 8-22, and 8-23. The machines provide additional leverage for bending heavier metal. In addition, special attachments permit complex shapes to be formed easily.

The bending sequence must be planned BEFORE cutting the metal. Do not forget to make allowances for the bends.

Metal formed in a bending machine has some "springback" (the metal tries to return to its original shape). Problems caused by this can be avoided by using a bending form with a radius slightly smaller than the required radius.

Springback varies with metal thickness. It may be helpful to make a sample bend. Use scrap metal of the required size to determine the correct form to use.

Be sure to study the instruction book supplied by the manufacturer of the machine being used.

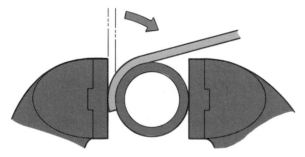

Fig. 8-19. Using a pipe held in a vise to form a scroll or curved section.

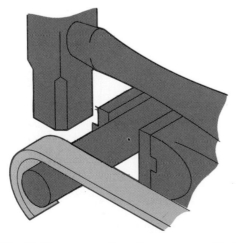

Fig. 8-20. Making a curved section over a rod held in a vise.

Fig. 8-21. The Di-Acro bender can be used to mechanically form rod and bar stock. (Di-Acro)

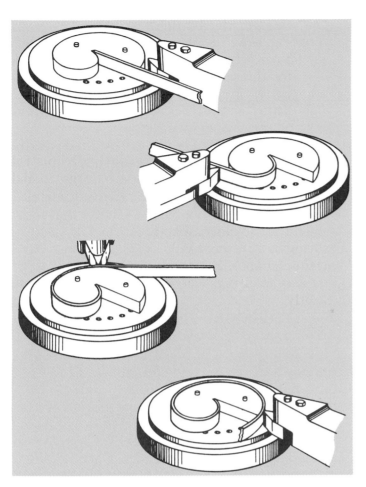

Fig. 8-22. Scrolls and other shapes of irregular radii can be formed with a bender. Use a collar having the same contour as the shape to be formed.

ASSEMBLY

Wrought ironwork may be assembled by riveting or welding.

FINISHING WROUGHT IRONWORK

Early wrought ironwork acquired a pleasing surface texture (finish) when the metal was forged to usable size. Today, with a large range of standard metal sizes available, it is no longer necessary to forge the material to size. In order to achieve the same surface look as forged metal, however, peening was developed. In PEENING, the work is struck over and over with the ball end of the hammer, Fig. 8-24. Today, peening is widely used. However, the "real" artisan still takes the time to forge the metal to size, Fig. 8-25.

Flat black lacquer or paint is applied for the final surface finish. It is easier to apply than the scorched linseed oil finish used by Colonial artisans.

Fig. 8-24. Close-up of a peened surface.

Fig. 8-23. The Metl-Former. (Swayne, Robinson and Co.)

Fig. 8-25. An attractive surface is made when metal is forged to size. It is not an artificial surface finish like a peened surface.

SAFETY

1. Remove all burrs and sharp edges from the metal before shaping it.
2. Wear your safety glasses when cutting, chiseling or grinding metal.
3. Have any cuts, bruises, or burns treated promptly.
4. Handle long sections of metal with extreme care so that persons working nearby will not be injured.
5. Keep your fingers clear of the moving parts of the bending machine.
6. Do not use a bending machine unless you know how to operate it safely. When in doubt, consult your instructor.
7. Do not use finishing materials near an open flame or in an area that is not properly ventilated. Store oily and solvent soaked rags in an approved closed container.

TEST YOUR KNOWLEDGE, Unit 8

Please do not write in the text. Place your answers on a separate sheet of paper.

1. The first wrought ironwork in America was made by _____.
2. Give two other names by which wrought ironwork is known.
3. Metal up to 1/4 in. thick can be bent _____. Heavier metal bends easier when it is _____.
4. There are many ways metal can be bent. List four of them.
5. Which two of the following points must be considered when making a bend in metal? (Choose all that apply.)
   a. Heavier metals must be used to allow extra metal for the bend.
   b. The length of the metal is reduced slightly by each bend and allowances must be made for this reduction.
   c. Bends should be made heated.
   d. Bends should be made with a hammer of the proper size.
   e. If several bends are to be made in the piece, the bending sequence must be planned very carefully.
6. What are some reasons for twisting metal?
7. The _____ is a curved section that looks like a loose clock spring.
8. What is springback?
9. How much will 25 feet of wrought iron cost if it is priced at 18.75 cents a foot?

TECHNICAL TERMS TO LEARN

acute angle
bending jig
blacksmith
carbon
circular
decorative
obtuse angle

ornamental ironwork
rod parter
scroll
sequence
springback
wrought

ACTIVITIES

1. Make a collection of illustrations showing wrought ironwork.
2. Secure catalogs from historic developments (Williamsburg, Cooperstown, Greenfield Village, Old Sturbridge Village, Mystic Seaport, etc.). Develop wrought iron projects that can be made in the school lab.
3. Demonstrate the proper and safe way to make bends on the Di-Acro bender.

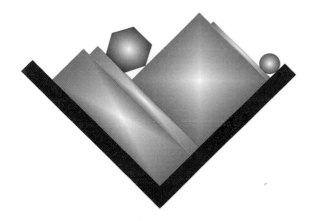

Unit 9
Fasteners

After studying this unit, you will be able to identify various types of fasteners. You will also be able to explain several uses for each type of fastener. And you will be able to demonstrate the safe use of epoxy adhesives.

FASTENERS (nuts, bolts, screws, rivets, etc.) may be thought of as small clamps that hold manufactured products together. They are made in many shapes, sizes, and styles, Fig. 9-1.

RIVETS

RIVETS are used for permanent assemblies. They cannot be removed without damaging the product. Rivets are made from many materials: soft iron, aluminum, copper, and brass, to name a few. And rivets come in many shapes and sizes, Fig. 9-2.

Fig. 9-1. Common fasteners found in the metals lab.

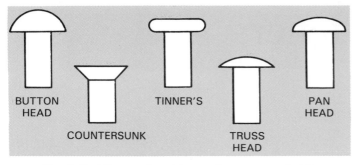

Fig. 9-2. Rivet styles. Many others are available.

Fig. 9-4. Rivet set.

SETTING A SOLID RIVET

Solid rivets can be set by hand or machine. The following steps should be followed when hand setting a rivet:

1. Select the type and length of rivet you will use. To find the proper rivet length for your work, add the total thickness of the material being joined to one and one-half times the rivet diameter. Wrought iron projects should be joined with soft iron rivets, aluminum projects with aluminum rivets, and so on.
2. Locate the point to be riveted. Drill or punch a hole of the proper size for the rivet. Countersink the hole if a countersunk flat head rivet is to be used. Remove all burrs and sharp edges.
3. Seat the rivet and draw the pieces together, Fig. 9-3A.
4. Flatten the rivet, Fig. 9-3B.
5. Form the rivet head, Fig. 9-3C, with a RIVET SET, Fig. 9-4.

If a rivet must be set flush with both work surfaces, each side of the hole must be countersunk. Place the work on a steel plate and flatten the rivet shank to fill the countersunk holes.

Button or round head rivets are held in the cup of a rivet set, while the head is formed on the shank.

BLIND RIVETS

BLIND RIVETS are used when joints are reachable from one side only. Special tools are used to put the rivets in place.

One type of blind rivet, and the tool needed to set it, are shown in Fig. 9-5. This set can be found in any hardware store. This rivet is set as shown in Fig. 9-6.

THREADED FASTENERS

Threaded fasteners permit the work to be taken apart and put back together without damage.

MACHINE SCREWS

MACHINE SCREWS, Fig. 9-7, are widely used in the metals lab. Many head styles are available with slotted or recessed heads.

Machine screws clamp parts together by screwing them into tapped holes. Square or hexagonal (six-sided) nuts may be used with the screws.

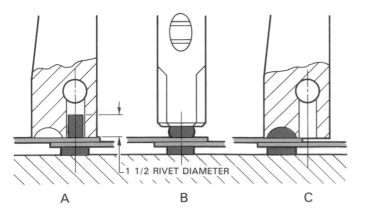

Fig. 9-3. Setting a rivet. A—Drawing the pieces together. B—Flattening the rivet. C—Forming the rivet head with a rivet set.

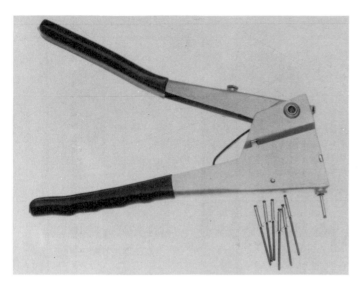

Fig. 9-5. Tool used to set one type of blind rivet.

MACHINE BOLTS

MACHINE BOLTS, Fig. 9-8, are used to assemble products that do not need close tolerance fasteners.

CAP SCREWS

CAP SCREWS, Fig. 9-9, are much like machine bolts. However, they are made more exactly than machine bolts. And, they are used in projects that require a higher quality fastener. Cap screws are available in a wide variety of head styles.

Nuts may be used with cap screws. However, common use of cap screws involves passing them through a CLEARANCE HOLE (a hole slightly larger than the screw) in one of the pieces. Then they are screwed into a threaded hole in the other piece, Fig. 9-10.

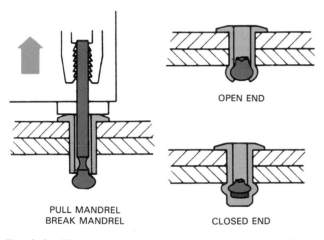

OPEN END

PULL MANDREL
BREAK MANDREL

CLOSED END

Fig. 9-6. This is how the blind rivet shown in Fig. 9-5 is set in place.

Fig. 9-8. Machine bolts.

Fig. 9-7. Three types of machine screw heads.

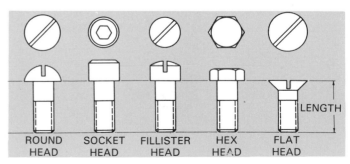

ROUND HEAD   SOCKET HEAD   FILLISTER HEAD   HEX HEAD   FLAT HEAD

LENGTH

Fig. 9-9. Cap screws.

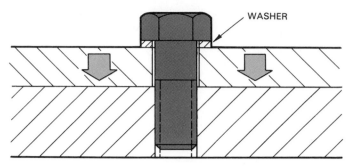

Fig. 9-10. This is how a cap screw joins work.

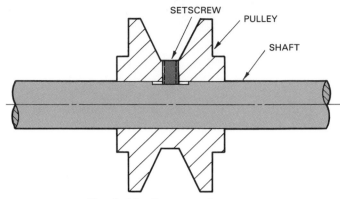

Fig. 9-12. Setscrew in use.

STUD BOLTS

STUD BOLTS, Fig. 9-11, are threaded at both ends. One end is screwed into a tapped hole, the piece to be clamped is fitted into place over the stud. A nut is screwed on the other end of the bolt, to clamp the two pieces together.

SETSCREWS

SETSCREWS, Fig. 9-12, have several uses. They are used to prevent pulleys from slipping on a shaft. They are used to hold collars in place on a shaft. And they are used to hold shafts in place on assemblies. Many types of setscrews are made, Fig. 9-13.

NUTS

NUTS may be standard square or hexagonal in shape. They are used with bolts having the same shape heads. They are also made for decorative and special applications, Fig. 9-14.

WASHERS

WASHERS permit a bolt or nut to be tightened without damage to the work surface.

STANDARD and LOCK WASHERS, Fig. 9-15, are commonly found in the metals lab.

KEYS

Keys prevent gears and pulleys from moving on shafts. Several key styles are shown in Fig. 9-16.

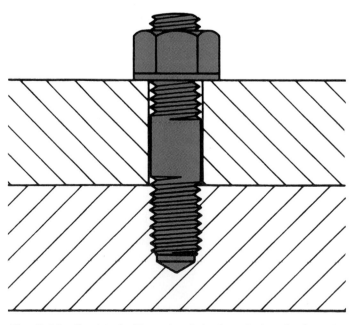

Fig. 9-11. Stud bolt. Note that it is threaded on both ends.

Fig. 9-13. Typical setscrews. These are only a few of the many types made.

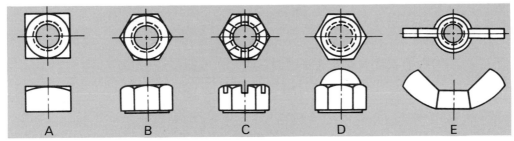

Fig. 9-14. Commonly used nuts. A—Square nut. B—Hex nut. C—Slotted nut. D—Acorn or cap nut. E—Wing nut.

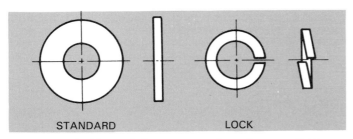

Fig. 9-15. Two of the many steel washers available.

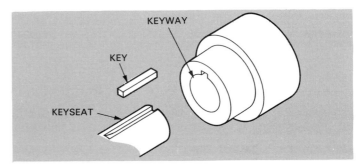

Fig. 9-17. Key, keyway, and keyseat.

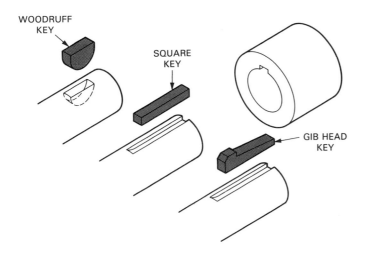

Fig. 9-16. Types of keys.

Fig. 9-18. Many types of adhesives are available that will bond metal to metal and metal to other materials. They are safe to use, if reasonable precautions are taken and manufacturer's instructions are strictly followed.

One-half of the key fits into a KEYSEAT on the shaft. The rest of the key fits into a KEYWAY in the hub of the gear or pulley. See Fig. 9-17.

EPOXY ADHESIVES

There are many adhesives used in metalworking. Many can be bought at any good hardware store, Fig. 9-18.

When using adhesives in metalworking, it is very important that the surfaces to be joined are cleaned correctly. Clean them with a fine abrasive paper. Then wipe with a solvent to remove all oil, grease, and dirt.

Before using an adhesive, carefully read the manufacturer's instructions on the container AND

FOLLOW THEM. If an adhesive must be mixed, blend the parts EXACTLY as specified.

CAUTION: ALL ADHESIVES CAN IMPAIR, OR CAUSE LOSS OF, SIGHT. SKIN IRRITATIONS MAY ALSO RESULT IF SKIN COMES IN CONTACT WITH SOME TYPES OF ADHESIVES. WEAR SAFETY GLASSES. KEEP ADHESIVES AWAY FROM YOUR EYES. WEAR DISPOSABLE PLASTIC GLOVES WHEN PREPARING AND USING ADHESIVES. WASH YOUR HANDS THOROUGHLY AFTER USING AN ADHESIVE. READ AND STRICTLY FOLLOW ALL MANUFACTURER'S INSTRUCTIONS BEFORE USING ANY ADHESIVE.

Adhesives are a good way to join metal to metal, or metal to other materials. They are safe to use if the manufacturer's instructions are closely followed, and safety precautions taken.

TEST YOUR KNOWLEDGE, Unit 9

Please do not write in the text. Place your answer on another sheet of paper.

1. Permanent assemblies are made with _____.
2. What can be done with threaded fasteners that cannot be done with rivets?
3. Which of the following are fasteners?
   a. Bolts.
   b. Nuts.
   c. Screws.
   d. All of the above.
4. What is similar to a machine bolt, but made to more exacting standards?

MATCHING QUESTIONS: Match each of the following terms with their correct definitions.

    a. Washer.   d. Stud bolt.
    b. Key.   e. Nut.
    c. Setscrew.

5. ____ Has several uses, one of which is to prevent a pulley from slipping on a shaft.
6. ____ Used with bolt that has the same shape head.
7. ____ Permits a bolt or nut to be tightened without damage to work surface.
8. ____ Is threaded on both ends.
9. ____ Prevents a gear or pulley from moving on a shaft.

TECHNICAL TERMS TO LEARN

acorn nut	lock washer
adhesive	machine bolt
blind rivet	machine screw
bolt	nut
cap screw	rivet
clearance hole	screws
epoxy	set screw
fasteners	solid rivet
key	washer
keyseat	wing nut
keyway	

ACTIVITIES

1. Prepare a collection of fasteners. Mount them on a display panel.
2. Research the manufacturing process used for threaded fasteners.
3. Obtain samples of several fasteners in common use.

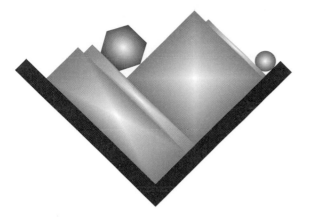

Unit 10
Sheet Metal

After studying this unit, you will be able to use the different methods of pattern development. You will also be able to cut and bend sheet metal using a number of tools. You will be able to demonstrate soldering procedure. And, you will be able to work safely with sheet metal and when soldering.

Great amounts of sheet metal are used in the manufacturer of metal objects such as aircraft, automobiles, furniture, and household appliances.

See Fig. 10-1. Sheet metal is also used by the building trades (air conditioning, heating, roofing, and prefabricated structures), Fig. 10-2.

Products made from sheet metal are given three-dimensional shape and rigidity. This is done by bending and forming the metal sheet into the required shape.

In many manufacturing jobs metal is cut to shape using a pattern as a guide. A PATTERN is a full-size drawing of the surfaces of the object.

Fig. 10-1. The automotive industry uses millions of tons of various sheet metals. The sheet metal used for body panels gains additional strength and rigidity as it is formed into a three-dimensional shape. The vehicle shown here is a concept model used to study the best way to manufacture production models. (Nissan)

Fig. 10-2. A building made from sheet metal.   (Stran-Steel)

It is stretched out as a single surface, Fig. 10-3. The pattern drawing is often called a STRETCH-OUT. It is made using a form of drafting called PATTERN DEVELOPMENT.

PATTERN DEVELOPMENT

There are two basic types of pattern development: parallel line development and radial line development.

PARALLEL LINE DEVELOPMENT is a technique used to make patterns of prisms and cylinders, Fig. 10-4.

RADIAL LINE DEVELOPMENT is used to make patterns of regular tapering forms such as cones and pryamids, Fig. 10-5.

Patterns developed from more complex geometric shapes are drawn using variations and combinations of basic pattern development techniques.

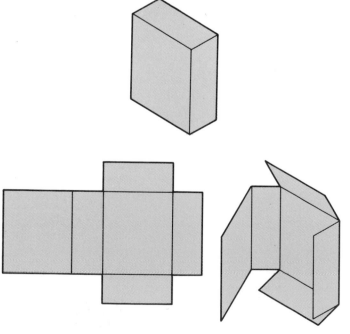

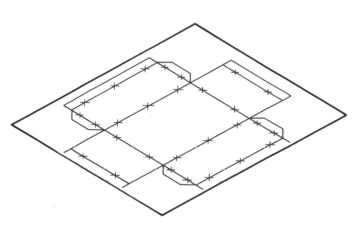

Fig. 10-3. A sheet metal layout or pattern. Folds are to be made as shown by the letter ''X.''

Fig. 10-4. A pattern made by parallel line development.

110

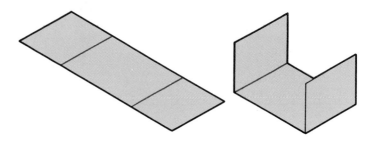

Fig. 10-6. A heavy solid line indicates a sharp fold.

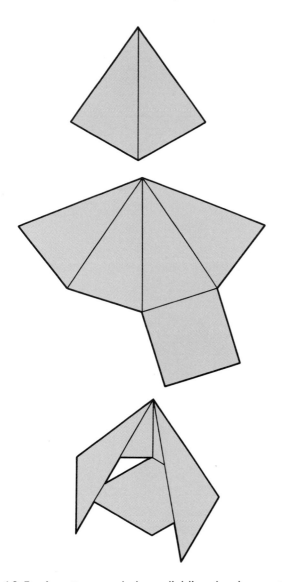

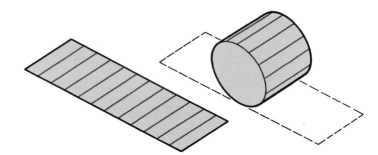

Fig. 10-7. Construction (light) lines or centerlines indicate a curved or circular surface.

Fig. 10-5. A pattern made by radial line development.

Lines used in pattern development give additional meaning to the drawing. A heavy solid line (visible object line) indicates a sharp fold or bend, Fig. 10-6. Curved surfaces are shown on the pattern by construction lines or centerlines, Fig. 10-7.

A word of caution. Be sure to allow enough material to make the various joints, hems, and seams. These are needed to join metal together and to add rigidity to thin metal sheets. This will be explained later in the unit.

Patterns may be developed on paper and transferred to the metal, or they can be drawn directly onto the metal. Carefully plan layouts made directly on metal so there will be a minimum of waste, Fig. 10-8.

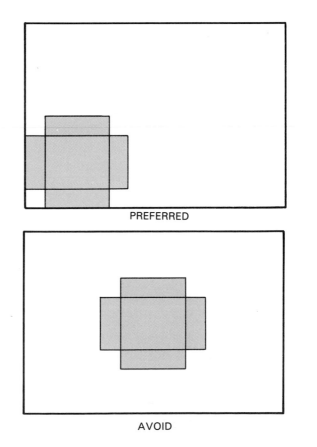

Fig. 10-8. Locate the pattern on the metal so there will be a minimum of waste.

111

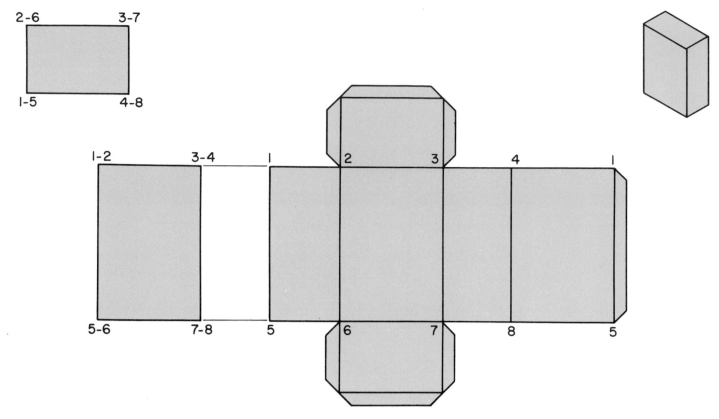

Fig. 10-9. Developing a pattern for a rectangular or square object.

PATTERN DEVELOPMENT OF A RECTANGULAR OBJECT

To develop rectangular or square objects, see Fig. 10-9. The steps to follow for this method are explained in the next section.

1. Draw front and top views.
2. The height of the pattern is the same as the height of the front view. Project light lines from the top and bottom of the front view. Number as shown.
3. Measure about 1 in. from the front view. Then draw a vertical line between the extended lines to locate 1 - 5.
4. Set the compass or divider from 1 to 2 on the top view. Transfer this distance to the extended lines to locate 2 - 6. Locate the other distances in the same manner.
5. Using 2 - 3 and 6 - 7 as one side draw the top and bottom.
6. Allow about 1/4 in. for seams. (This amount may vary, depending on the size of the object). Go over all outlines and folds with heavy lines (object lines).

PATTERN DEVELOPMENT OF CYLINDRICAL OBJECTS

Cylindrical objects, Fig. 10-10, are developed using the following method.

1. Draw the front and top views. Divide the top view into twelve equal parts and number as shown.
2. The height of the pattern is the same as the height of the front view. Project light lines from the top and bottom of the front view.
3. Measure about 1 in. from the front view. Draw a vertical line between the extended lines to locate 1.
4. Set the compass or divider from 1 to 2 on the top view. Transfer this distance to the extended lines to locate points 1, 2, 3, etc., as shown. Draw light vertical lines at each point.
5. Draw the top and bottom tangent to the extended lines.
6. Allow material for the seams. Go over all outlines with heavy lines (object lines). The lines that depict curves or circular lines may be drawn in color, or left as light lines.

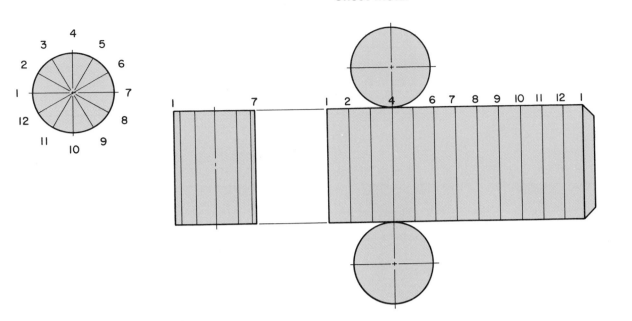

Fig. 10-10. Pattern development for a cylinder.

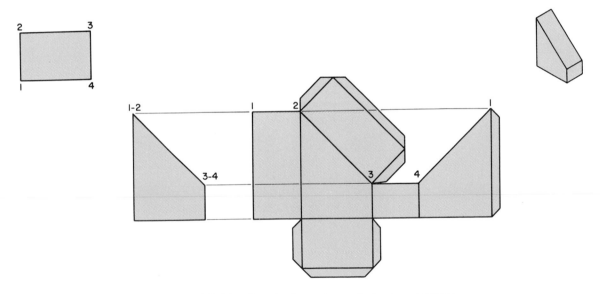

Fig. 10-11. Developing a truncated prism pattern.

PATTERN DEVELOPMENT OF A TRUNCATED PRISM

A truncated prism is cut off at an angle to its base, Fig. 10-11. Development of this shape is made as follows.

1. Draw front and top views. Number views as shown.
2. Proceed as in the previous examples of parallel line development.
3. Mark off and number the folding points. Project point 1 on the front view, to line 1 of the stretchout. Repeat with points 2, 3, and 4.
4. Connect the points 1 to 2, 2 to 3, 3 to 4, and 4 to 1.
5. Draw the top and bottom (if needed) in position.
6. Allow material for seams. Go over the outline and folds with visible object lines.

113

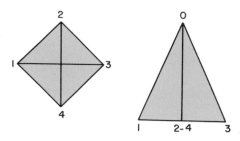
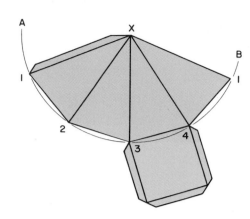
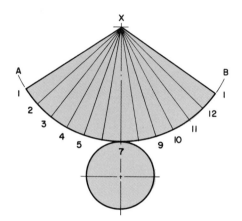
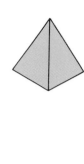

Fig. 10-12. Developing a pattern for a pyramid.

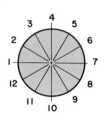
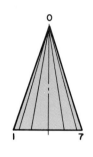

Fig. 10-13. Developing a pattern for a cone.

PATTERN DEVELOPMENT OF A PYRAMID

The pattern for a pyramid, Fig. 10-12, is developed as follows.

1. Draw front and top. Number as shown.
2. Locate centerline X of the stretchout.
3. Set the compass to a radius equal to 0 - 1 on the front view. Using X of the stretchout as the center, draw arc AB.
4. Draw a vertical line through centerline X and arc AB.
5. Set the compass from 1 to 2 on the top view. Then, using the place where the vertical line intersects the arc as the starting point, step off two divisions on each side of the line (points 1-2-3-4-1 on the stretchout).
6. Connect the points and draw the bottom in place (if needed). Go over the outline and folds with object lines.

PATTERN DEVELOPMENT OF A CONE

Developing the pattern for a cone, Fig. 10-13, is done as follows.

1. Draw front and top views. Divide the top view into twelve equal parts and number as shown.
2. Locate centerline X of the stretchout.
3. Set the compass from 0 to 1 on the front view. Using X as the centerline, draw arc AB.
4. Draw vertical construction line through centerline X and the arc.
5. Set the compass from 1 to 2 on the top view. Using the place where the vertical line intersects the arc as the starting point, step off six divisions on both sides of the line (points 1-2-3-4 etc. on the pattern).
6. Go over the outline with object lines. The lines that represent the curved portion are drawn in color, or are left as construction lines.

WORKING WITH SHEET METAL

METALS USED IN SHEET METAL

TIN PLATE (a mild steel with a tin coating), GALVANIZED STEEL (a mild steel with a zinc coating), and COLD FINISHED STEEL SHEET are most often used in sheet metal work in the school shop.

These metals are available in standard thicknesses and sizes. The thickness can be measured with a micrometer or SHEET METAL GAUGE. See Fig. 10-14.

CUTTING SHEET METAL

SNIPS are used a great deal for cutting metal sheets. They are made in a number of sizes and styles, Fig. 10-15.

Large sheet metal sections are cut on SQUARING SHEARS, Fig. 10-16. When a quantity of sheet metal must be cut on the job, smooth, clean cuts can be made with an electric tool, Fig. 10-17.

Do not cut wire, band iron, or steel rod with snips or squaring shears. This will damage the cutting edges.

KEEP YOUR HANDS CLEAR OF THE BLADE AND YOUR FOOT FROM BENEATH THE FOOT PETAL WHEN USING SQUARING SHEARS.

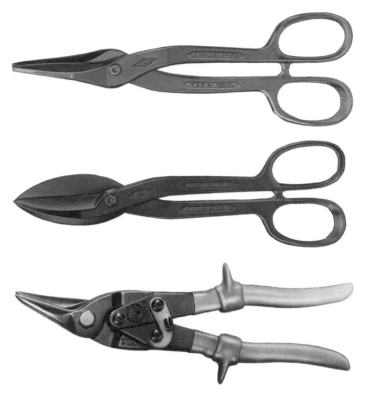

Fig. 10-15. Typical snips used to cut sheet metal. Top. Circular pattern snips. Center. Straight pattern snips. Bottom. Aviation snips, left cut.   (Diamond Tool Co.)

Fig. 10-14. Measuring sheet metal thickness with a sheet metal gauge.

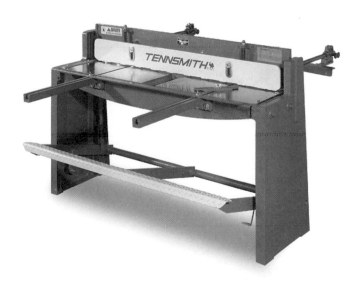

Fig. 10-16. Foot powered squaring shears. (Tennsmith, Inc.)

MAKING SMALL DIAMETER HOLES IN SHEET METAL

In addition to drilling, small holes can be made in sheet metal with a HOLLOW PUNCH, Fig. 10-18, and a SOLID PUNCH, Fig. 10-19.

BENDING SHEET METAL

Sheet metal is often bent into three-dimensional shapes. The bending makes thin metal rigid.

Some hand and machine bending techniques are described in the following paragraphs.

BENDING METAL BY HAND

Sheet metal may be bent by fitting it between two blocks of hardwood or pieces of angle iron, Fig. 10-20. Press the metal over by hand. Square up the bend with a wooden mallet.

USE EXTREME CARE WHEN HANDLING SHEET METAL TO AVOID PAINFUL CUTS FROM THE SHARP EDGES OF THE MATERIAL.

Bending can also be done using a STAKE, Fig. 10-21.

Fig. 10-18. Using a hollow punch to cut holes in sheet metal.

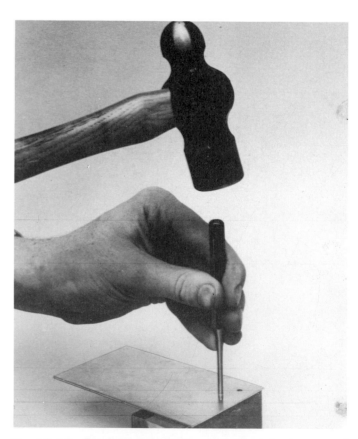

Fig. 10-19. Small diameter holes can be made in sheet metal with a solid punch. Use a bar of lead or end-grain wood block to support the work while making the hole.

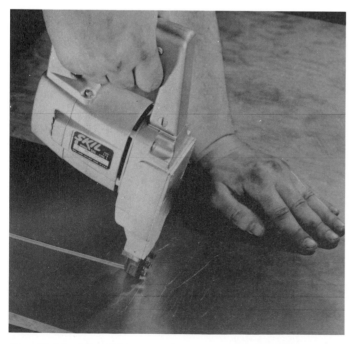

Fig. 10-17. The nibbler cuts sheet metal cleanly and smoothly, with neither cut edge bending. (Skil Corp.)

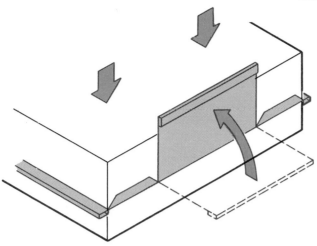

Fig. 10-20. Using blocks of wood to make a bend in sheet metal.

Fig. 10-22. Box and pan brake. (Tennsmith, Inc.)

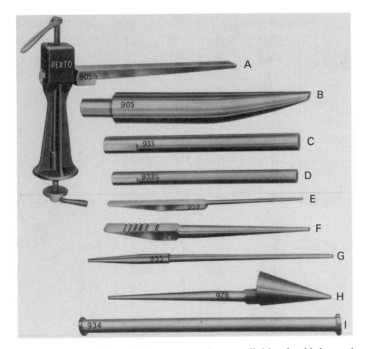

Fig. 10-21. A few of the many stakes available. A—Universal stake holder fitted with the rectangular end of a beakhorn stake. B—Beakhorn stake. C—Conductor stake, large end. D—Conductor stake, small end. E—Needle case stake. F—Creasing stake with horn. G—Candle mould stake. H—Blowhorn stake. I—Double seaming stake.   (Pexto)

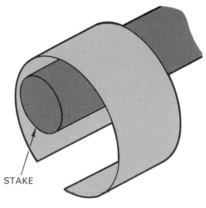

STAKE

Fig. 10-23. Circular shapes can be formed by hand, over a stake or metal rod.

BENDING SHEET METAL ON BOX AND PAN BRAKE

The BOX AND PAN BRAKE, Fig. 10-22, makes accurate bends mechanically. Its upper jaw is made of a number of blocks that are of different widths. These can be positioned or removed to permit all four sides of a box to be formed.

MAKING CIRCULAR AND CONICAL SHAPES

Circular and conical shapes can be made by hand, over a stake, Fig. 10-23. However, it is difficult to make the curves smooth and accurate.

Most cylindrical shapes can be formed quickly and accurately on a SLIP ROLL FORMING MACHINE, Fig. 10-24. The rolls can be adjusted to fit different thicknesses of metal and to form the desired curve, See Fig. 10-25.

COMMON SHEET METAL SEAMS, HEMS, AND EDGES

Various SEAMS are used to join sheet metal sections, Fig. 10-26. Sheet metal seams are often finished by soldering.

HEMS, Fig. 10-27, are used to straighten lips of sheet metal objects. These are made in standard fractional sizes — 3/16, 1/4, etc.

The WIRE EDGE, Fig. 10-28, gives additional strength and rigidity to sheet metal edges.

FOLDS are basic to the making of seams, edges, and hems. Folds are made on a BAR FOLDER, Fig. 10-29.

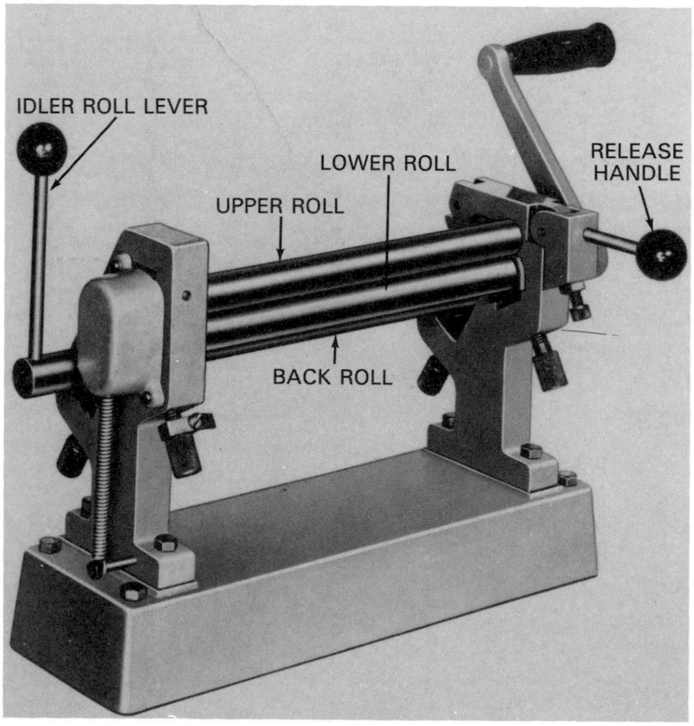

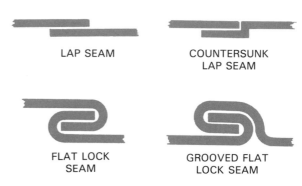

LAP SEAM

COUNTERSUNK LAP SEAM

FLAT LOCK SEAM

GROOVED FLAT LOCK SEAM

Fig. 10-26. Typical sheet metal seams.

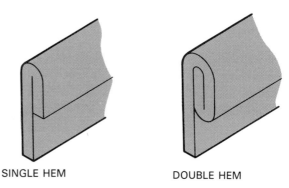

SINGLE HEM

DOUBLE HEM

Fig. 10-27. Hems are made on sheet metal to provide added strength to the metal.

Fig. 10-24. Slip roll forming machine.

IDLER ROLL LEVER

LOWER ROLL

RELEASE HANDLE

UPPER ROLL

BACK ROLL

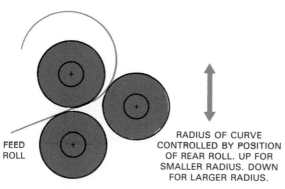

FEED ROLL

RADIUS OF CURVE CONTROLLED BY POSITION OF REAR ROLL. UP FOR SMALLER RADIUS. DOWN FOR LARGER RADIUS.

Fig. 10-25. The rolls can be adjusted to fit sheet metal of different thickness and to form various curvatures.

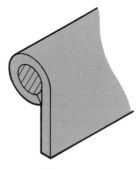

Fig. 10-28. Wire edge.

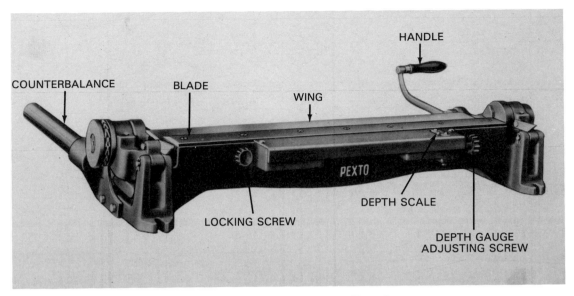

Fig. 10-29. Bar folder. (Pexto)

The width of the folded edge is set on the DEPTH GAUGE of the machine. The sharpness of the folded edge can vary. It will be sharp for a hem or seam, and rounded for a wire edge. This sharpness is determined by the position of the WING, Fig. 10-30. The machine can also be adjusted to make bends of 45 deg. and 90 deg.

Folds for seams are made on a bar folder. After the sections are fitted together, the seams may be locked with a HAND GROOVER, Fig. 10-31.

A COMBINATION ROTARY MACHINE, Fig. 10-32, can be used to make a wire edge. See Fig. 10-33. The machine can be fitted with a variety of rolls to perform operations such as CRIMPING, Fig. 10-34.

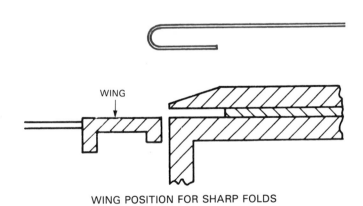

WING POSITION FOR SHARP FOLDS

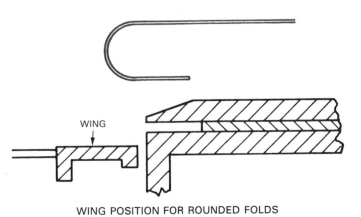

WING POSITION FOR ROUNDED FOLDS

Fig. 10-30. Setting the wing of the bar folder to make different size folds.

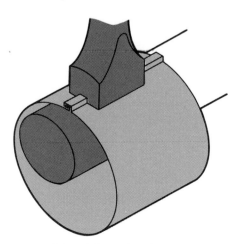

Fig. 10-31. Using a hand groover.

SHEET METAL SAFETY

1. Treat cuts and bruises immediately, no matter how minor.
2. Remove all burrs formed during the cutting operation, before doing further work on the metal.
3. Clean the work area with a brush. NEVER brush metal with your hands.
4. Use sharp tools.
5. Keep your hands clear of the blade on the squaring shears.
6. Do not run your hand over metal that has just been cut or drilled. Painful cuts from the burrs may result.
7. Place scrap pieces of sheet metal in the scrap box.
8. Do not use tools that are not in first-class condition—hammer heads should be tight, files should have handles, machines should have guards.
9. Always wear goggles when working in the shop.

SOLDERING

To SOLDER properly the steps outlined here should be followed.

1. A 50-50 solder alloy (50 percent tin-50 percent lead) is most commonly used.
2. The correct FLUX must be applied. A NON-CORROSIVE flux such as RESIN works best on tin plate and brass.

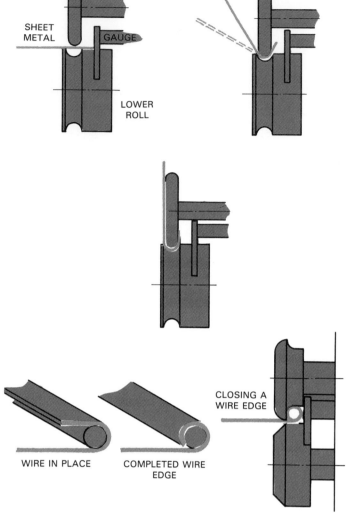

Fig. 10-33. Top. Turning a wire edge with a rotary machine. Bottom. Closing the wire edge.

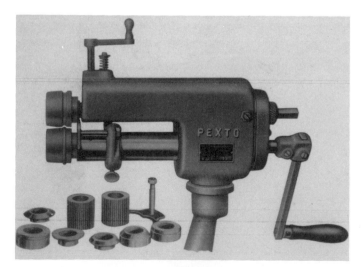

Fig. 10-32. Combination rotary machine with extra forming rolls.

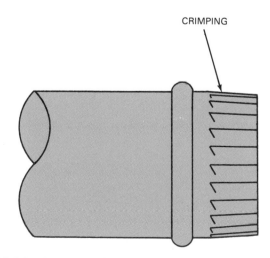

Fig. 10-34. Crimping makes one end of the metal pipe smaller. It will fit easily into the end of another pipe of the same diameter.

3. The SOLDERING COPPER must furnish sufficient heat for the job.
4. The surfaces being soldered must be clean.

SOLDERING DEVICES

Heat for soldering is often applied with an ELECTRIC SOLDERING COPPER, Fig. 10-35. Also used is a SOLID SOLDERING COPPER, Fig. 10-36, that is heated in a SOLDERING FURNACE.

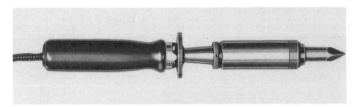

Fig. 10-35. Electric soldering copper.
(American Electrical Heater Co.)

Fig. 10-36. Solid soldering copper.

Fig. 10-37. Properly tinned soldering copper.

A soldering copper must be TINNED, Fig. 10-37, before it will solder properly. TINNING (coating the tip of the soldering copper with solder) is done by first cleaning the copper tip with a file. Heat the tip until it will melt solder freely. Rub it on a sal ammoniac block on which a few drops of solder have been melted, Fig. 10-38. This will clean the tip and cause the solder to adhere. Remove excess solder by rubbing the tip over a clean cloth.

SOLDERING SHEET METAL

Clean the area to be soldered and apply flux. Place the pieces on a heat-proof surface.

Heat and tin the soldering copper. Hold the seam together with the tang of the file or stick of wood, Fig. 10-39. Tack it with small amounts of solder. Apply the solder directly in front of the soldering copper tip rather than on it.

Keep the seam pressed together with a file tang. Lay the soldering copper FLAT on the work. Start moving the copper slowly toward the far end of the joint, as the solder melts and begins to flow.

Clean the soldered seam with hot water to remove all traces of flux.

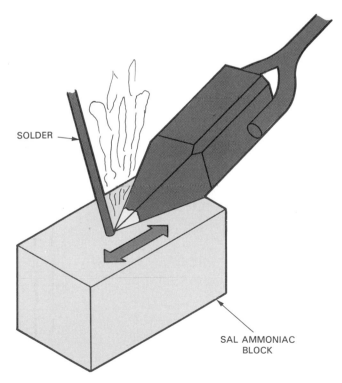

SOLDER

SAL AMMONIAC BLOCK

Fig. 10-38. Tinning soldering copper on a sal ammoniac block.

121

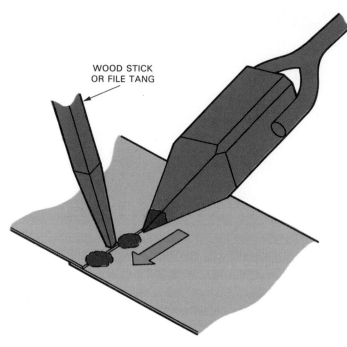

WOOD STICK OR FILE TANG

Fig. 10-39. Hold sheet metal pieces together with a wooden stick or file tang until the molten solder "freezes."

SOLDERING SAFETY

1. Wear safety glasses when soldering.
2. Wash your hands carefully after soldering to remove all traces of flux.
3. Have burns treated promptly.
4. Avoid touching just-soldered joints.
5. Use care when storing the soldering copper after use. Improper or careless storage can result in serious burns or a fire.

TEST YOUR KNOWLEDGE, Unit 10

Please do not write in the text. Place your answers on a separate sheet of paper.

1. Sheet metal is cut to shape using a _____ as a guide.
2. Name the two basic types of pattern development.
3. A heavy, solid line on a pattern indicates a:
   a. Fold or bend.
   b. Flat surface.
   c. Curve.
   d. Cut.

4. What three steels are most often used for sheet metal work in the school shop?

MATCHING QUESTIONS: Match each of the following terms with their correct definitions.

a. Hollow punch.   c. Squaring shears.
b. Sheet metal gauge.   d. Snips.

5. ____ Used to cut large sheet metal sections.
6. ____ Used to make small holes in sheet metal.
7. ____ Used to cut sheet metal.
8. ____ Used to measure sheet metal thickness.
9. What is the purpose of a hem, a seam, and a fold?
10. A soldering copper must be _____ before it will solder properly.

TECHNICAL TERMS TO LEARN

bar folder	radial
bending	rectangular
box and pan brake	rigidity
cone	rotary machine
crimping	sal ammoniac
curved	seam
cylindrical	shape
flux	sheet metal
forming	sheet metal gauge
hand groover	ships
hem	squaring shears
parallel	stake
pattern	stretchout
prism	three-dimensional
punch	tinning
pyramid	truncated

ACTIVITIES

1. Collect illustrations that show objects made from sheet metal. Prepare a bulletin board using the pictures.
2. Use an empty cereal box to demonstrate how a pattern is used.
3. Prepare posters on sheet metal safety.
4. Demonstrate the correct way to use tin snips.
5. Demonstrate the correct way to use the box and pan brake.
6. Prepare examples of hems, folds, and a wire edge.
7. Demonstrate how to tin a soldering copper.
8. Prepare examples of properly soldered seams.

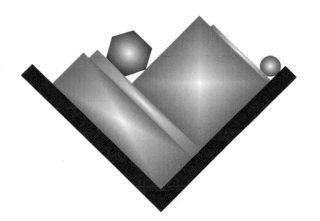

Unit 11
Art Metal

After studying this unit, you will be able to demonstrate the annealing and the pickling processes. You will also be able to form bowls and trays using the beating down and raising methods. In addition, you will be able to explain how to hard solder joints and demonstrate how to safely work with art metal.

Hand skills can easily be developed in art metal since most of the work is done by hand, Fig. 11-1. Machines are seldom used.

There are four classifications of art metal. They are hollow ware, Fig. 11-2; flatware, Fig. 11-3; strip work, Fig. 11-4, and jewelry making, Fig. 11-5.

Fig. 11-1. Artisans in the James Geddy Silversmith Shop. They use 18th century methods to pound out colonial design articles from silver bars. (Colonial Williamsburg)

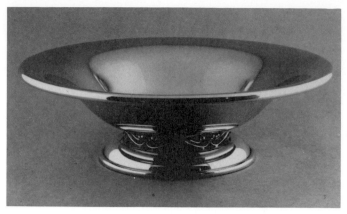

Fig. 11-2. This beautiful bowl is an example of hollow ware. (Shirley Pewter Shop, Williamsburg, VA)

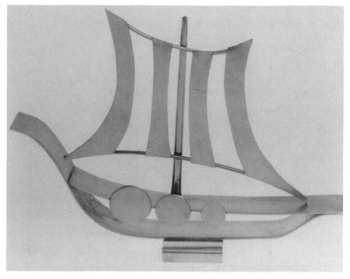

Fig. 11-4. Strip work model of early Norse ship.

Fig. 11-3. This salad server set is typical of flatware work.

Fig. 11-5. Jewelry making is one type of art metalwork.

ANNEALING AND PICKLING METAL

As metal is worked, it becomes hard and brittle. Therefore, it must be ANNEALED (softened by heating) from time to time. Then further shaping and forming can be done. Without annealing, the metal may crack.

PICKLING is closely related to the annealing process. Annealing causes an oxide (a coating like rust) to form on the metal. The oxide must be removed or it will mar the surface of the metal when any other work is done on it.

Abrasives can be used to remove this oxide. However, it is a time-consuming job. A simpler method is to heat the piece and plunge it into a dilute solution of sulphuric acid. HAVE YOUR INSTRUCTOR DEMONSTRATE THIS OPERATION FOR YOU. WHEN YOU DO IT YOURSELF, WEAR A FACE SHIELD.

Anneal copper and sterling silver by heating to a dull red. Quench the hot metal in water or pickling solution.

Brass, bronze, and German silver are heated to a dull red. Then they are allowed to cool slightly before plunging into water or pickling solution.

The metal may be heated with a torch or soldering furnace.

CAUTION: ASK YOUR INSTRUCTOR TO MIX THE PICKLING SOLUTION. THE ACID IT CONTAINS IS DANGEROUS. IT CAN CAUSE SERIOUS BURNS IF NOT PROPERLY HANDLED.

CUTTING AND PIERCING METAL

The metals used in art metal can be cut using hacksaws, snips, squaring shears, etc. However, these tools are not useful for cutting the decorative, internal designs often used in art metal. These cuts should be done with a JEWELER'S SAW, Fig. 11-6. The operation is known as PIERCING. Use a solid support when sawing, Fig. 11-7. The openings made with a jeweler's saw are cleaned with JEWELER'S FILES, Fig. 11-8.

HAMMERS AND MALLETS

Many different HAMMERS, Fig. 11-9, are used to shape and form metal. They are made of steel. The hammer faces should be kept polished.

MALLETS, Fig. 11-10, are made of hardwood, plastic, rawhide, and rubber. They are used to form softer metals. After forming with a mallet, the piece is finished with a hammer.

Fig. 11-8. Jeweler's files.

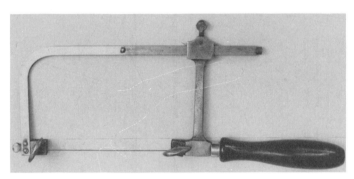

Fig. 11-6. Jeweler's saw.

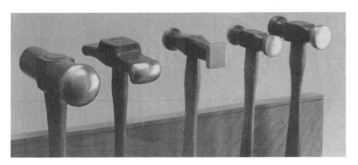

Fig. 11-9. Common types of art metal hammers.

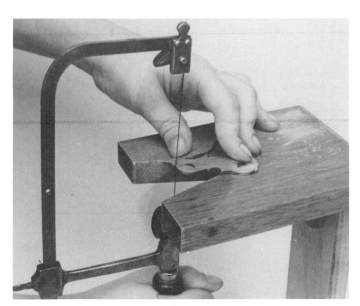

Fig. 11-7. Support work solidly when cutting with a jeweler's saw.

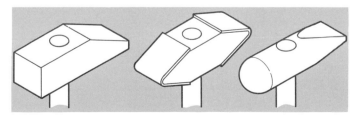

Fig. 11-10. Art metal mallets. Left. Hardwood forming mallet. Center. Leather faced forming mallet. Right. Round end forming mallet.

FORMING BY BEATING DOWN

Shallow trays and plates are often formed by beating down portions of the metal. They are beat over either a hardwood stake or form.

Both methods for beating down require a pattern, Fig. 11-11. Guide lines indicate the part to be beat down.

Hold the metal over the stake with the guide lines about 1/8 in. from the edge of the stake, Fig. 11-12. Use a forming hammer to beat down the metal. Rotate the blank slightly after each blow until the correct depth is reached. A template may be used to check the progress of the forming operation.

A wooden form block, Fig. 11-13, may also be used to shape the piece.

Position the metal over the block. Fasten it at the corners with small nails or screws. Use a mallet when working on soft metals (pewter and soft aluminum). For use on other metals, select a hammer with about the same contour (shape) as the form block sides.

Start working at the outer edges and slowly work toward the center.

The work will have a tendency to warp slightly and must be flattened. This is done on a flat, clean surface. Use a wooden block and a mallet to hammer the surface flat, Fig. 11-14.

Trim, planish (smooth by hammering lightly), and polish the piece.

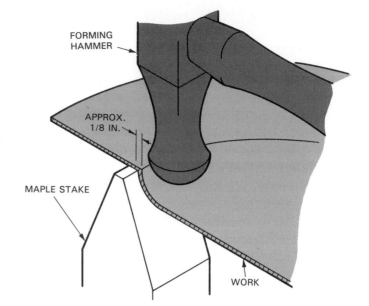

Fig. 11-12. Hold the metal on the stake.

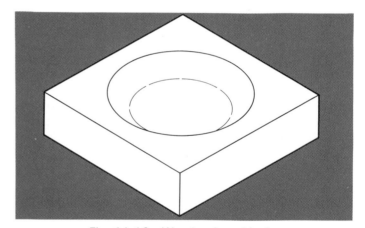

Fig. 11-13. Wooden form block.

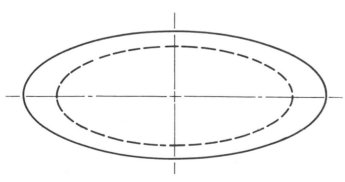

Fig. 11-11. Pattern with guide lines showing the portion to be formed.

Fig. 11-14. One method that may be used to flatten work that has been warped in the forming operation.

PLANISHING

PLANISHING is an operation done to make a metal surface smooth. See Fig. 11-15. It is done with a planishing hammer. Only a hammer with a mirror smooth face should be used.

Lay the hammer blows on evenly. Rotate the piece so that no two blows fall in the same place. Use a stake with a shape that is near to the desired shape.

FORMING BY RAISING METAL

A bowl may be raised by one of several methods. The design of the project will determine the method to be used.

Shallow pieces can be raised using a block of hardwood that has a shallow depression in one of the end grain sides, Fig. 11-16.

Slowly form the metal into the depression. When the desired depth has been reached, clean the piece and planish it on a MUSHROOM STAKE, Fig. 11-17. Trim, Fig. 11-18, smooth the edges, and polish.

Fig. 11-19 shows how to determine the diameter of the metal blank.

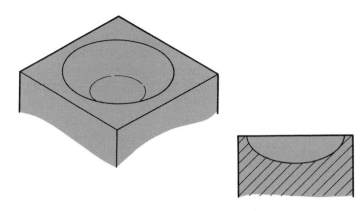

Fig. 11-16. A forming block.

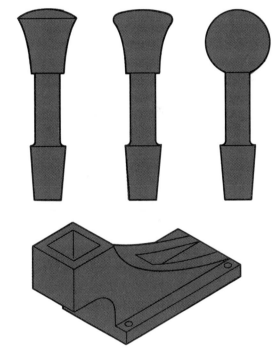

Fig. 11-17. Mushroom stakes with stake holder.

Fig. 11-15. Note how the planish marks enhance the appearance of this brass mug. (Henry Kauffman)

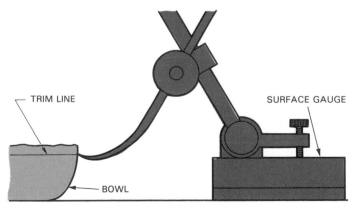

Fig. 11-18. Using a surface gauge to mark a line for trimming.

127

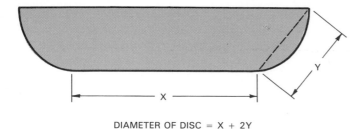

DIAMETER OF DISC = X + 2Y

Fig. 11-19. Calculating the diameter of the metal disc.

Fig. 11-21. Starting to shape a bowl on a sandbag.

Bowls can be raised using a sandbag and mallet. Scribe concentric circles (several circles having the same center) on a disc of the proper size, Fig. 11-20. Place the disc on the sandbag. Raise the edge opposite the one to be struck with the mallet, Fig. 11-21. Shape the bowl by striking the disc a series of blows around the outer circle. Continue working toward the center until the desired depth is obtained. Planish on a stake, trim to height, and polish.

A form can also be raised over a stake. Forms of considerable height can be obtained using this method, Fig. 11-22.

Cut the disc and scribe the concentric circles. Start the raising on the sandbag. Anneal the piece and place it on a RAISING STAKE, Fig. 11-23. Begin hammering by going around the scribed circles.

Fig. 11-22. Forms of varying heights can be raised over a stake.

To continue the raising operation, hold the work on the stake. The hammer blows must land just ABOVE the point where the metal touches the stake, Fig. 11-24.

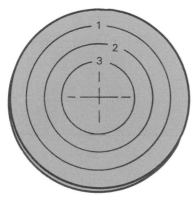

Fig. 11-20. Concentric circles on metal disc.

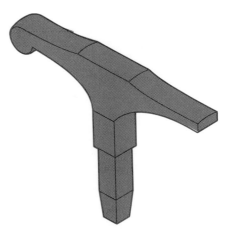

Fig. 11-23. Stake used for raising work.

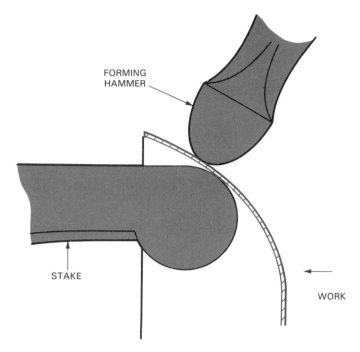

Fig. 11-24. Land hammer blows just above the point where the bowl rests on the stake.

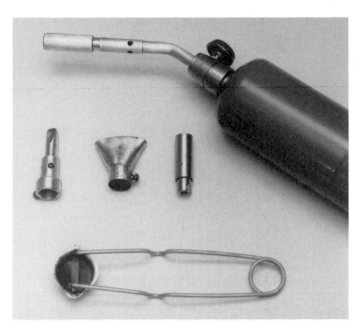

Fig. 11-25. A gas torch with a few of the tips available.

Continue the operation. Anneal when needed, until the desired height is reached.

Pickle, trim, planish, and polish.

HARD SOLDERING

HARD SOLDERING, or SILVER SOLDERING, produces a joint much stronger than soft solder. It is often used in art metal. A torch, Fig. 11-25, must be used to reach the temperatures needed. These range from 800 to 1400°F (430 to 760°C), depending on the silver solder used. The solder is available in sheet and wire form.

Most nonferrous metals (copper, brass, and silver) can be joined by this method.

When silver soldering, follow these steps:

1. Carefully fit the pieces together, Fig. 11-26.
2. Clean the metal. This may be done with abrasive cloth or by pickling.
3. Fluxing the metal. A proper flux can be made by mixing borax and water to form a thick paste.
4. Support the joint, Fig. 11-27. The pieces must not slip during the heating. Use clamps or binding wire.

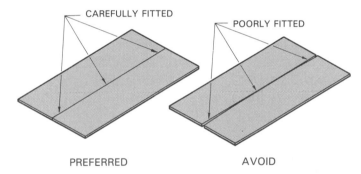

Fig. 11-26. The pieces to be hard soldered (silver soldered) must be carefully fitted together.

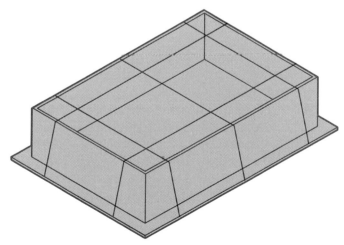

Fig. 11-27. Support the joints to be soldered. This can often be done by wiring them together with soft iron wire.

129

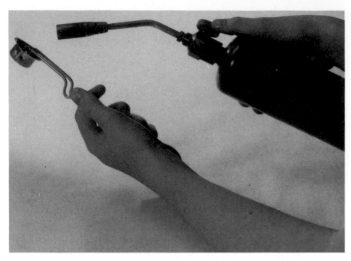

Fig. 11-28. The recommended way to light the torch. NEVER USE MATCHES!

5. Cut the solder into pieces about 1/16 in. long. Dip them in flux and place them on the joint with a small brush or tweezers. CAUTION: WHEN LIGHTING THE TORCH HOLD IT SO IT IS POINTING AWAY FROM YOU. USE A SPARK LIGHTER, Fig. 11-28.

Keep the torch in motion while heating the joint. This will assure a uniform heat. As the solder melts, it will be drawn into the joint.

If different size pieces are to be soldered, apply more heat to the larger section.

6. Remove all traces of flux and oxides from the joint using abrasive cloth or pickling.

SAFETY

1. Remove all burrs and sharp edges from metal before attempting to work with it.
2. Use caution when handling hot metal. Do not place it where it can start a fire.
3. Have your instructor mix the pickling solution. It contains acid and may cause serious and painful burns.
4. Wear a face shield when using the pickling solution. Wear your regular goggles when working in the lab.
5. Use the pickling solution in a well ventilated area. Do not breathe the fumes.
6. Stand to one side when plunging hot metal into the pickling solution.

7. Clear the soldering area of solvents and other flammable material before soldering.
8. Have cuts, bruises, and burns treated promptly.

TEST YOUR KNOWLEDGE, Unit 11

Please do not write in the text. Place your answers on a separate sheet of paper.

1. List the four classifications of art metal.
2. Why must metal be annealed?
3. Cleaning metal by heating it and then plunging it into a dilute solution of acid is called _____.
4. Name two methods used to form bowls and trays.
5. _____ is the name of the operation done to make a metal surface smooth.
6. _____ soldering produced a stronger joint than _____ soldering.
7. List several safety rules that must be observed when working with art metal.

TECHNICAL TERMS TO LEARN

annealing	mallet
art metal	pickling
beating down	piercing
borax	planish
flatware	raising
gas torch	silver soldering
hard soldering	stake
hollow ware	strip work
jewelry making	surface gauge

ACTIVITIES

1. Collect illustrations from magazines, newspapers, and catalogs to use for art metal project.
2. Visit a museum to study examples of art metal made by well-known artisans.
3. Demonstrate hard soldering.
4. Design and craft an outstanding artisan award for the Industrial Arts department of your school.
5. Prepare posters on art metal safety.
6. Secure samples of the various types of art metalwork. Describe them to your class. Stress what you consider to be their good points of design.

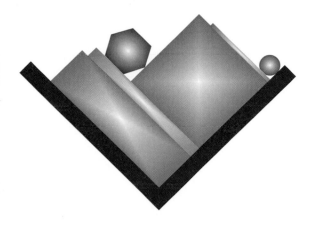

Unit 12
Metal Finishes

After studying this unit, you will be able to discuss various types of abrasives and identify the best uses for each type. You will also learn about several types of finishing techniques and how they are done.

Finishes are applied to metal for two reasons. They protect surfaces from corrosion and enhance surface appearance, Fig. 12-1. Many types of finishes are available.

No matter what type of finish is applied, the surface of the metal must be clean.

HAND POLISHING WITH ABRASIVES

Any hard, sharp material that can be used to wear away another material is an abrasive. For many purposes, manufactured abrasives (aluminum oxide, silicon carbide, etc.) are superior to natural abrasives (emery and iron oxide).

Fig. 12-1. Aircraft is painted to enhance its appearance and make it easier to see in the air. The top of the fuselage is painted white, or any light color tint, to reflect the heat of the sun. (Canadair Limited)

Abrasive grains are bonded to a cloth or paper backing. The coarseness (grain size) of an abrasive is identified by number. The higher the number, the smaller the grain and the finer the finish it will produce. Aluminum oxide and silicon carbide abrasive sheets used in the metals lab usually range from 1 extra coarse to 8/0 extra fine.

Properly filed work can be polished using only a fine-grain abrasive cloth. However, if there are deep scratches, it is best to start with coarse-grain cloth. Change to medium-grain cloth, then finish with a fine-grain abrasive. A few drops of machine oil will speed the operation.

Support the abrasive cloth by wrapping it around a wood block or using a sanding block, Fig. 12-2. Applying pressure, move the abrasive back and forth in a straight line. Rub it parallel to the long edge of the work, if possible.

DO NOT POLISH MACHINED SURFACES.

BUFFING

BUFFING, Fig. 12-3, provides a bright mechanical finish. It is commonly applied to nonferrous metals like aluminum, copper, brass, pewter, and silver. Copper base alloys then must be covered by a clear plastic or lacquer coating to protect the polished surface. Buffing is done on a BUFFER-POLISHER, Fig. 12-4

Remove all scratches with an abrasive before buffing.

Fig. 12-3. Buffing produces a highly polished surface on metal.

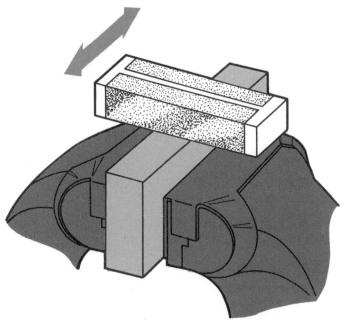

Fig. 12-2. When polishing, wrap the abrasive cloth around a block of metal or wood, or use a sanding block.

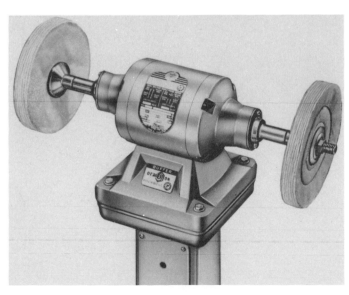

Fig. 12-4. This buffing machine is typical of those found in the school metal shop. (Delta-Rockwell)

132

To buff, use a fairly stiff wheel. Charge it with tripoli or pumice. This will remove the small scratches formed by the abrasive cloth.

Final polishing is done with a loose flannel wheel and polishing compound.

Buffing should be done with the work held below the centerline of the wheel, Fig. 12-5.

WIRE BRUSHING

WIRE BRUSHING produces a smooth, satin sheen on the metal. Wire size will determine the smoothness of the finish: the finer the wire the smoother the finish.

PAINTING

PAINTING, Fig. 12-6, is another finishing technique. Many colors and finishes are available. They can be applied to the project by spraying, Figs. 12-7 and 12-8, brushing, rolling, or dipping.

MACHINED SURFACES SHOULD NOT BE PAINTED.

COLORING WITH HEAT

This type of finish may be applied to steel. Clean the metal and slowly heat it. Watch the colors as they appear. Plunge the metal into cool water when the correct color is reached. Protect the resulting finish with clear lacquer.

Fig. 12-6. The painted finish on this factory-built structure is attractive and also reduces maintenance and repair costs.   (Stran-Steel)

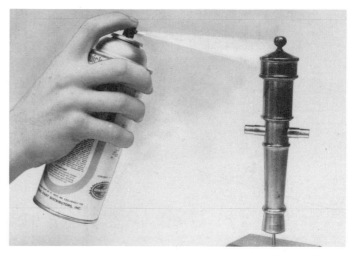

Fig. 12-7. A wide selection of paint colors are available in pressurized spray cans. Project being finished is a barrel for a model cannon.

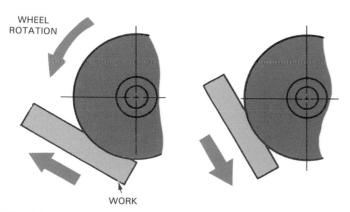

Fig. 12-5. Work below the centerline of the buffing wheel. Left. Pull the metal toward you when starting the operation. If the work is pulled from your grasp, your hands will not be pulled into the wheel. Right. For the final high polish, pass the work lightly into the wheel.

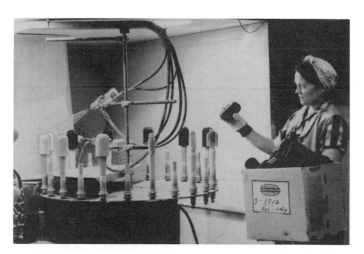

Fig. 12-8. An automatic rotary machine sprays porcelain enamel on metal parts for lighting fixtures. The machine boosted production 300 percent over hand spraying and had fewer rejects.   (DeVilbiss Co.)

Fig. 12-9. Metal parts being hot dipped to produce a zinc (galvanized) coating on the parts. (American Hot Dip Galvanized Assoc., Inc.)

OTHER METAL FINISHES

There are several other finishes that can be applied to metal. However, some of these finishes require special equipment for application. This equipment is not readily available in a school lab.

HOT DIPPING, Fig. 12-9, is applied to steel by dipping the metal into molten aluminum, zinc, tin, or lead. GALVANIZED STEEL is an example of this finishing technique.

ANODIZING is a process used to form a protective layer of aluminum oxide on aluminum parts. The anodized coating can be dyed a wide range of colors. The color becomes a part of the surface of the metal.

ELECTROPLATING is a finishing technique in which a metal coating is applied to or deposited on a metal surface by an electrical current, Fig. 12-10.

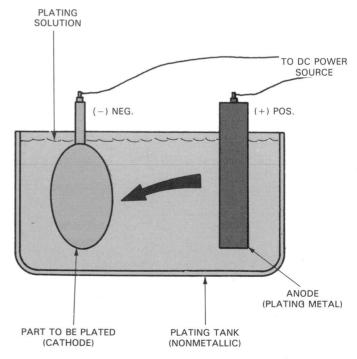

PLATING SOLUTION

TO DC POWER SOURCE

(−) NEG.

(+) POS.

ANODE (PLATING METAL)

PART TO BE PLATED (CATHODE)

PLATING TANK (NONMETALLIC)

Fig. 12-10. Electroplating is the process by which one metal is coated with another by electricity.

TEST YOUR KNOWLEDGE, Unit 12

Please do not write in the text. Place your answers on a separate sheet of paper.

1. Finishes are applied to:
   a. Protect surfaces from corrosion.
   b. Enhance surface appearance.
   c. Both a and b.
   d. None of the above.
2. Hard, sharp material that can be used to wear away another material is an _____.
3. Aluminum oxide and silicon carbide are examples of _____ _____.
4. Name two types of natural abrasives.

MATCHING QUESTIONS: Match each of the following terms with their correct definitions.
   a. Painting.   c. Buffing.
   b. Wire brushing.
5. ____ Produces a smooth, satin sheen.
6. ____ Technique in which finishes are applied by spraying, rolling, brushing, or dipping.
7. ____ Produces a bright mechanical finish.
8. Galvanized steel is an example of a finish applied by the _____ _____ finishing technique.
9. Finishing technique in which a metal coating is applied to a metal surface.
   a. Electroplating.
   b. Anodizing.
   c. Hot dipping.
   d. None of the above.

TECHNICAL TERMS TO LEARN

abrasive
aluminum oxide
anodizing
appearance
application
buffer-polisher
buffing
corrosion
electroplating

emery
hot dipping
iron oxide
painting
polish
silicon carbide
surface finish
wire brushing

ACTIVITIES

1. Collect samples of metals with different types of finishes. Why was each finish used?
2. Look around your metals lab. Note the different types of finishes used on the tools and machinery. How many types can you identify?
3. Prepare samples of metal sheet with different kinds of painted finishes. Devise a method for testing the durability of each finish.
4. Demonstrate electroplating. SAFETY NOTE: Before attempting to demonstrate electroplating, have your instructor check over the equipment and its wiring harness. Be sure they are in safe operating condition. Wear safety glasses and rubber or disposable plastic gloves when electroplating.

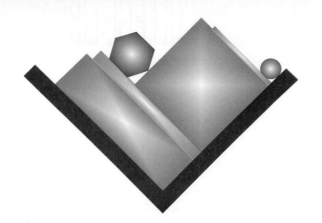

Unit 13
Hand Forging

After studying this unit, you will be able to explain and demonstrate various forging processes. You will be able to identify various forging tools and demonstrate their use.

Metal is forged by heating it to less than the melting point, then using pressure to shape the hot metal. Heating makes the metal PLASTIC (more easily shaped).

In HAND FORGING, pressure is applied by using a hammer, Fig. 13-1. Forging improves the physical characteristics (strength, toughness) of the metal.

Fig. 13-1. Master blacksmith hand forges useful ironwork articles at the Deane Forge at Colonial Williamsburg, just as his 18th century predecessors did.   (Colonial Williamsburg)

EQUIPMENT FOR HAND FORGING

FORGE

Metal is heated in a FORGE, Fig. 13-2. The forge may be gas- or coal-fired. Gas is preferred because it is cleaner to use.

Lighting a gas forge

1. Open the forge door. Check to be sure the gas valve is closed.
2. Start the air blower and open the air valve slightly.
3. Apply the lighter and slowly turn on the gas. CAUTION: STAND TO ONE SIDE AND DO NOT LOOK INTO THE FORGE WHEN YOU LIGHT IT.
4. After the gas has ignited, adjust the gas and air valves for the best combination.

ANVIL

Heated metal is shaped on an ANVIL, Fig. 13-3. The HORN, which is conical in shape, is used to form circular sections.

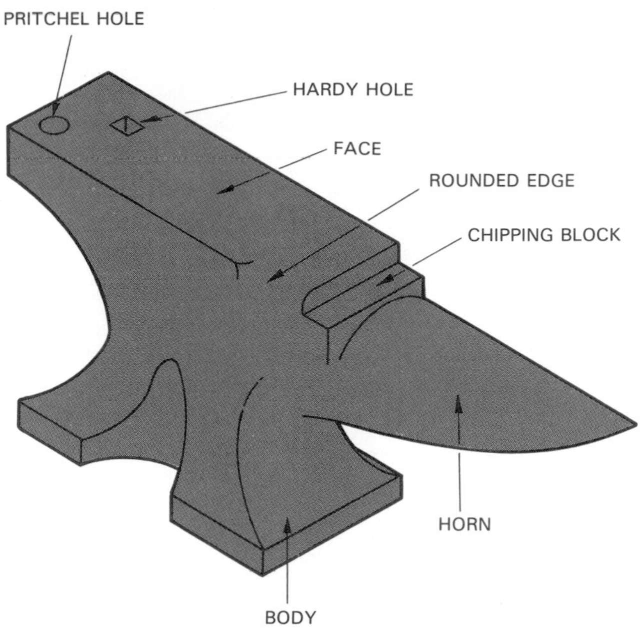

Fig. 13-3. The parts of an anvil.

Various anvil tools can be mounted in the HARDY HOLE. The PRITCHEL HOLE is used to punch holes or to bend small diameter rods.

ANVIL TOOLS

Many anvil tools are available. You will probably use the HARDY, Fig. 13-4, to cut metal. It is inserted in the Hardy hole. The metal to be cut is placed on the cutter and struck with a hammer, Fig. 13-5, until it is cut. If metal thicker than 1/2 in. is to be cut, it should be cut hot.

Fig. 13-2. A gas-fired forge. (Johnson Gas Appliance Co.)

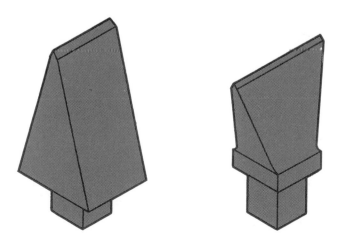

Fig. 13-4. Hardies.

137

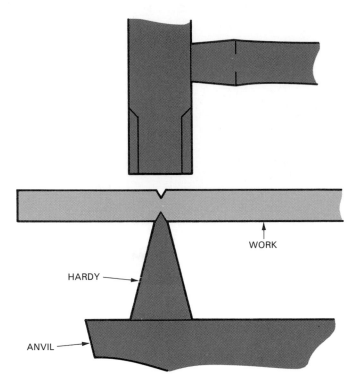

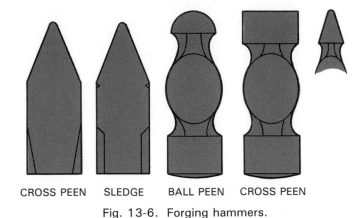

CROSS PEEN   SLEDGE   BALL PEEN   CROSS PEEN

Fig. 13-6. Forging hammers.

Fig. 13-5. When using the Hardy to cut metal, nick it on both sides. Then bend it back and forth until it breaks. Metal thicker than 1/2 in. should be cut while hot.

STRAIGHT LIP

HAMMERS

Many different types of HAMMERS are available for hand forging. See Fig. 13-6. Use a 1 1/2 - 2 lb. hammer for light work. A 3 lb. hammer will be satisfactory for heavy work. Do not "choke-up" on the hammer handle — you may injure yourself.

CURVED LIP

Fig. 13-7. Typical forging tongs.

TONGS

Tongs are used to hold the hot metal while it is being forged, Fig. 13-7. Use tongs that are best suited for the work being done, Fig. 13-8.

FORGING

It is necessary to heat metal to the correct temperature before it can be forged. Mild steel should be heated to a BRIGHT RED. Carbon steel should not be heated beyond a DULL RED. Avoid WHITE HEAT, where sparks fly from the piece.

CAUTION: WEAR A FACE SHIELD WHEN FORGING. THE SCALE THAT FLIES ABOUT IS HOT. GLOVES SHOULD BE WORN, TOO.

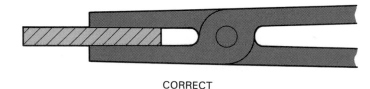

CORRECT

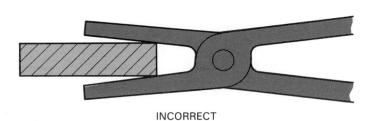

INCORRECT

Fig. 13-8. Use the correct tongs for the work to be forged.

DRAWING OUT METAL

The forging operation employed to stretch or lengthen metal is called DRAWING OUT. Fig. 13-9 shows the sequence recommended for drawing out round stock. Square stock is drawn out in much the same way. Round stock can be pointed as shown in Fig. 13-10.

BENDING METAL

There are several ways to make a bend in stock. Unless a small size rod is to be bent, all require the metal to be heated to a red heat.

1. Bend the metal over the anvil face and edge, Fig. 13-11.
2. Place the stock in the Hardy or Pritchel hole and bend it over, Fig. 13-12. Square the bend over the anvil edge.
3. Curved sections are made on the anvil horn, Fig. 13-13.

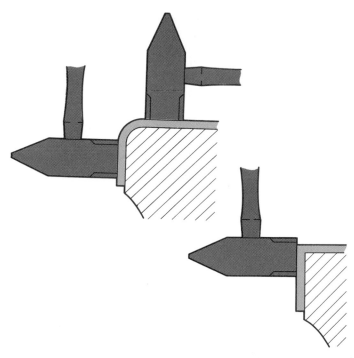

Fig. 13-11. Making a right angle bend over the edge of an anvil.

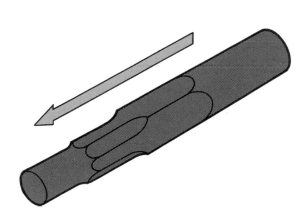

Fig. 13-9. Drawing out sequence.

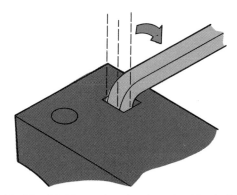

Fig. 13-12. Making a bend using the Hardy hole.

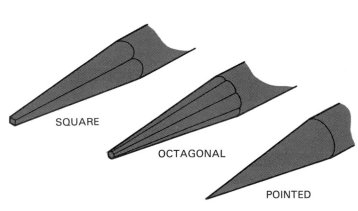

SQUARE

OCTAGONAL

POINTED

Fig. 13-10. Steps in drawing out a point.

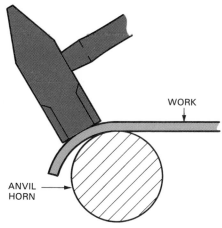

WORK

ANVIL HORN

Fig. 13-13. Circular shapes can be made over the anvil horn.

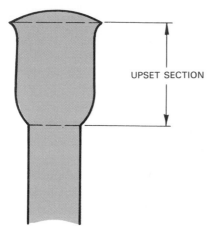

Fig. 13-14. Upsetting decreases the length of the metal but increases the thickness.

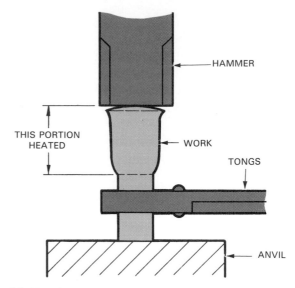

Fig. 13-15. Short pieces can be upset on the face of the anvil.

UPSETTING METAL

UPSETTING is an operation that increases the thickness of the metal at a given point, Fig. 13-14. This also shortens the metal.

1. Heat the metal to a red heat.
2. Short work is upset using the anvil face. See Fig. 13-15.
3. Mount the work in a vise if it is short enough. The heated end extends above the vise jaws and is hammered to increase its size.

If the metal becomes hard to work because of cooling, reheat it and continue the operation.

TEST YOUR KNOWLEDGE, Unit 13

Please do not write in the text. Place your answers on a separate sheet of paper.

1. Explain how metal is forged.
2. Heating makes the metal _____.
3. Where is metal heated?

MATCHING QUESTIONS: Match the following terms with their correct definitions.
a. Tongs.   d. Pritchel hole.
b. Anvil.   e. Hardy hole.
c. Horn.
4. ____ Used to punch holes or to bend small diameter rods.
5. ____ Used to shape heated metal.
6. ____ Used to mount various anvil tools.
7. ____ Used to hold hot metal while it is being forged.
8. ____ Conical shape used to form circular sections of metal.
9. _____ _____ is an operation that lengthens or stretches metal.

TECHNICAL TERMS TO LEARN

anvil	horn
drawing out	physical characteristics
forge	plastic
forging	pressure
gas forge	Pritchel hole
Hardy	tongs
Hardy hole	upsetting

ACTIVITIES

1. Prepare posters on forging safety.
2. Demonstrate the proper way to light a gas forge.
3. Hand-forge a cold chisel.
4. Secure products that have been forged.

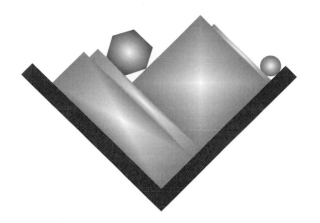

Unit 14
Casting Metals

After studying this unit, you will be able to explain various casting processes. You will also be able to demonstrate the proper way to make a sand mold. Finally, you will be able to demonstrate safety in the foundry area.

Casting metals is one of the oldest processes in metalworking. Objects are made by pouring molten metal into a mold. The mold is a cavity the shape and size of the object to be cast. It is made of material suitable for holding the molten metal until it cools. The part made by this process is called a CASTING. In industry, castings are made in a FOUNDRY.

PERMANENT MOLD CASTING

Permanent mold castings are made in metal molds. These molds are not destroyed when removing the casting. That is why they are called permanent molds. The process creates accurate parts with fine surface finishes. Many nonferrous metals (metals other than iron) can be cast this way.

The molds used to cast fishing sinkers and toy soldiers are examples of castings made in permanent molds, Fig. 14-1. Many of the pistons that are used in automobiles are made by this technique.

Fig. 14-1. Familiar items made by the permanent mold casting process.

Fig. 14-2. Die casting machines are the most complex equipment used in any major casting procedure. They provide economical production of large quantities at very high output rates. Die casting develops superior finish quality and surface detail. Shown is a crane setting die in machine.
(Aluminum Company of America)

DIE CASTING

Die casting is a variation of the permanent mold process. It is very useful in the metalworking industry. The metal is forced into the mold or die under pressure, Figs. 14-2 and 14-3. Castings produced have smoother finishes, finer details, and are more accurate than regular sand castings, Fig. 14-4. Large quantities of die castings can be produced quickly on automatic casting machines.

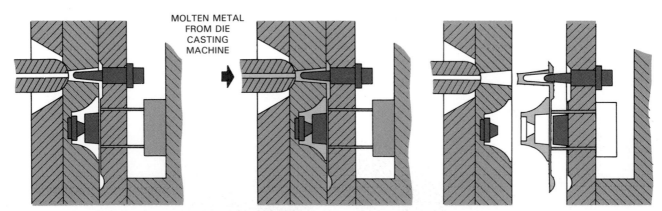

Fig. 14-3. Left. Two sections of a die, closed and locked to receive the "shot" of metal to form a casting. Center. The cavity of the die is now completely filled. Note the metal in the overflow well at the bottom of the cavity. This provides an outlet for the air trapped in the die cavity. Right. The die is opened to permit ejection of the casting. Note the two pins which free the casting from the die.

Fig. 14-4. Typical items made by the die casting method.

INVESTMENT OR LOST WAX CASTING

Investment or lost wax casting is a foundry process used when very accurate castings of complex shapes or detailed designs must be produced, Figs. 14-5 and 14-6.

In the investment casting process, Fig. 14-7, patterns of wax or plastic are placed (invested) in a refractory mold. A refractory mold will withstand high heat. When the mold has hardened, it is placed in an oven. Then the mold is heated until the wax or plastic pattern is burned out (lost). This leaves a cavity in the mold the shape of the pattern. Molten metal is forced into the mold. It is allowed to cool and the mold is broken apart to remove the casting.

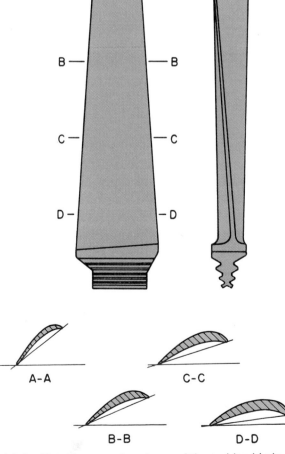

Fig. 14-5. Jet engine turbine blade made by the investment or lost wax casting process.

Fig. 14-6. Note the complex shape of the turbine blade. This made it expensive to manufacture by any other method.

143

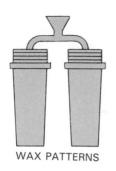

WAX PATTERNS

PATTERNS INVESTED IN
REFRACTORY MOLD

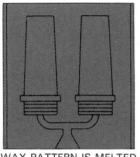

WAX PATTERN IS MELTED
OR BURNED FROM MOLD

MOLD IS POURED

Fig. 14-7. The investment casting process.

Fig. 14-9. Pattern halves are ready to receive the specially prepared mix used in shell molding.

SHELL MOLDING

Shell molding, a fairly new foundry process, is a variation of sand casting. The molds are made in the form of thin sand shells, Fig. 14-8.

A metal pattern attached to a steel plate is fitted into the molding machine. Then the pattern is heated to the required temperature. A measured amount of mix is deposited on the pattern, Fig. 14-9. The mix consists of thermosetting resin and sand. The heat causes the thermosetting resin to take a permanent shape. The thin soft shells are then cured until the desired hardness is obtained.

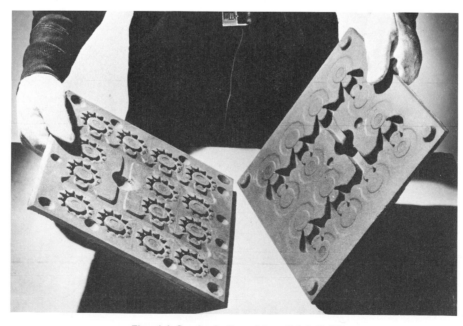

Fig. 14-8. A shell mold.   (Link-Belt)

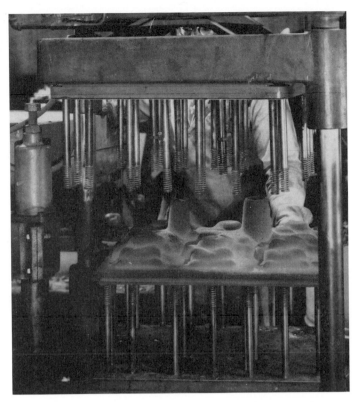

Fig. 14-10. Joining the two mold sections together.

Upon cooling, the pattern is removed. The shell halves are bonded together using a special adhesive, Fig. 14-10. Like regular sand molds, shell molds must be destroyed to remove the casting.

Shell molding is well suited for mass-production. The casting produced, Fig. 14-11, has a superior finish. It also has greater accuracy than castings made by the regular sand casting technique.

CASTING IN THE METALS LAB

While several foundry processes just discussed can be used in the metals lab, SAND CASTING is most often used. It is also the most widely used of the industrial casting techniques.

The sand casting technique used in a school lab is similar to that used in industry. The major differences are casting size and the quantity of castings produced.

Sand castings are cast in MOLDS made of a mixture of sand and clay, Fig. 14-12. The sand is used in the moist stage and is called a GREEN SAND MOLD. The mold is made by packing the sand in a box called a FLASK, around a pattern of the shape to be cast. Parts of the flask separate to allow easy removal of the pattern.

A GATING SYSTEM is used to get the molten metal to the mold cavity. This consists of vertical openings, called SPRUES. These are connected to the mold cavity by grooves called GATES and RUNNERS.

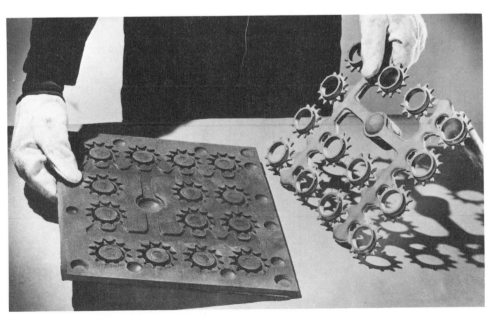

Fig. 14-11. Shell molding is well suited to mass production of quality castings.

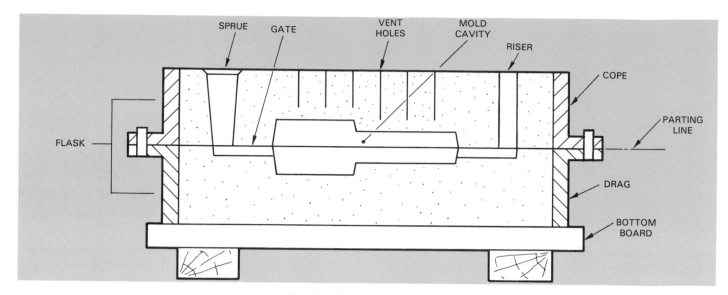

Fig. 14-12. Sand mold.

Metals shrink as they cool. Therefore, steps must be taken to supply additional metal to the parts of the casting that cool last. If this is not done, hollow parts may occur in the finished casting. These reservoirs of extra metal are called RISERS or FEEDERS.

Sand molds can be used only once. They are destroyed in removing the finished casting.

SAND MOLD CONSTRUCTION

In order to produce a sound casting in the metals lab, a certain sequence of operations should be followed. This sequence is outlined in the following sections.

PATTERNMAKING

The cavity in the sand mold is made with a PATTERN, Fig. 14-13. A SIMPLE PATTERN is made in one piece. SPLIT PATTERNS are patterns with two or more parts. They are used to make castings of more complex shape.

A pattern must have DRAFT, Fig. 14-14. Draft is a slight taper. It permits the pattern to be lifted from the sand without damaging the mold.

The pattern must be made slightly larger than the casting. This is because metal shrinks as it cools from the molten state. The amount of shrinkage will depend on the metal being cast.

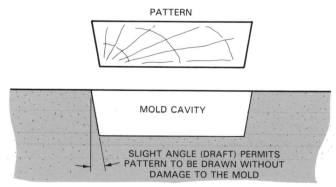

Fig. 14-13. The split pattern (top) is used to make castings of more complex shape than the simple pattern (bottom).

Fig. 14-14. A slight angle (draft) on the pattern permits it to be easily drawn from the mold.

Openings or hollow spaces are made in a casting with a sand CORE, Fig. 14-15. Cores are made from a special sand mix. When baked, this is hard enough to withstand casting pressures.

MOLDING SAND

Good molding sand is essential for good castings. Two types of sand are commonly used in the school foundry. NATURAL SAND is used as it is dug from the ground. SYNTHETIC SAND has an oil binder and no water is needed.

If a synthetic sand is not used, the sand must be TEMPERED. This means it is dampened enough to be workable. Water is added and serves as a binder. Mix the sand until it is uniformly moist. Test this by squeezing a handful of sand and breaking it in half. Properly tempered sand will break cleanly and retain the imprint of your fingers, Fig. 14-16. The sand will cling to your hand if it is too moist, and will crumble if too dry.

Fig. 14-16. Notice how cleanly properly tempered sand breaks.

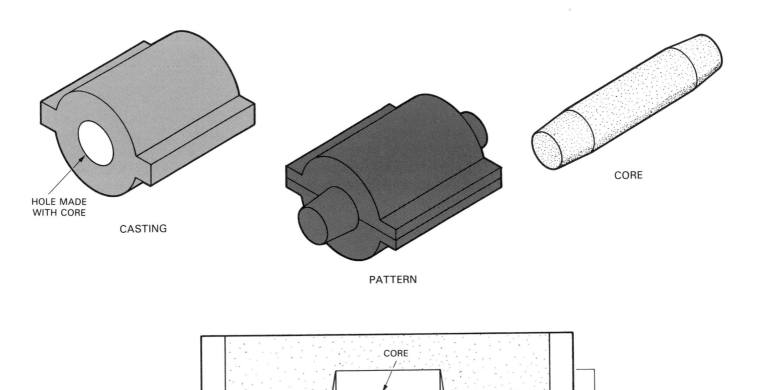

HOLE MADE WITH CORE

CASTING

PATTERN

CORE

CORE

FLASK

PARTING LINE

MOLD CAVITY

Fig. 14-15. Cores are made of a special sand and are used to produce openings in castings.

147

PARTING COMPOUND

PARTING COMPOUND is a waterproofing material. A light dusting will prevent moist sand from sticking to the pattern or to the mold faces.

TOOLS AND EQUIPMENT

Typical foundry tools and equipment are shown in Fig. 14-17. Their uses are explained in the next section.

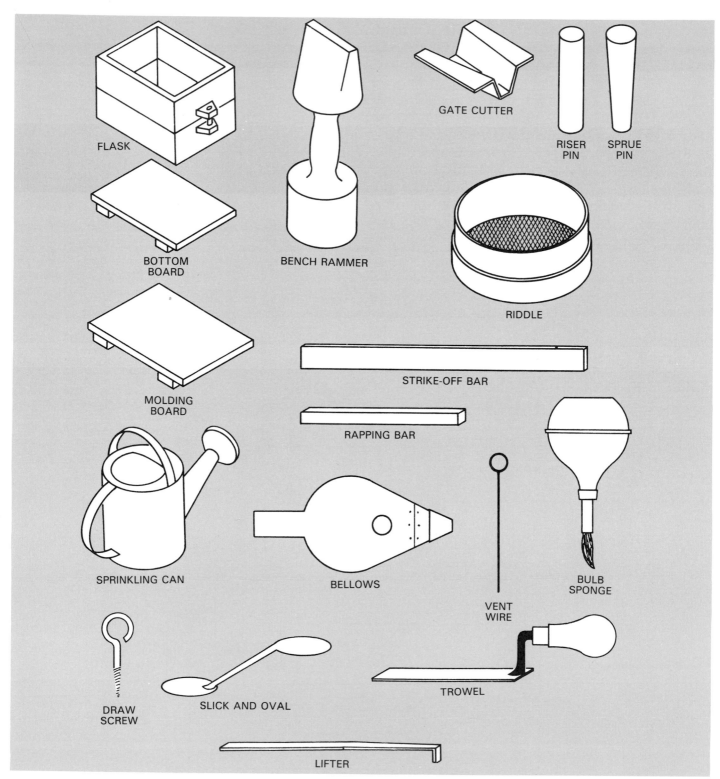

Fig. 14-17. Tools used in the foundry.

SIMPLE MOLD CONSTRUCTION AND CASTINGS

1. Temper the molding sand.
2. Clean and wax the pattern.
3. Position the drag on the molding board. Aligning pins are down. Place the pattern in position, Fig. 14-18. The flat back of the pattern is placed on the board.

4. Lightly dust the pattern with parting compound.
5. Fill the riddle with sand. Then sift a 1 to 1 1/2 in. (25 to 38 mm) layer of sand over the pattern, Fig. 14-19.
6. Using your fingers, pack the riddled sand around the pattern. Roughen the surface of the packed sand and fill the drag with unriddled sand.
7. Pack loose sand around the pattern and inside edges of the drag. Do this with the PEEN edge of the bench rammer, Fig. 14-20. Be careful not to hit the pattern. Ram the sand firmly enough around the pattern to give a good, sharp impression.
8. Add extra sand if necessary to fully pack the drag. Use the BUTT end of the rammer for the final packing.
9. Use a strike-off bar to remove excess sand, Fig. 14-21.

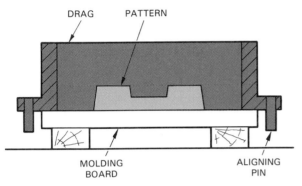

Fig. 14-18. Position the drag on the molding board and position the pattern near its center.

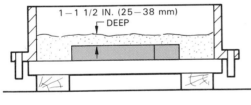

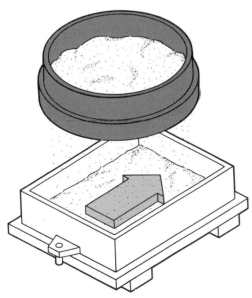

Fig. 14-19. Riddle (sift) about 1 to 1 1/2 in. deep layer of sand over the pattern.

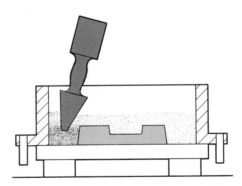

Fig. 14-20. Ram the mold using the peen end of the bench rammer. Be careful not to hit the pattern!

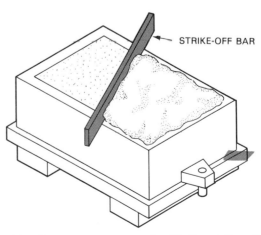

Fig. 14-21. Remove excess sand with the strike-off bar.

10. Place the bottom board on top of the drag and roll (turn) it over.
11. Remove the molding board to expose the pattern. Examine the surface of the sand and, if necessary, smooth and level it with a trowel or slick.
12. Place the cope on the drag. Place the sprue and riser pins in the drag about 1 in. (25 mm) away from each end of the pattern, Fig. 14-22. Place the riser pin near the heaviest section of the pattern.
13. Dust parting compound over the face of the mold. This will prevent the surfaces from sticking together when the cope and drag are separated.
14. Riddle, ram, and strike off the sand in the cope, as was done before.
15. Gases are generated when the molten metal is poured into a mold. To permit these gases to escape, vent the mold with a VENT WIRE, Fig. 14-23. The vent holes should almost touch the pattern.
16. Remove the sprue and riser pins from the mold. Smooth the edges of the holes with your fingers.
17. Carefully lift the cope from the drag. Place it away from the immediate work area.
18. Use a bulb sponge to moisten the sand around the pattern. This will lessen the chance that the sand will break up when the pattern is DRAWN (removed from the sand).

19. Withdraw the pattern from the mold. Insert a draw screw into the back of the pattern. Tap the screw lightly with a rapping bar to loosen the pattern. DO NOT HIT THE BACK OF THE PATTERN!
20. Draw the pattern from the mold. Repair any defects using a slick and/or trowel, Fig. 14-24.

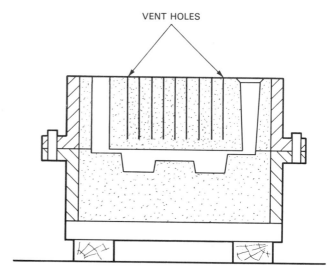

Fig. 14-23. Gases are generated when molten metal contacts moist sand. Vent holes permit the gas to escape, without damaging the mold.

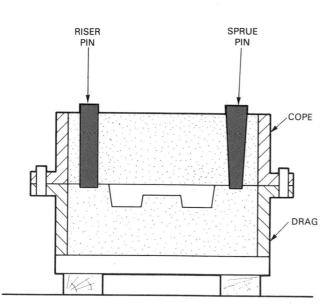

Fig. 14-22. Place the riser and sprue holes about 1 in. from the pattern. Place the riser hole near the heaviest section of the pattern.

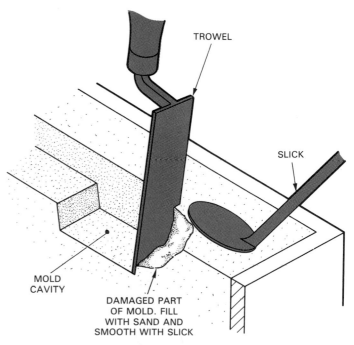

Fig. 14-24. Defects in the mold can often be repaired using a trowel and slick.

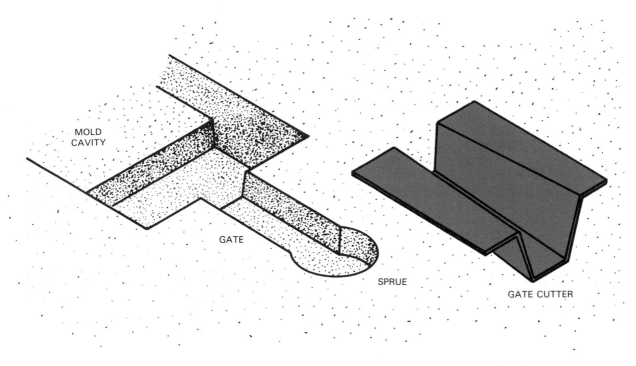

Fig. 14-25. The gate cutter is used to make the trough that connects the riser and sprue holes to the mold cavity.

21. Cut a gate from the mold cavity to the sprue hole and riser hole. Use a gate cutter, Fig. 14-25. The gate is about 1/2 in. (13 mm) wide and 1/2 in. deep. Smooth the gate surfaces with your finger or a slick.
22. Use a bellows to remove any loose sand that has fallen into the mold cavity.
23. Replace the cope on the drag.
24. Move flask to the pouring area. Allow to dry for a short time before pouring.
25. Pour metal into sprue hole carefully and rapidly. Hold the ladle or crucible close to the mold. Large molds are weighted or clamped.

This will prevent the molten metal from lifting the cope, allowing molten metal to flow out of the mold at the parting line.
26. Allow the casting to cool, then break the sand from around it.

The mold is made in much the same manner when a split pattern is used. The pattern is made in two parts. One is fitted with aligning pins, the other has holes to receive the pins. The pattern half with the holes is rammed in the drag, Fig. 14-26. The other half is placed in position after the drag has been rolled.

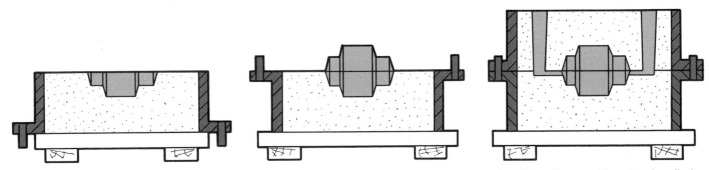

Fig. 14-26. Ramming up a split pattern. Left. Pattern half with aligning holes is rammed up first. Center. After drag is rolled over, the pattern half with the aligning pins is put in space. Right. Mold is rammed up in conventional manner.

FOUNDRY SAFETY

1. Wear protective clothing and goggles when pouring and handling molten metal.
2. UNDER NO CIRCUMSTANCES SHOULD MOIST OR WET METAL BE ADDED TO MOLTEN METAL. A VIOLENT EXPLOSION MIGHT RESULT.
3. Place hot castings in an area where they will not cause accidental burns.
4. Stand to one side of the mold as you pour. Never stand directly over it. Steam is generated when molten metal meets moist sand and this may burn you. Also, molten metal may spurt from the mold if the sand is too moist.
5. Keep the foundry area clean.
6. Do not wear oily or greasy shop clothing when working with hot metals.
7. Follow instructions carefully. DO NOT TAKE CHANCES. WHEN YOU ARE NOT SURE WHAT MUST BE DONE, ASK YOUR INSTRUCTOR.
8. Never look into the furnace when it is lighted. Stand to one side. DO NOT LIGHT THE FURNACE UNTIL YOU HAVE LEARNED THE PROPER WAY TO DO IT.

MELTING METALS

A CRUCIBLE FURNACE, Fig. 14-27, is used to melt aluminum and brass. A SOLDERING FURNACE, Fig. 14-28, is used to melt metals with lower melting temperatures. This includes lead, pewter and zinc alloys (garalloy).

A CRUCIBLE, Fig. 14-29, is used to hold aluminum and brass being melted. It is removed from the furnace with CRUCIBLE TONGS, Fig. 14-30. It is placed in a CRUCIBLE SHANK for pouring.

TURN OFF THE GAS BEFORE REMOVING THE CRUCIBLE FROM THE FURNACE.

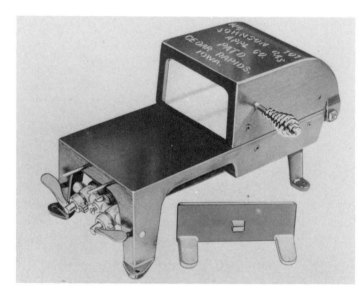

Fig. 14-28. Gas fired soldering furnace. (Johnson Gas Appliance Co.)

Fig. 14-27. A gas fired melting furnace with safety features. (Johnson Gas Appliance Co.)

Fig. 14-29. Crucible used to hold melted aluminum and brass.

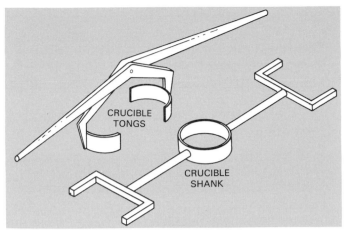

Fig. 14-30. Tongs are used to lift crucible out of furnace. The crucible is placed in the shank for pouring into the mold.

MELTING POINTS OF METAL			
Tin	232°	Garalloy	444°
Lead	327°	Aluminum	659°
Zamak	391°	Brass	956°
Zinc	420°	Copper	1084°

Fig. 14-31. Chart of melting temperatures. Temperatures are given in both degrees Fahrenheit and Celsius.

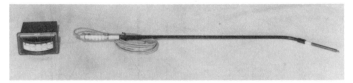

Fig. 14-32. The pyrometer probe is placed in the molten metal. The temperature is read on the direct reading gauge.

POURING METAL

Gather enough metal before lighting the furnace. Do not mix different metals. Use a crucible large enough to hold sufficient metal for the casting. Place it in the furnace.

Light the furnace. Allow the metal to come to pouring temperature, Fig. 14-31. Use a PYROMETER, Fig. 14-32, to check the temperature of the metal. If additional metal must be added to the crucible, prevent metal with moisture from being added to the molten metal. WATER AND MOLTEN METAL REACT VIOLENTLY WHEN THEY COME IN CONTACT. BE SURE TO KEEP THEM APART.

When the metal has been heated to the correct temperature, turn off the furnace. Be careful not to overheat or poor casting will result. Remove the crucible from the furnace with tongs. Place it in a crucible shank for pouring. Add FLUX and skim the SLAG or DROSS (impurities in the metal) from the surface of the metal.

Pour the molten metal rapidly. STAND TO ONE SIDE OF THE MOLD AS YOU POUR. NEVER STAND DIRECTLY OVER IT. Stop when the riser and sprue holes are full.

Allow the casting to cool, then shake out (break up) the mold.

After cutting off the sprue and riser, the casting is ready for machining and finishing.

TEST YOUR KNOWLEDGE, Unit 14

Please do not write in the text. Place your answers on a separate sheet of paper.
1. Name four casting processes that are used in industry.
2. What casting process is used most often in school labs?

MATCHING QUESTIONS: Match each of the following terms with their correct definitions.

a. Flask.   f. Parting line.
b. Cope.   g. Core.
c. Drag.   h. Draft.
d. Green sand mold.   i. Bench rammer.
e. Mold.   j. Crucible.

3. ____ Container used to hold metal that is being melted.
4. ____ Box into which sand is packed to make mold.
5. ____ Mold made with moist sand.
6. ____ Opening in sand into which molten metal is poured to produce casting.
7. ____ Bottom half of mold.
8. ____ Tool used to pack sand into flask.
9. ____ Top half of flask.
10. ____ Enables pattern to be removed from sand without damaging mold.
11. ____ Point at which flask comes apart.
12. ____ Inserts that make holes and other openings in casting.
13. List four safety precautions that must be observed when casting metals.

TECHNICAL TERMS TO LEARN

bellows
bottom board
casting
cope
core
crucible
crucible furnace
crucible shank
crucible tongs
die casting
draft
drag
dross
flask
foundry
gate
gate cutter

investment casting
lifter
lost wax casting
mold
parting compound
pattern
permanent mold
pyrometer
riddle
runner
sand casting
shell molding
slag
slick and oval
sprue
trowel

ACTIVITIES

1. Secure samples of various types of castings. Prepare a display board showing the processes followed to make each casting. Use photographs and drawings to show the steps.
2. Prepare a series of posters on safe foundry operations.
3. Demonstrate to the class the proper way to make a sand mold. Follow all safety rules.
4. Certain products require molds made of plaster. Research and prepare a paper on the plaster casting process.
5. Visit a local dentist or a dental supply shop to see how the dental profession uses investment castings. Give a short report to the class on what you observed.
6. Make arrangements to visit a foundry that uses the shell molding technique. If possible, secure a casting made by the process. Compare it with a sand mold casting. How do surface finish and dimensional accuracy compare between the two?

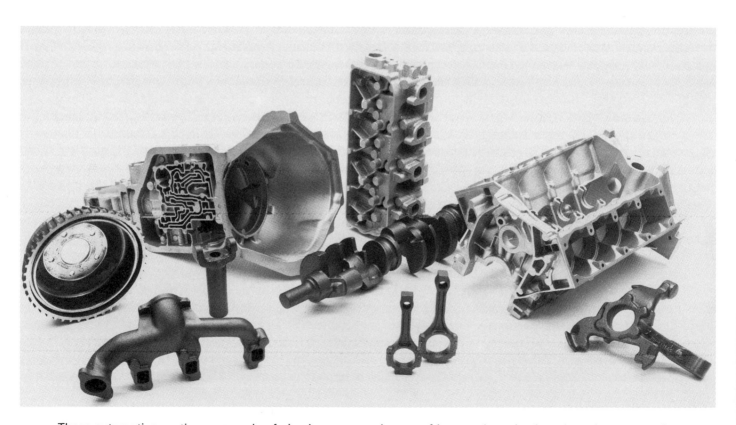

These automotive castings are made of aluminum, several types of iron, and an aluminum/grey iron composite.

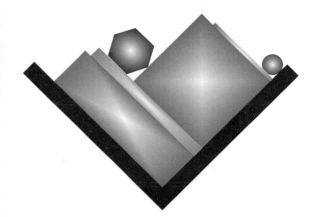

Unit 15
Welding and Brazing

After studying this unit, you will be able to explain the three most common types of welding. In addition, you will be able to demonstrate various welding techniques in a safe manner.

WELDING is a process used to fabricate metal parts. Two or more pieces of metal are joined into a single unit. This is done by heating them to a temperature that is high enough to cause them melt and combine. A permanent joint is produced by this method.

Welding is a very important facet of modern fabricating techniques, Fig. 15-1. There are almost 100 welding and allied processes classified by the American Welding Society (AWS). Three of the most common are ARC, OXYACETYLENE, and SPOT WELDING. Welding makes it possible for most metals and their alloys to be welded to themselves, and often to each other.

Fig. 15-1. Worker puts finishing touches on a crosshead for a large press. This was made in the weldment shop of Bethlehem Steel Corporation.

With careful design, welded standard stock metal shapes are often simpler, lighter and, in many cases, stronger and less expensive than their cast counterparts, Fig. 15-2. Welding is often used to make combinations of castings, forgings, and stock steel shapes.

WELDING SAFETY

When welding, be sure you are shielded from the direct rays of the arc. The rays can cause serious and often permanent eye damage. Protect your eyes, face, and neck with an arc welding helmet. NEVER USE GAS WELDING GOGGLES FOR THIS PURPOSE.

Keep your sleeves down. Wear gauntlet type leather gloves to protect your arms and hands from "arc burn" and molten weld metal.

Wear a leather apron or jacket to protect your clothing from weld "spatter." WEAR LEATHER (NOT CANVAS) SHOES WHEN WELDING.

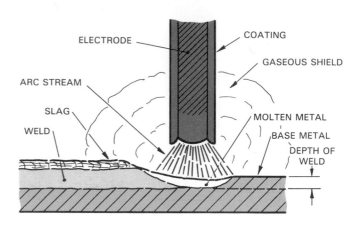

Fig. 15-3. Close-up showing electric welding procedure.

ARC WELDING

Arc welding is a joining technique that uses an electric current to generate the necessary heat for making welds. The tip of the electrode (welding rod) and a small portion of the work become molten in the intense heat, Fig. 15-3.

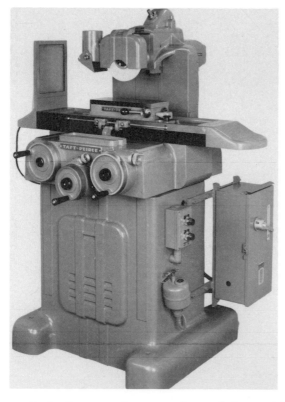

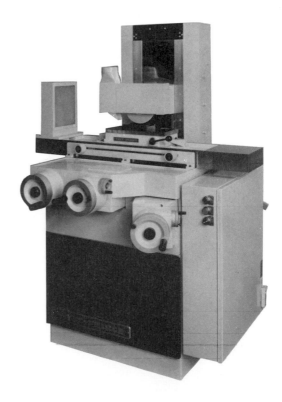

Fig. 15-2. Compare the two surface grinder models shown. Left. This grinder was made from castings. Right. This grinder has been redesigned to make use of welded and cast components.

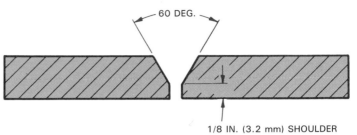

Fig. 15-5. A joint prepared for welding.

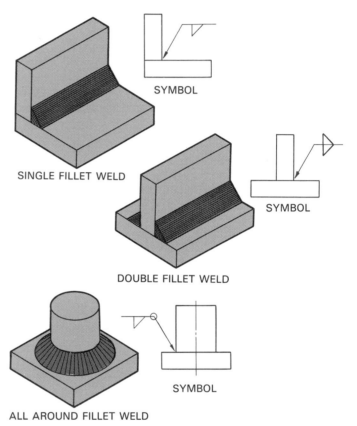

Fig. 15-6. Common welding symbols.

Fig. 15-4. Arc welding equipment. (Lincoln Electric)

Typical arc welding equipment is illustrated in Fig. 15-4.

PREPARING WORK FOR WELDING

Metal to be welded must be clean. It may also be necessary to shape the joint in a V. In this way, electrode metal will fuse with the parent metal better, Fig. 15-5.

Special welding symbols are used on drawings. These show the type and size of the weld to be made. Fig. 15-6 shows some of these symbols and their meaings.

STRIKING THE ARC

Adjust the machine to the correct power setting. Your instructor will help you do this. Your instructor will also help you select the correct type and size of welding rod.

Clamp the electrode into the holder, Fig. 15-7. It should be at a 90 deg. angle to the jaws.

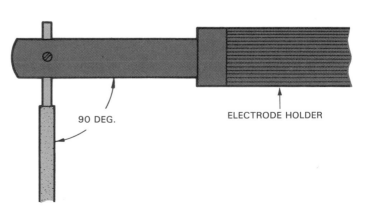

Fig. 15-7. Clamp electrode into the holder as shown.

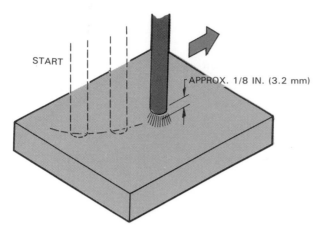

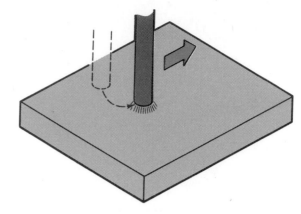

Fig. 15-8. The scratching method of striking an arc. The striking end of the electrode is dragged across the work. It might be compared to striking a match. To prevent the electrode from "freezing" to the work, withdraw the rod from the work immediately after contact has been made.

Fig. 15-9. In the tapping method of striking an arc, the electrode is brought straight down on the work. It is immediately raised on contact to provide the proper arc length.

There are two methods for striking an arc. They are the SCRATCHING method, Fig. 15-8, and the TAPPING method, Fig. 15-9.

Once the arc is established, a short arc (1/16 to 1/8 in. or 1.6 to 3.2 mm) should be held.

DEPOSITING WELD METAL

The electrode should be held at a right angle to the bead. And it should be slightly tilted in the direction of travel, Fig. 15-10.

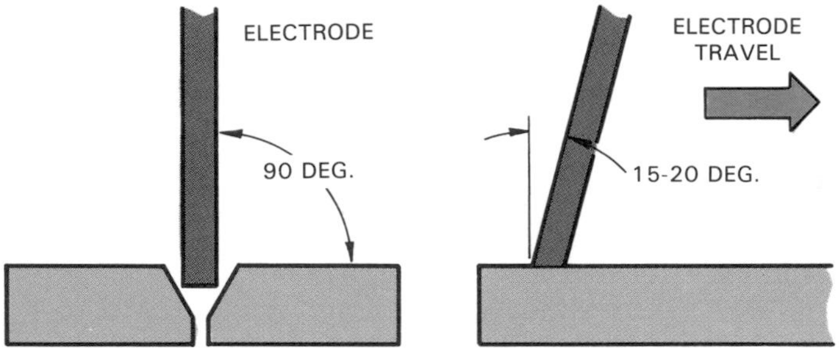

Do not watch the arc. Watch the puddle of molten metal directly behind the arc and the ridge that forms as the molten metal becomes solid. Move the electrode at a uniform speed. The correct speed will be reached when the ridge formed is about 3/8 in. (9.5 mm) behind the arc. Fig. 15-11 shows weld characteristics.

Fig. 15-10. The electrode is held at a slight angle, in the direction of travel.

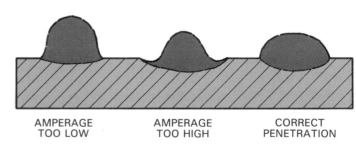

Fig. 15-11. Arc weld characteristics.

ARC WELDING SAFETY

1. NEVER arc weld or watch arc welding without using a shield. Gas welding goggles or sun glasses are NOT satisfactory.
2. Wear goggles when chipping slag.
3. Wear suitable clothing for welding.
4. Do not weld where solvent or paint fumes may collect. Remove all flammable materials from the welding area.
5. Weld only in a well-ventilated area.
6. Treat any cut or burn promptly.
7. Safety goggles should be worn under your welding shield for additional safety.
8. Do not attempt to weld a container until you have determined whether flammable liquids were stored in it. Have containers steam cleaned or fill them with water before welding.
9. Use care when handling metal that has just been welded. A serious burn could result from touching a weld that has not cooled.

OXYACETYLENE WELDING

Oxyacetylene welding equipment, Fig. 15-12, includes two heavy cylinders. One is for acetylene gas, one is for oxygen. PRESSURE REGULATORS reduce high pressure in the cylinders to usable pressures. They also help maintain constant pressure at the torch. One gauge of the regulator shows cylinder pressure. The other shows pressure being delivered to the torch.

HOSES are used to transfer the acetylene and oxygen from cylinders to the torch. The oxygen hose is green and the acetylene hose is red.

The TORCH mixes the gases in the proper proportion for welding.

A WRENCH is used to fit the various connections. A SPARK LIGHTER is used to light the torch. NEVER USE MATCHES! A pair of suitable GOGGLES and heavy GLOVES are need for eye and hand protection.

FLUXES

FLUXES are needed when gas heating certain metals to be joined with solder, or by welding or brazing. The flux promotes better fusion of the metals. It also prevents harmful oxides (which cause poor weld joints) from forming.

WELDING PROCEDURE

Attach the torch tip best suited for the job (check with your instructor). Be sure it is clean, and that both valves on the torch are closed.

Open the oxygen tank valve SLOWLY until the oxygen high pressure gauge shows tank pressure. Then open the valve as far as it will go.

Turn the handscrew until the oxygen low pressure gauge shows the pressure needed for the torch tip being used. Your instructor will provide this information.

Do the same with the acetylene cylinder. However, the cylinder valve should be opened only 1 to 1 1/2 turns—do not open it any more.

Stand to one side of the regulator valves when the cylinder valves are opened.

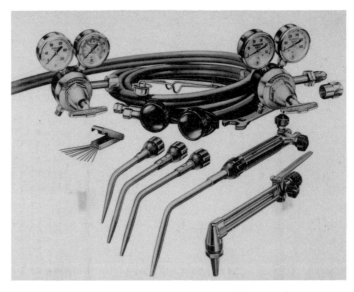

Fig. 15-12. Typical oxyacetylene welding equipment. (Marquette Mfg. Co.)

Have your instructor show you how to light and adjust the welding flame. Your instructor should also check you as you prepare the equipment for welding.

The following steps are done to close down the welding unit:

1. Shut off the acetylene valve ON THE TORCH, then on the oxygen valve.
2. Turn off each tank valve.
3. Open both valves on the torch to drain the gases from the hoses.
4. When both gauges read "zero," turn the adjusting screws on both regulators all the way out.
5. Close the torch valves. Store the torch and hoses.

MAKING A WELD

Welds can be made with or without the use of welding rod. Either type of weld is made holding the torch at a 45 deg. angle to the work, Fig. 15-13. Position the torch until the cone of the flame is about 1/8 in. (3.2 mm) from the metal. When the metal begins to melt into a puddle, move the torch along in a zigzag motion, Fig. 15-14. Do this as fast as the metal will melt and form a smooth, uniform weld. The BACKHAND GAS WELDING TECHNIQUE is shown in Fig. 15-15.

If welding rod is used, preheat it. Do this by bringing it to within 3/8 in. (9.5 mm) of the flame.

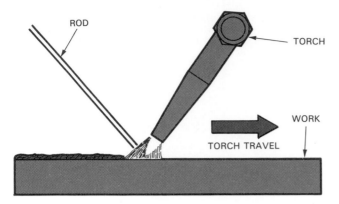

Fig. 15-13. Torch position for gas welding.

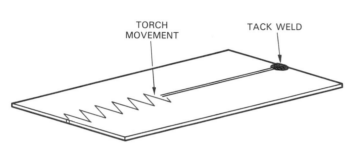

Fig. 15-14. When the metal begins to melt into a puddle, move the torch along in a zigzag motion. The tack weld at the end will prevent the pieces from spreading apart when they expand from being heated.

Fig. 15-15. Producing a smooth, strong, uniform weld using the backhand gas welding technique. The weld is made from left to right and the welding rod is between the completed weld and the torch flame. For a forehand weld, the weld is made from right to left. The flame is then between the completed weld and the welding rod.

Preheating causes the rod to melt faster when it is dipped into the molten weld pool as more metal is needed.

Enough rod should be added to raise the metal in the joint into a slight crown.

A number of typical weld joints are illustrated in Fig. 15-16.

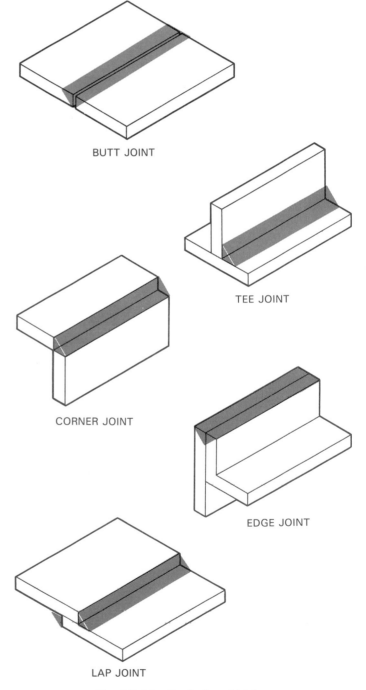

BUTT JOINT

TEE JOINT

CORNER JOINT

EDGE JOINT

LAP JOINT

Fig. 15-16. Typical weld joints.

BRAZING

There are two major differences between welding and brazing. The base metal is not melted in brazing and a nonferrous rod is used as filler metal. However, gas welding equipment is used.

The joint must be clean. Heat the pieces until they are red hot. Then heat the end of the brazing rod and dip it in flux. Hold the rod just ahead of the flame. Allow it to melt and flow into the joint. An example of a properly deposited brazed joint is shown in Fig. 15-17.

OXYACETYLENE WELDING SAFETY

1. Do not attempt to gas weld until you have received instructions on how to use the equipment.
2. Wear appropriate welding goggles.
3. Remove all flammable material from the welding area.
4. Never light the torch with matches. Use a spark lighter, Fig. 15-18.
5. Never attempt to blow dirt from your clothing with gas pressure. Your clothing may become saturated with oxygen and/or acetylene and may explode if a spark comes in contact with it.
6. For added protection, dress properly for the job. Wear welder's gloves.
7. Use care when handling work that has just been welded to avoid serious burns.
8. It is only necessary to turn off the torch if your work needs to be repositioned. However, the entire unit should be turned off when the job is completed. Carefully hang up the torch.

SPOT WELDING

A SPOT WELD is a resistance type weld in which the metals to be welded are clamped between two electrodes. An electric current is passed between the two electrodes. Resistance to the current of the metal between the electrodes heats the metal at the point of contact and fuses the pieces together, Fig. 15-19. Filler metal is not required. Spot welding can be done with a portable unit, Fig. 15-20, or a larger stationary unit, Fig. 15-21.

Fig. 15-17. An example of a properly deposited brazed joint.

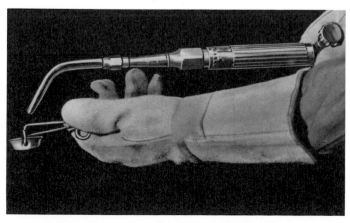

Fig. 15-18. Hold the torch away from the cylinders and pointed away from the body when lighting it. Note the gloves worn by the welder.   (Linde Co.)

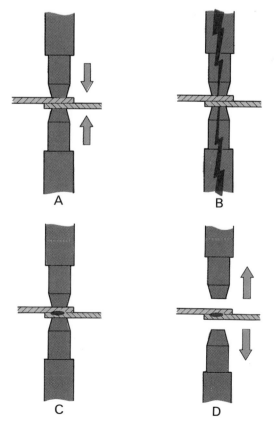

Fig. 15-19. Making a spot weld. A—Squeeze time. B—Weld time. C—Hold time. D—Off time.

Fig. 15-20. Portable spot welding unit. (Marc-L-Tec, Inc.)

Fig. 15-21. Spot welding unit. A safety screen behind the operator protects other workers from sparks. (LORS Machinery, Inc.)

TEST YOUR KNOWLEDGE, Unit 15

Please do not write in the text. Place your answers on a separate sheet of paper.

1. Define welding.

2. List the three welding techniques most commonly used.
3. There are two methods for striking an arc. Name them.
4. List five safety precautions to follow when arc welding.
5. Gases used in oxyacetylene welding are stored in:
   a. Hoses.
   b. Regulators.
   c. Cylinders.
   d. None of the above.
6. _____ prevents oxidation from forming during the welding operation.
7. There are two major differences between welding and brazing. Name them.
8. A _____ _____ is a resistance type weld.

TECHNICAL TERMS TO LEARN

amperage	spot weld
arc burn	tack weld
brace	tank
butt joint	tee joint
edge joint	torch
electrode	weld
fillet weld	welding
joint	welding rod
lap joint	weld splatter
oxyacetylene	weld symbol
pressure regulator	

ACTIVITIES

1. Prepare a series of posters on arc welding safety.
2. Prepare a series of posters on oxyacetylene welding safety.
3. Secure samples of work that have been arc welded, gas welded, brazed, and spot welded.
4. Acquire drawings that show parts that must be fabricated by welding.
5. Make a chart showing the various welding symbols used on drawings.

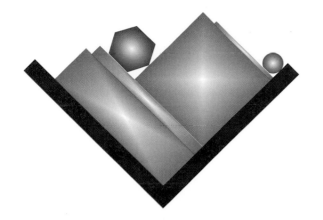

Unit 16
Heat-Treating Metals

After studying this unit, you will be able to explain the reasons metal is heat-treated. You will also be able to demonstrate the annealing and case hardening heat-treating processes. You will also be able to harden and temper a steel object.

Metals must be soft in order to be machined. And they must be hard and tough enough to perform the job they were designed to do, without quickly dulling or wearing down, Fig. 16-1. These conditions are difficult to meet in the manufacture of many metal products. However, through the science of heat-treating, this is made possible. Most metals and alloys can be heat-treated.

The heat treatment of metal includes a number of processes. All of them involve CONTROLLED heating and cooling of a metal to obtain certain changes in metal properties. This might mean a change in properties such as toughness and hardness.

Fig. 16-1. Many parts on this tractor and scraper must be heat-treated to prevent them from wearing out rapidly. (Caterpillar Tractor Co.)

One heat-treating process, ANNEALING, can be used to soften some metals for easier machining. Another, CASE HARDENING, can be used to produce a very hard surface on steel. This makes the metal more resistant to wear.

Heat-treating is done by heating the metal to a predetermined temperature, Figs. 16-2, 16-3, and 16-4. Then, the metal is cooled rapidly (QUENCHING) in water, oil, or brine (salt water). Another heating (at a lower temperature) and cooling cycle may be needed to give the metal the desired degree of hardness and toughness.

Fig. 16-2. Electric heat-treating furnace.

HEAT-TREATING IN THE SCHOOL SHOP

There are any heat-treating processes used by industry. The following techniques can be used in the metals lab. However, before attempting any of them, the following safety precautions are recommended.

Heat-treating requires the handling of very hot metals. Wear a face shield and leather or fabric gloves. Have any burns, no matter how small, treated promptly.

ANNEALING

Some metals become hard as they are worked. They will fracture or be very difficult to work if not softened. ANNEALING reduces the hardness of metals, making them easier to machine and shape.

Annealing is done by heating the metal in a HEAT-TREATING FURNACE, Fig. 16-5. The metal is then cooled slowly in an insulating material such as ashes or vermiculite. The temperature of the furnace and the amount of time the metal is to be heated depends on the kind of metal being annealed and its size. Your instructor can help you find this information in a machinist's handbook.

Fig. 16-4. The parts have reached the required temperature and are now ready for quenching. Note how the parts are placed in the furnace so they will be uniformly heated.

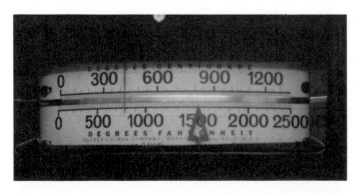

Fig. 16-3. The thermocouple is set to the desired temperature of 1500°F (815°C). The furnace has reached a temperature of 800°F (425°C).

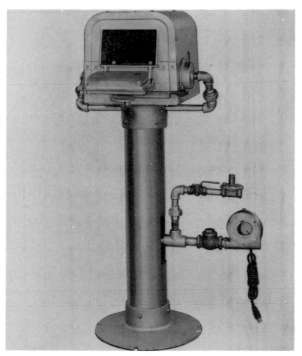

Fig. 16-5. A gas fired heat-treating furnace.
(Johnson Gas Appliance Co.)

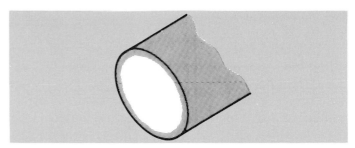

Fig. 16-6. Only a thin outer case is hardened when steel is case hardened. The center portion of the metal remains soft.

Fig. 16-7. A pyrometer is used to measure the high temperatures needed in the heat treatment of metals.
(Johnson Gas Appliance Co.)

Metals like copper, silver, and nickel silver require a different annealing operation than those used with steel. These metals are heated and cooled very rapidly by quenching in water.

Some aluminum alloys are annealed by letting the heated metal cool in still air.

CASE HARDENING

CASE HARDENING adds carbon to low carbon steel so that a hard shell or case is produced, Fig. 16-6. The interior of the metal remains soft. The process is used on steel parts that need a tough, hard wearing surface, such as gears and roller bearings.

Case hardening is done by heating the steel to a bright red at 1650-1700°F (900-925°C). Use a PYROMETER, Fig. 16-7, to measure the temperature. Then roll or sprinkle case hardening compound on the heated piece until a coating of uniform thickness forms, Fig. 16-8. These powders add carbon to the surface of the steel.

Reheat according to the instructions furnished with the carburizing compound used. Then quench in clean, cold water.

Fig. 16-8. Heated piece being rolled in case hardening compound, until a coating forms. It is then reheated and plunged in the quenching fluid, according to manufacturer's instructions.

HARDENING

HARDENING is done by heating carbon or alloy steel to a certain temperature (CRITICAL TEMPERATURE), and cooling it rapidly in water, oil, or brine. The exact temperature is determined by the type of steel being hardened. This information can be found in a machinist's handbook.

Be careful when quenching the heated metal. Dip it into the liquid, Fig. 16-9. NEVER drop it in. This may cause sections of the metal to cool faster than other sections causing it to warp or crack.

Properly hardened steel will be "glass hard" and too brittle for most uses. Hardness can be tested by trying to file the metal with an old file. The file will not cut if the steel has been properly hardened.

Another heat-treating process must be performed to make this brittle steel usable. This process is called tempering.

TEMPERING

Tempering should be done immediately after hardening. The brittle steel may crack if exposed to sudden changes in temperature.

The process removes some of the hardness and reduces brittleness.

To do tempering, use the following steps.

1. Polish the piece to be tempered with abrasive cloth.
2. Reheat to the correct tempering temperature. Use a COLOR SCALE, Fig. 16-10, as a guide. Quench the steel when the proper color has been reached.
3. Temper small tools by placing them on a steel plate that has been heated red hot. Have the point of the tool extending beyond the edge of the plate, Fig. 16-11. Quench the piece when the correct color has reached the tool point.

Degrees Fahrenheit	Temper Colors	Tools
380	Very Light Yellow	Tools that require maximum hardness — lathe centers, lathe tools
420	Light Straw	Drills, taps, milling cutters
460	Dark Straw	Tools that need both hardness and toughness — punches
500	Bronze	Cold chisels, hammer faces
540	Purple	Screwdrivers, scribers
580	Dark Blue	Wrenches, chisels
620	Pale Blue	Springs, wood saws

Fig. 16-10. Tempering colors and temperatures for 0.95 percent carbon steel.

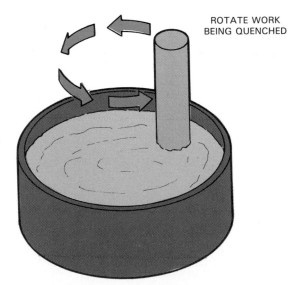

Fig. 16-9. General work can be quenched using a rotary motion. However, long, slender work should be plunged into the quenching fluid in an up-and-down motion. This prevents the piece from warping.

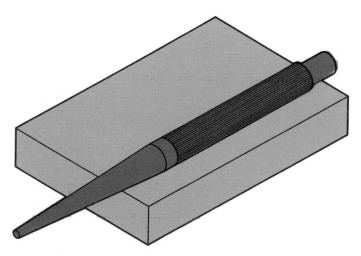

Fig. 16-11. Tempering small tools on a steel plate. It is important that the tool be in continuous motion during the tempering process, or portions of the tool will get overheated.

SAFETY

1. Wear goggles and the proper protective clothing: leather or fabric gloves and an apron (never one that is greasy or oil-soaked).
2. Heat-treating involves metal heated to very high temperatures. Handle it with the appropriate tools.
3. Never look at the flames of the furnace unless you are wearing tinted goggles.
4. Do not light the furnace until you have been instructed in its operation. If you are not sure how it should be done, ask for further instructions.
5. Be sure the area is properly ventilated and that all solvents and flammable materials have been removed.
6. Do not stand over the quenching bath when immersing heated metal.

TEST YOUR KNOWLEDGE, Unit 16

1. Heat treating involves the _____ heating and cooling of metals.
2. A heat treating process used to soften metals for easier machining.
   a. Tempering.
   b. Case hardening.
   c. Annealing.
   d. None of the above.
3. Define quenching.
4. Case hardening:
   a. Adds carbon to low carbon steel.
   b. Is used on steel parts that need a hard wearing surface
   c. Leaves the interior of the metal soft.
   d. All of the above.
5. A _____ is used to measure the temperature of the furnace.
6. List live safety precautions to follow when heat treating metals.

TECHNICAL TERMS TO LEARN

anneal	harden
brittle	heat treat
carburize	pyrometer
case harden	quench
color scale	temper
cooling	temper colors
critical temperature	toughness
glass hard	

ACTIVITIES

1. Secure examples of products that have been heat-treated.
2. Demonstrate how to anneal a piece of metal.
3. Demonstrate how to case harden a piece of low carbon steel.
4. Demonstrate how to harden and temper a center punch made from carbon steel.
5. Prepare a poster on one of the safety precautions that should be observed when heat-treating metals.

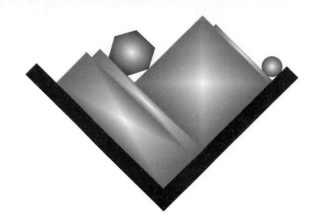

Unit 17
Machining
Technology

After studying this unit, you will be able to explain the functioning of and uses for a lathe. You will be able to set up a lathe for operation and safely perform several lathe operations. In addition, you will be able to discuss the uses and operations of shapers, milling machines, bands saws, grinding machines, and planing machines. And, finally, you will be able to explain the manual and computer-assisted operations of numerical control systems.

Many kinds of machine tools are found in the MACHINE SHOP. MACHINE TOOLS are power-driven tools. They may be used to mass produce accurate, uniform parts used in such machinery as printing presses, washers, and trains, Fig. 17-1.

THE LATHE

The lathe is a versatile machine tool. Rotating work is shaped by a cutting tool that is fed against

Fig. 17-1. Many machined parts are used in the construction of this high-speed train, which tilts as it goes around curves at speeds of up to 100 miles per hour. (Siemens)

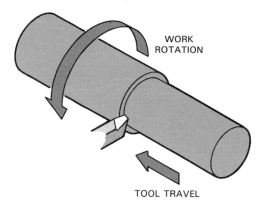

Fig. 17-2. The lathe rotates work against a cutting tool that shapes the work.

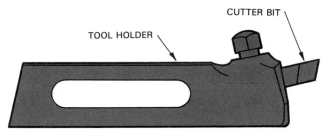

Fig. 17-4. Lathe tool holder and cutter bit.

the work, Fig. 17-2. The major parts of the lathe are shown in Fig. 17-3.

PREPARATION FOR LATHE OPERATION

Become familiar with the names and locations of the lathe parts. Learn what each does. Get permission to operate the various handwheels and levers of the lathe with the power off. The parts should move freely. There should be no binding.

Lubricate the lathe before operating it. Use the lubricants specified by the manufacturer.

Clean the lathe after each working period. Use a paint brush to remove the chips. NEVER USE YOUR HANDS. Wipe all surfaces with a soft cloth.

LATHE CUTTING TOOLS AND TOOL HOLDERS

The lathe cuts metal with a small piece of metal called a CUTTER BIT, Fig. 17-4. The cutter bit used in the school lab is usually made from a special alloy steel. This special alloy is called HIGH SPEED STEEL (HSS).

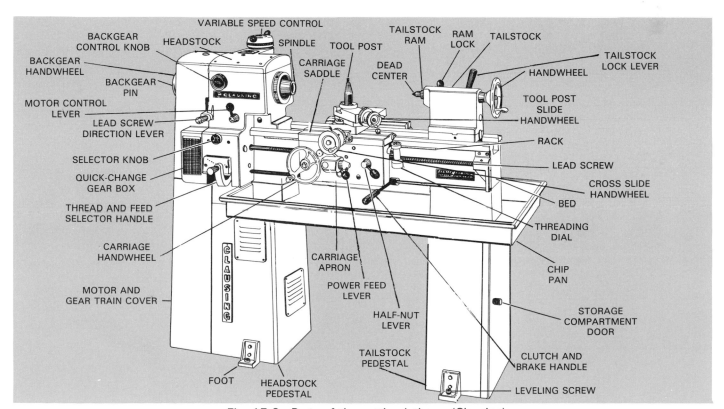

Fig. 17-3. Parts of the cutting lathe.   (Clausing)

169

On most lathes the cutter bit must be supported in a TOOL HOLDER. Tool holders are made in STRAIGHT, RIGHT-HAND, and LEFT-HAND shapes, Fig. 17-5. This allows for many different machining operations.

CUTTING TOOL SHAPES

Some cutting tools used for general turning are shown in Fig. 17-6. The cutter bit should have a keen, properly shaped cutting edge. The type of work and the kind of metal being cut will determine the shape to grind the cutter bit.

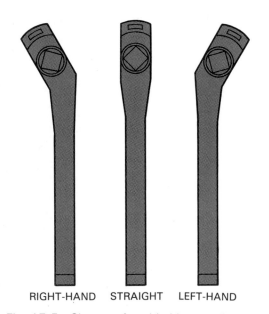

RIGHT-HAND   STRAIGHT   LEFT-HAND

Fig. 17-5. Shapes of tool holders available.

ROUGHING CUTS are used to reduce the work diameter quickly to an approximate size (about 1/32 in. oversize).

The FINISH CUT brings the work to EXACT size. The work surface is also made smooth.

The cutter bit is sharpened on a BENCH or PEDESTAL GRINDER, Fig. 17-7. The bench grinder is a grinder fitted to a bench or table. The grinding wheel mounts directly to the motor shaft. The pedestal grinder is usually larger than the bench grinder. It is equipped with a base (pedestal) that is fastened to the floor.

CAUTION: Always wear goggles when doing *any* grinding. Be sure all guards and safety shields are in place and securely fastened. Because it is not always possible to check the wheels on a grinder each time it is used, stand to one side of the machine when it is first turned on, until it reaches operating speed. This will keep you clear of flying pieces if a wheel shatters.

Before attempting to grind a cutter bit, be sure the tool rest is positioned with a space of about 1/16 in. (1.6 mm) between the tool rest and the grinding wheel, Fig. 17-8.

HOLDING WORK ON THE LATHE

Different kinds of work require different methods of holding. Two methods are the 3-jaw universal and the 4-jaw independent chucks.

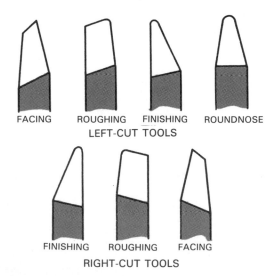

FACING   ROUGHING   FINISHING   ROUNDNOSE
LEFT-CUT TOOLS

FINISHING   ROUGHING   FACING
RIGHT-CUT TOOLS

Fig. 17-6. Common cutting tools used on the lathe.

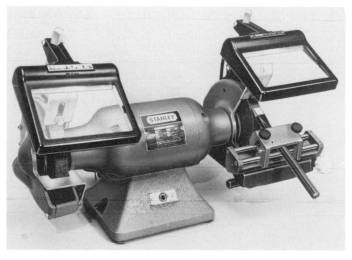

Fig. 17-7. Typical bench grinder.   (The Stanley Works)

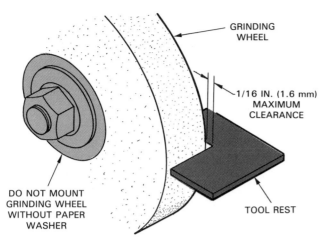

GRINDING WHEEL

1/16 IN. (1.6 mm) MAXIMUM CLEARANCE

DO NOT MOUNT GRINDING WHEEL WITHOUT PAPER WASHER

TOOL REST

Fig. 17-8. Adjusting the tool rest for grinding.

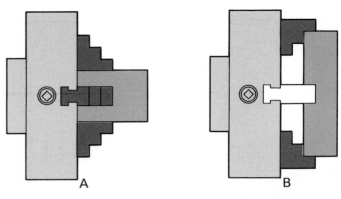

A

B

Fig. 17-10. A—Chuck jaw position for holding small diameter work. B—Chuck jaw position for holding large diameter work.

CHUCKS, Fig. 17-9, are the easiest and fastest method for mounting work for turning. They fit on the headstock spindle.

The jaws on a 3-JAW UNIVERSAL CHUCK all operate at one time. The jaws automatically center round or hexagon shaped stock.

Each jaw on a 4-JAW INDEPENDENT CHUCK operates individually. This permits irregular shaped work (square, rectangular, octagonal, etc.) to be centered.

The jaws can be reversed on the 4-jaw independent chuck to hold large diameter work, Fig. 17-10. Another set of jaws must be used with the 3-jaw universal chuck.

At first, it may appear difficult to center work on a 4-jaw chuck. Practice is all that is required.

Center the work approximately by using the concentric rings on the chuck face as a guide, Fig. 17-11. Final centering can be done using a piece

Fig. 17-9. Top. The 3-jaw universal chuck. Bottom. 4-jaw independent chuck. (L.W. Chuck Co.)

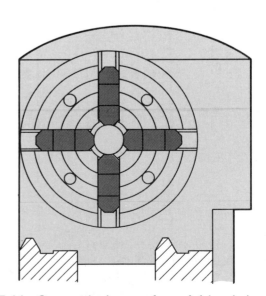

Fig. 17-11. Concentric rings on face of 4-jaw independent chuck can be used for approximate centering of work.

Fig. 17-12. Using chalk to center work in a 4-jaw chuck. The chalk mark indicates the ''high point.'' Loosen jaw opposite chalk mark, then tighten jaw on chalked side. Use small, gradual adjustments rather than large adjustments.

Fig. 17-13. The facing operation. Note that a right-hand tool holder is used.

of chalk, Fig. 17-12. Rotate the chuck slowly and bring the chalk into contact with the work. Slightly loosen the jaw(s) opposite the chalk mark. Then tighten the jaw(s) on the side where the chalk mark is located. Continue the operation until the work is centered. The cutter bit may be used instead of chalk if the work is large enough.

REMOVE THE CHUCK KEY FROM THE CHUCK BEFORE TURNING ON THE LATHE.

FACING STOCK HELD IN A CHUCK

FACING is the term used when the end or face of the stock is machined square. The tool is positioned on center and mounted as shown in Fig. 17-13. The cut can be made in either direction.

PLAIN TURNING

Plain turning, Fig. 17-14, is done when the work must be reduced in diameter. One precaution must be noted. If the work projects from the chuck more than a few inches, it should be center drilled and supported with the tailstock center, Fig. 17-15. (Also refer to Fig. 17-27.) This will prevent the work from springing away from the cutting tool while it is being machined.

Fig. 17-14. Plain turning. A left-hand tool holder is used for most operations of this type. It will permit maximum clearance between the rotating chuck and the lathe carriage.

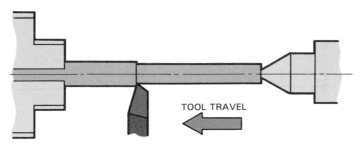

Fig. 17-15. Support long or slender work with the tailstock center. Otherwise the resulting "chatter" will cause a poor surface finish or work cut on a slight taper.

DO NOT FORGET TO LUBRICATE THE TAILSTOCK (DEAD) CENTER. Without lubrication, the tailstock center will heat up and "burn" off. Lubrication can be done with either a dab of white lead mixed with machine oil or a commercially prepared center lubricant.

The first operation in plain turning is called ROUGH TURNING. The diameter is reduced to within 1/32 in. (0.8 mm) of desired size. Use a LEFT-HAND TOOL HOLDER to hold the cutter bit. Position the tool holder as shown in Fig. 17-16.

CAUTION: Before starting the lathe, be sure that the cut can be made without danger of the rotating chuck striking the tool holder or compound, Fig. 17-17.

The tool cutting edge should be on center or slightly over center. There should not be too much overhang, Fig. 17-18. Too much overhang will cause a rough surface.

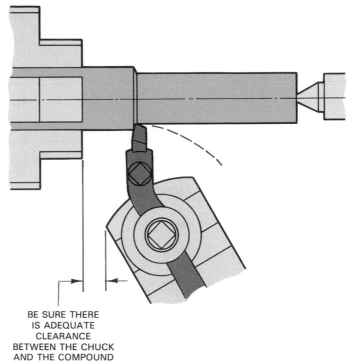

BE SURE THERE IS ADEQUATE CLEARANCE BETWEEN THE CHUCK AND THE COMPOUND

Fig. 17-17. Be sure the cut can be made without danger of the rotating chuck striking the tool holder or compound. Also, position the tool holder so it will swing clear of the work if the tool holder slips in the tool post.

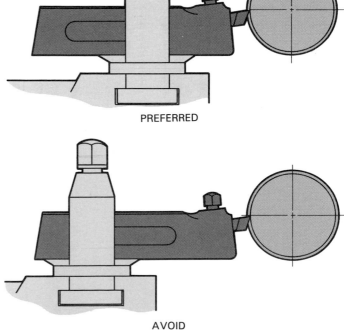

PREFERRED

AVOID

Fig. 17-18. Top. Recommended tool set up for general turning. Bottom. Avoid excessive overhang of the tool holder.

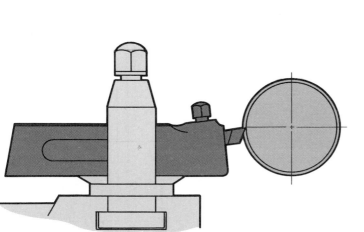

Fig. 17-16. Tool set up for general turning.

Adjust the lathe to the proper CUTTING SPEED (revolutions per minute) and FEED. Feed is the distance the cutter travels in each revolution of the work. This will vary for different metals. See Fig. 17-19.

Make the cut from the tailstock to the headstock. Feed the cutter into the stock. Engage the power feed and make a trial cut about 1/4 in. (6.4 mm) wide. Check with a micrometer or caliper. You may have to make several roughing cuts to get the work to the desired size plus 1/32 in. (0.8 mm).

Insert a finishing tool. INCREASE the speed but DECREASE the feed. Again, make a trial cut 1/4 in. (6.4 mm) wide. DO NOT CHANGE THE CROSS FEED SETTING. Measure the diameter with a micrometer. If, for example, the work is 0.006 in. (0.152 mm) oversize, the cross feed micrometer dial is fed in 0.003 in. (0.076 mm) and another cut is taken. Check your lathe. Some of them are set differently. The cross micrometer dial must be fed in 0.006 in., if 0.006 in. of metal is to be removed. When the correct diameter is reached, finish the cut.

CUTTING TAPERS

The lathe center is cut on a TAPER. There are several ways to cut tapers on the lathe. The COMPOUND REST method of turning tapers, Fig. 17-20, is perhaps the easiest. However, the taper length is limited to the compound rest travel. The base of the compound is graduated in degrees.

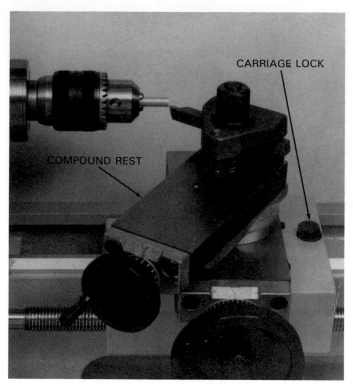

Fig. 17-20. Turning a taper using the compound rest method.

When cutting tapers, the compound rest is ALWAYS set parallel to the desired taper.

Long tapers can be cut using the TAILSTOCK SET-OVER METHOD, Fig. 17-21. To calculate the amount of set-over use the following formulas It is suggested that all fractions be converted to decimal fractions.

When taper per inch is known:

$$\text{SET-OVER} = \frac{\text{Total length of piece} \times \text{Taper per inch}}{2}$$

When taper per foot is known:

$$\text{SET-OVER} = \frac{\text{Total length of piece} \times \text{Taper per foot}}{24}$$

When the dimensions of the tapered section are known:

$$\text{SET-OVER} = \frac{\text{Total length of piece} \times (\text{Major Dia.-Minor Dia.})}{2 \times \text{Length of Taper}}$$

The set-over can be measured by using center points to determine amount of tailstock set-over, Fig. 17-22. Or it can be measured by measuring the distance between the witness marks on base of the tailstock. See Fig. 17-23.

STOCK DIA. IN IN.	MACHINE STEEL	BRASS	ALUMINUM
3/8	960	1700	2100
1/2	720	1200	1600
5/8	576	960	1280
3/4	500	800	1066
1	360	600	800
1 1/4	288	480	640
1 1/2	240	400	540
2	180	300	400

Fig. 17-19. Speeds for various metals are given in revolutions per minuts (rpm) and are only approximate. Increase or decrease as necessary for the job being machined. Feeds will vary from 0.010-0.020 in. for roughing cuts to 0.002-0.010 in. for finishing cuts.

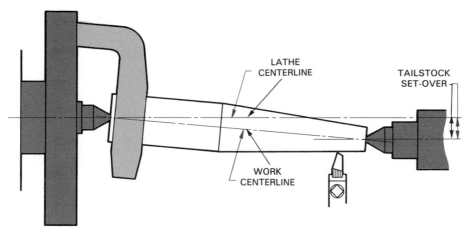

Fig. 17-21. Tailstock set-over method of turning tapers. The work must be mounted between centers.

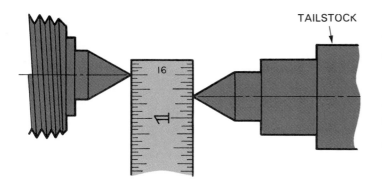

Fig. 17-22. The set-over can be measured using the center points.

When using the tailstock set-over method of turning tapers the work must be mounted between centers, Fig. 17-24.

You can also cut tapers by using the taper attachment that is built on many lathes, Fig. 17-25. The taper attachment is a guide that provides an accurate way to cut tapers.

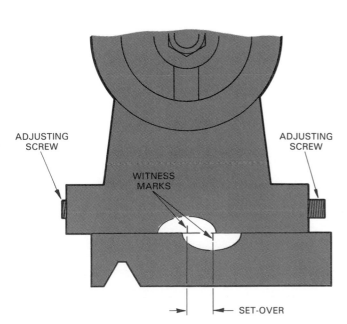

Fig. 17-23. The distance between the witness marks on the base of the tailstock can be used to measure the amount of set-over.

Fig. 17-24. Working between centers.

Fig. 17-25. Lathe fitted with taper attachment. When using this method to cut tapers, the work can be mounted in a chuck or between centers. (Clausing)

Fig. 17-27. Locating stock center using a 3-jaw chuck.

TURNING BETWEEN CENTERS

Much lathe work is done with the stock mounted between centers. Refer back to Fig. 17-24. The chuck is replaced with a FACEPLATE. A LIVE CENTER and a SLEEVE are inserted in the headstock spindle. A DEAD CENTER is placed in the tailstock. The work is connected to the faceplate with a DOG.

Before stock can be turned between centers, it is necessary to drill a CENTER HOLE in each end. Center holes are drilled with a COMBINATION DRILL AND COUNTERSINK, Fig. 17-26.

There are several ways to locate the center of the stock. An easy way is to mount the work in a 3-jaw chuck. Face the stock to length and drill the center holes, Fig. 17-27. Fig. 17-28 shows a properly drilled center hole.

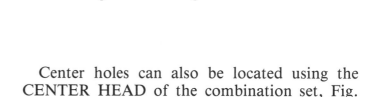

TOO SHALLOW   TOO DEEP   CORRECT

Fig. 17-28. Drilling center holes.

Center holes can also be located using the CENTER HEAD of the combination set, Fig. 17-29. The holes are drilled on a drill press.

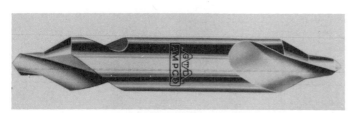

Fig. 17-26. Combination drill and countersink (center drill). (Greenfield Tap and Die)

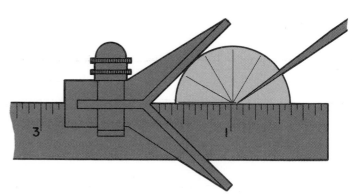

Fig. 17-29. Using the center head to locate the center of round stock.

The centers must run true for accurate work. Check center alignment by bringing them together, Fig. 17-30. A tapered piece will result if they are not aligned. Adjust the tailstock to bring them into alignment.

Clamp a dog on one end of the stock. Place a dab of lubricant (white lead and oil) in the center hole at the tailstock end. Insert the centers in the holes. Adjust the tailstock center until it is snug enough to prevent the lathe dog from "clattering." Check the adjustment from time to time. The heat generated during the machining operation causes the work to expand. The tailstock center may burn off if too much heat is generated.

The cutting operation is identical to machining work mounted in a chuck.

If the work must be reversed to machine its entire length, protect the section under the dog setscrew. This can be done by inserting a piece of soft copper or aluminum sheet, Fig. 17-31.

DRILLING

Drilling can be done on the lathe. Hold the work in a chuck and mount a Jacobs chuck in the tailstock to grip the drill, Fig. 17-32.

Start the hole by first "spotting" the work with a center drill.

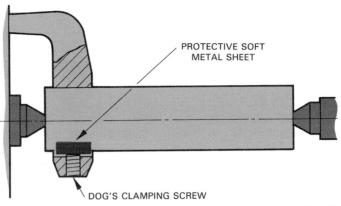

Fig. 17-31. Protect the work by placing a piece of soft aluminum or copper under the clamping screw.

Fig. 17-32. Drilling with a straight shank drill held in a Jacobs chuck.

Drills up to 1/2 in. (12.7 mm) diameter are held in a Jacobs chuck. Larger drills are mounted in the tailstock spindle. See Fig. 17-33.

Fig. 17-30. Center alignment can be checked by bringing centers together.

Fig. 17-33. Drilling on the lathe.

177

A PILOT HOLE (small hole) should be drilled first if a drill larger than 1/2 in. (12.7 mm) diameter is used, Fig. 17-34.

BORING

BORING, Fig. 17-35, is done when a hole is not a standard drill size. It is also done when a very accurate hole diameter is needed.

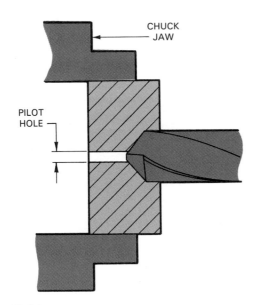

Fig. 17-34. A pilot hole permits larger drills to cut easier and faster.

A BORING TOOL HOLDER supports the BORING BAR. A cutter bit is inserted in the boring bar. The hole is first drilled slightly smaller than the desired diameter. The cutter tool is set on center with the boring bar parallel to the centerline of the hole. The cutting is done in much the same manner as is external turning.

KNURLING

KNURLING, Fig. 17-36, is the operation that presses horizontal or diamond shaped serrations on the circumference of the stock. This provides a gripping surface.

To knurl, complete the following steps.

1. Mark off the section to be knurled.
2. Put the lathe into back gear (slow speed) with a fairly rapid feed.
3. Position the knurling tool so that both knurl wheels bear evenly and squarely on the work.
4. Start the lathe and feed the knurl slowly into the work until a pattern develops. Engage the automatic feed. USE A LUBRICANT ON THE KNURL WHEELS. After the knurl has traveled the desired distance, reverse the spindle rotation. Apply additional pressure. Repeat until a good knurl has formed.

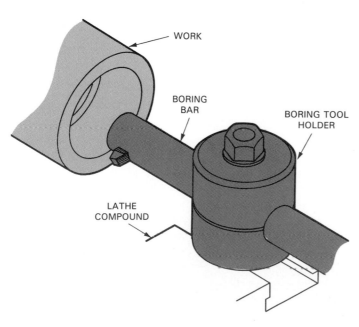

Fig. 17-35. Boring on the lathe.

Fig. 17-36. Knurling.

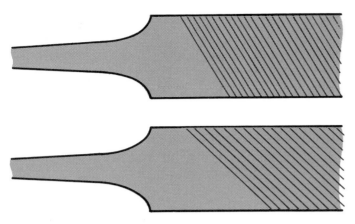

Fig. 17-37. Top. Long angle lathe file. Bottom. Compare the long angle file with this regular file.

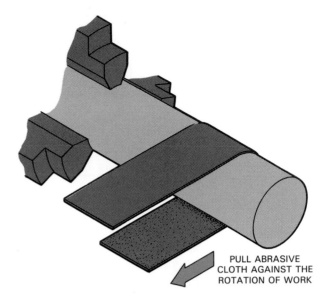

PULL ABRASIVE CLOTH AGAINST THE ROTATION OF WORK

Fig. 17-39. Polishing on the lathe. Do not forget to protect the lathe ways from the abrasive particles.

FILING AND POLISHING ON THE LATHE

If the cutter bit is properly sharpened, there should be no need to file or polish work machined on the lathe. However, due to lack of experience, you may wish to smooth and polish a machined surface.

A fine mill file or long angle lathe file should be used, Fig. 17-37. Take long even strokes across the rotating work. Hold the file as shown in Fig. 17-38. Keep the file clean and free from chips.

A filed surface can be made very smooth by using fine grades of abrasive cloth, Fig. 17-39. Apply oil to the abrasive cloth. Keep the cloth moving across the rotating work. It is recommended that you protect the lathe ways from the abrasive particles.

CUTTING THREADS ON THE LATHE

Threads can be cut on many lathes, Fig. 17-40. The cutter bit is sharpened to the shape of the desired threads. The feed and speed selector controls are adjusted to move the carriage and cutter bit across the work. This is done at a rate that will cut the desired number of threads per inch.

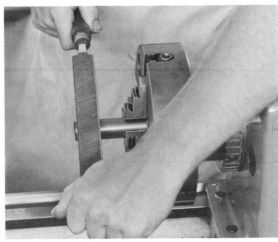

Fig. 17-38. Left. Left-hand filing method. The hands are clear of the rotating chuck. Right. Right-hand filing method. The left hand and arm must be over the revolving chuck.

Fig. 17-40. Cutting threads.

Cutting threads is a complex operation. The correct way to set up the lathe and how threads are cut is best demonstrated by your instructor.

THE SHAPER

The SHAPER, Fig. 17-41, is another machine tool often found in the metals lab. It is used primarily to machine flat surfaces, Fig. 17-42. However, a skilled machinist can use it to cut curved and irregular shapes, grooves, and keyways. The cutting tool is mounted to a ram which moves back and forth across the work.

Mount the work in the vise or clamp it directly to the worktable. The table can be moved horizontally or vertically. The head, to which the tool holder is mounted, can be pivoted for angular cutting, Fig. 17-43.

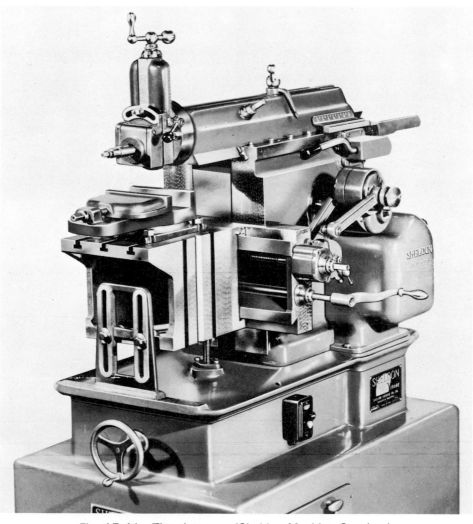

Fig. 17-41. The shaper. (Sheldon Machine Co., Inc.)

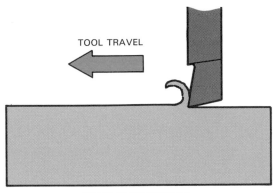

TOOL TRAVEL

WORK IS STATIONARY

Fig. 17-42. How a shaper works.

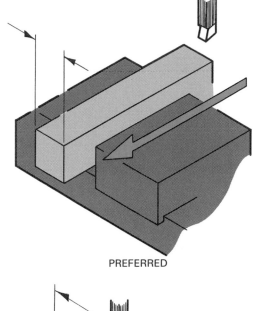

PREFERRED

Fig. 17-43. The head of the shaper can be pivoted to make angular cuts.

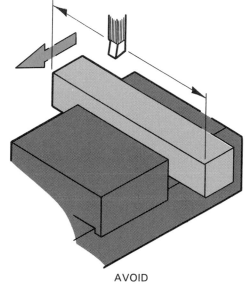

AVOID

Fig. 17-44. Mount the work in the shaper so that the work will be done in the least amount of time.

Position the work so that the cut can be made in the shortest possible time, Fig. 17-44. The STROKE (length of cut) is adjustable and should be positioned as shown in Fig. 17-45.

SHAPER SAFETY

KEEP YOUR FINGERS CLEAR OF THE CUTTING TOOL WHILE ADJUSTING THE STROKE.

DO NOT RUB YOUR FINGERS ACROSS THE WORK WHILE THE TOOL IS CUTTING.

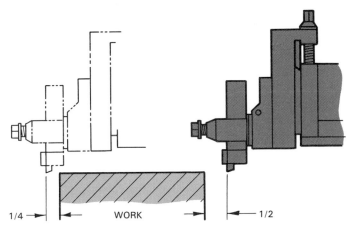

1/4   WORK   1/2

Fig. 17-45. The correct way to position the shaper stroke for cutting.

THE MILLING MACHINE

Industry uses many types of milling machines. The VERTICAL MILLING MACHINE, Fig. 17-46, and the HORIZONTAL MILLING MACHINE, Fig. 17-47, are most commonly found in the school shop.

Milling machines operate by a rotating multi-toothed cutter being fed into a moving piece of work, Fig. 17-48. The horizontal milling machine uses a cutter mounted on an arbor to smooth and shape metal, Fig. 17-49. The cutter used on a vertical milling machine is called an END MILL. It clamps in the machine spindle, Fig. 17-50.

Special attachments permit gears to be cut on the horizontal milling machine, Fig. 17-51.

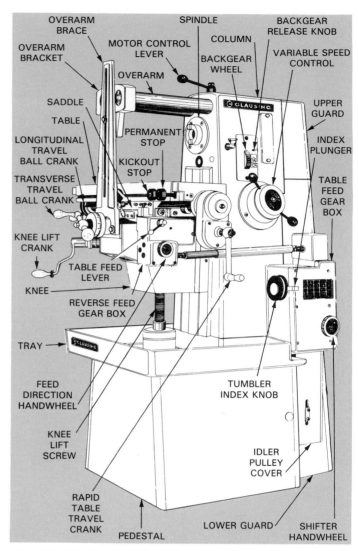

Fig. 17-47. Horizontal milling machine with the parts identified.

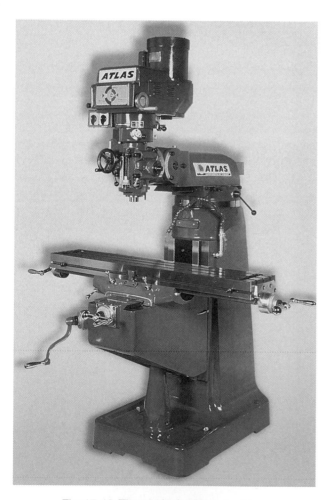

Fig. 17-46. The vertical milling machine.
(Clausing Industrial, Inc.)

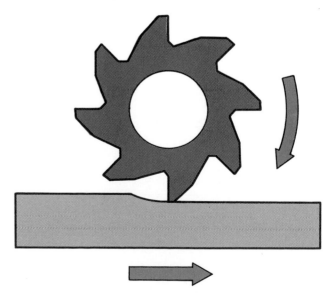

Fig. 17-48. How the milling machine works.

Fig. 17-49. The cutter on the horizontal milling machine is mounted on an arbor.

One kind of unique machine tool is a combination LATHE/MILLING MACHINE, Figs. 17-52 and 17-53. It is being used often in school labs. The machine tool is used for basic milling and lathe operations.

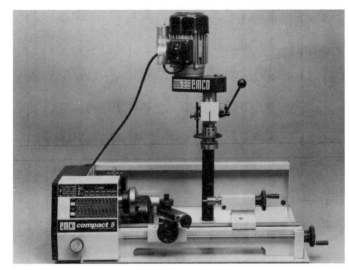

Fig. 17-52. The lathe/miller is a combination lathe and milling machine.

Fig. 17-50. An end mill is used to do the cutting in a vertical milling machine.

Fig. 17-51. Cutting gears. A dividing head is used to space the gear teeth.

Fig. 17-53. The milling machine portion of the lathe/miller.

Milling machines range in size from a small vertical milling machine similar to those used in the school metals lab, to the 50 hp horizontal milling machine shown in Fig. 17-54.

OTHER MACHINE TOOLS

Many other types of machine tools are used by industry. Several of the more widely used machines are discussed in the following sections.

BANDSAW

The cutting operation done on the METAL CUTTING BANDSAW, Fig. 17-55, is called BAND MACHINING. The bandsaw used one-piece saw blade to do the cutting. Cutting can be at any angle or in any direction, Fig. 17-56.

PRECISION GRINDING MACHINES

There are many types of precision grinding machines. The SURFACE GRINDER, Fig. 17-57, is frequently found in school labs. The machine uses a grinding wheel to produce a smooth, accurate surface on material regardless of its hardness. The work moves back and forth under the grinding wheel, Fig. 17-58. There are two types of surface grinders: the planer type and the rotary type.

Fig. 17-54. This 50 hp horizontal milling machine can make a cut 1/8 in. (3 mm) deep by 12 in. (305 mm) wide. (National Machine Tool Builders Assoc.)

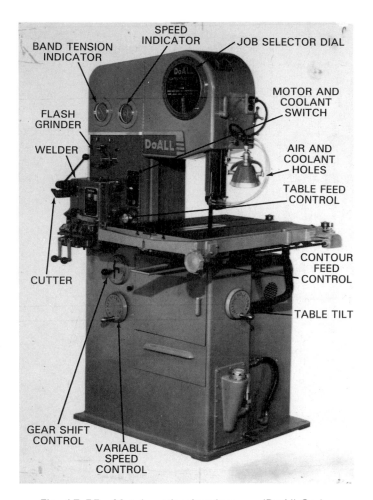

BAND TENSION INDICATOR — SPEED INDICATOR — JOB SELECTOR DIAL — FLASH GRINDER — WELDER — CUTTER — MOTOR AND COOLANT SWITCH — AIR AND COOLANT HOLES — TABLE FEED CONTROL — CONTOUR FEED CONTROL — TABLE TILT — GEAR SHIFT CONTROL — VARIABLE SPEED CONTROL

Fig. 17-55. Metal cutting bandsaw. (DoAll Co.)

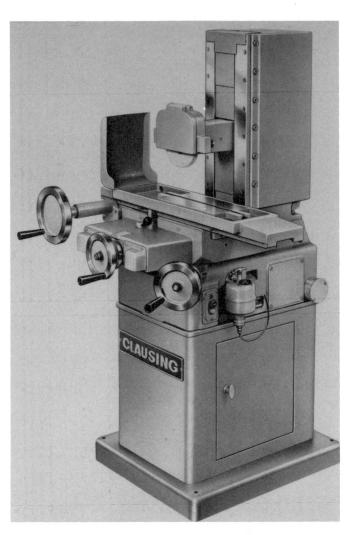

Fig. 17-57. Surface grinder.

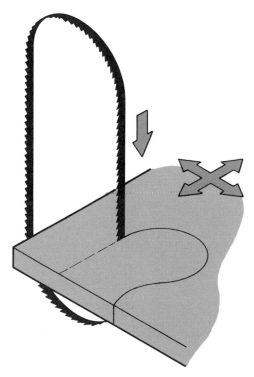

Fig. 17-56. Bandsaws machine at any angle or in any direction. The length of the cut is unlimited.

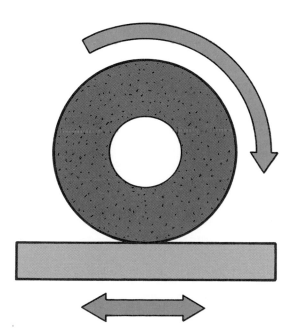

Fig. 17-58. How the surface grinder works.

PLANING MACHINES

PLANING MACHINES produce horizontal, vertical, and angular surfaces on metal much like the shaper does.

On the PLANER, Fig. 17-59, the material is mounted on the worktable. The worktable moves back and forth under the cutting tool, Fig. 17-60. The planer is used to machine large, flat surfaces.

The BROACH employs a long tool with many cutting surfaces. Each tool is slightly higher than the one before. Each increases in size to the exact size required, Fig. 17-61.

The tool is pushed or pulled across the surface to be machined, Fig. 17-62. Many flat surfaces on automobile engines are broached.

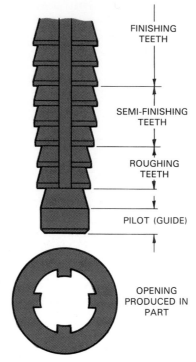

Fig. 17-61. Typical broaching tool.

Fig. 17-59. Large planer. Note the size of the work being machined. (G.A. Gray Co.)

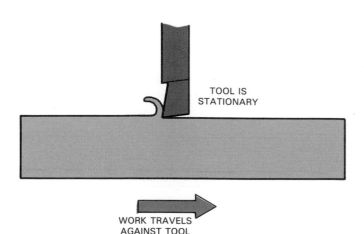

Fig. 17-60. How the planer works.

Fig. 17-62. Vertical broach machining the flat surface of compressor cylinder heads.
(National Machine Tool Builders Assoc.)

COMPUTER-ASSISTED MANUFACTURING TECHNOLOGY

The machine tools in your school metals lab are operated by hand controls. You must turn a handwheel or pull a lever to move the work against the cutting tool, Fig. 17-63. Some of the machines may also have power feed mechanisms.

Fig. 17-63. Some of the machines in your school's metals lab may be similar to those shown. They are operated by hand controls.

Fig. 17-64. This robot, or automated cell, consists of several machines with robotics material handling. Parts to be machined are delivered and finished parts are removed automatically as required. The cell becomes a fully automatic process through the application of robotics and other forms of automation. (Kearney & Trecker Corp.)

Industry employs many machine tools that perform machining operations automatically, even to the point where robots load and unload the machine, Fig. 17-64. In the system known as CAM (COMPUTER-ASSISTED MANUFACTURING or COMPUTER-AIDED MANUFACTURING), a computer "tells" the machine the required sequence of operations to be performed, the distance and direction the work is to travel, where the cutting is to be done, and the cutting tool(s) to be utilized.

NUMERICAL CONTROL or NC is the most widely used form of CAM. NC is an automated manufacturing technique. Coded numerical instructions, Fig. 17-65, direct a machine tool

PREPARATORY FUNCTIONS (G-CODES)*	
CODE	FUNCTION
G00	Rapid traverse (slides move only at rapid traverse speed).
G01	Linear interpolation (slides move at right angles and/or at programmed angles).
G02	Circular interpolation CW (tool follows a quarter part of circumference in a clockwise direction).
G03	Circular interpolation CCW (tool follows a quarter part of circumference in a counterclockwise direction).
G04	Dwell (timed delay of established duration. Length is expressed in X or F word).
G33	Thread cutting.
G70	Inch programming.
G71	Metric programming.
G81	Drill.
G90	Absolute coordinates.
G91	Incremental coordinates.

*G-CODES may vary on different NC machines.

MISCELLANEOUS FUNCTIONS (M-CODES)*	
CODE	FUNCTION
M00	Stop machine until operator restart.
M02	End of program.
M03	Start spindle—CW.
M04	Start spindle—CCW.
M05	Stop spindle.
M06	Tool change.
M07	Coolant on.
M09	Coolant off.
M30	End program and rewind tape.
M52	Advance spindle.
M53	Retract spindle.
M56	Tool inhibit.

*M-CODES may vary on different NC machines.

Fig. 17-65. Example of NC codes for preparatory and miscellaneous functions.

through a sequence of operations that produce the required part shape. Known as a PROGRAM, the coded instructions consist of ALPHANUMERIC DATA (letters, numbers, punctuation marks, and special characters). Each identifies a specific machine function.

NC coded information is sometimes punched into paper or mylar tape, Fig. 17-66, or recorded on magnetic tape, but is more often placed directly into the on-board machine tool computer, Fig. 17-67, as electronic data. Information also may be transmitted to a machine's control unit through a computer network.

The programmed information instructs small electric motors on the machine tool, called SERVOS, telling them how far and in what direction the work is to move. It also tells the cutting tool when to start cutting, how deep to cut, and when to stop cutting.

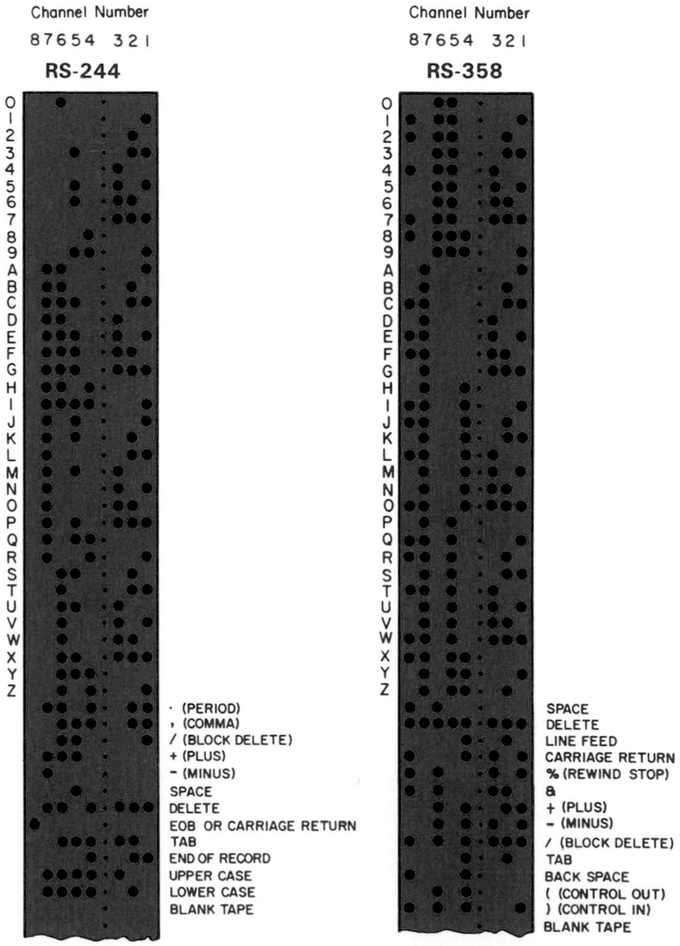

Present day NC equipment may be CNC (COMPUTER NUMERICAL CONTROL) or DNC (DIRECT NUMERICAL CONTROL). With CNC, one computer typically controls one machine tool, Fig. 17-68. DNC involves a single master computer linked to several machine tools and able to operate them simultaneously or individually.

Fig. 17-67. CNC milling/drilling machine found in many school metals labs. Note the on-board computer. (Dyna Electronics, Inc.)

Fig. 17-66. Tape used on manually programmed NC machines and the tape code.

Fig. 17-68. With CNC, one computer controls one machine. The machinist can input the program directly into the computer. (Bridgeport Machines, Inc.)

CARTESIAN COORDINATE SYSTEM

Work positioning is by the CARTESIAN COORDINATE SYSTEM, Fig. 17-69. This system is the basis of all NC programming. The X and Y axes are horizontal work movements. The Z axis is vertical work or cutting tool movement.

There are two basic NC movement systems:
- POINT-TO-POINT. There is no concern what path the tool taes when moving from Point A to Point B. See Fig. 17-70. This system is widely used for drilling, spot welding, and punching openings in sheet metal. A computer is not usually needed to prepare point-to-point programs.
- CONTOUR OR CONTINUOUS PATH. Employed to profile and contour complex two- and three-dimensional workpieces, Fig. 17-71. The cutting tool remains in continous contact with the work. Programming cannot be accomplished without a computer.

CNC technology has reached the position where a engineer can, using COMPUTER GRAPHICS, design a part and see how it will work with other parts in the assembly, Fig. 17-72. When the design has been confirmed, the computer is directed to figure out the tool paths required to make it. The tool paths are converted into a detailed sequence of commands. These commands are carried out by the proper CNC machine tool.

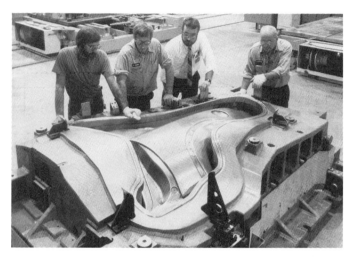

Fig. 17-71. Auto body dies (shaped steel blocks used to form body sections) are machined using contour or continuous path NC milling. (Buick)

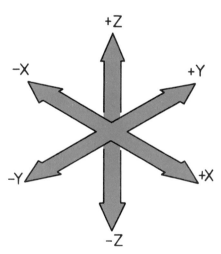

Fig. 17-69. The Cartesian Coordinate System is the basis of all NC programming.

Fig. 17-72. Computer-Aided Design (CAD) starts by defining the outline of the part being designed. After being analyzed for correct specifications, cutter paths for NC machining are defined and modified to meet production needs. Here, an engineer is starting to design an engine mount for a new type of aircraft. (Beech Aircraft Corp.)

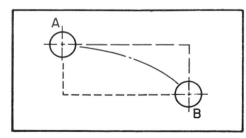

Fig. 17-70. In point-to-point NC systems, there is no concern what path is taken when the tool moves from Point A to Point B.

The technology that makes all of this possible is called CAD/CAM (COMPUTER-AIDED DESIGN/COMPUTER-AIDED MANUFAC-TURING).

ROBOTICS

Industrial ROBOTS, Fig. 17-73, are found in numerous applications. They can be programmed to do many types of jobs: loading and unloading machines, drilling, spot welding, Fig. 17-74, inspection, and paint spraying. They are often found performing jobs that would be tedious or hazardous for human workers. Many varieties of robots have been developed, Fig. 17-75.

MACHINING SAFETY

1. Have ALL guards in place before operating any machine.
2. When using the drill press, clamp the work to the table or mount it solidly in a vise. Otherwise, the work might spin rapidly, possibly causing injury.
3. Never reverse a machine before it comes to a full stop. This could cause the chuck on some lathes to spin off.
4. ALWAYS remove the chuck key from the lathe or drill press chuck.
5. Stop machines before making measurements and adjustments.
6. Do not operate portable electric power tools in areas where thinners and solvents are used and stored. A serious fire or explosion might result.
7. Roll up sleeves and remove dangling jewelry before starting to work.
8. ALWAYS WEAR GOGGLES.
9. Get cuts, burns, and bruises properly treated.
10. Do not wear loose clothing when operating machines. The clothing may get caught in a machine, causing injury.

Fig. 17-74. Industrial robots like these are usually found performing jobs that are tedious or hazardous for human workers. (Ford Motor Co.)

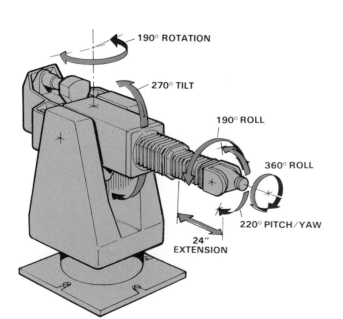

Fig. 17-73. A typical industrial-type robot. It has great freedom of movement.

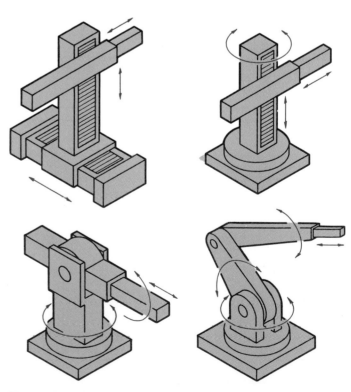

Fig. 17-75. A few variations of industrial robot design. Their movements are planned for performing specific jobs.

TEST YOUR KNOWLEDGE, Unit 17

Please do not write in the text. Place your answers on a separate sheet of paper.

1. _____ _____ are power driven tools
2. What object is used to clean a lathe?
   a. Hands.
   b. Paint brush.
   c. Screwdriver.
   d. Vacuum.
3. What is a cutter bit?
4. A _____ cut is used to quickly reduce the diameter of work.
5. What operation would you perform to reduce work in diameter?
   a. Facing.
   b. Boring.
   c. Plain turning.
   d. None of the above.
6. Cutting speed is measured in:
   a. Revolutions per minute.
   b. Revolutions per second.
   c. Revolutions per hour.
   d. None of the above.
7. Feed is the distance the _____ travels in each revolution of work.
8. List two ways to cut tapers on the lathe.
9. What machine tool is used to machine flat surfaces?
10. What cutting operation is done on the band saw?
11. In numerical control, work is positioned using the _____ _____ _____.

MATCHING QUESTIONS: Match each of the following terms with their correct definitions.
a. Boring.   d. Center head.
b. Boring tool holder.   e. Center holes.
c. Knurling.
12. ____ Operation that presses horizontal or diamond-shaped serrations on stock.
13. ____ Drilled with a combination drill and countersink.
14. ____ Done when a hole is not standard flat size.
15. ____ Used to locate center holes.
16. ____ Supports a boring bar.
17. Name two types of milling machines.
18. List five safety rules to follow in the machine shop.
19. It costs $25 to manufacture a part using conventional machine tools. The cost is reduced by 20 percent when the part is manufactured on a CNC machine tool. How much money will be saved using a CNC machine tool to produce 500 parts?

TECHNICAL TERMS TO LEARN

band saw
band machining
bench grinder
boring
broach
Cartesian coordinate
  system
center hole
computer-aided
  drafting
computer-aided
  manufacturing
computer graphics
Computer Numerical
  Control (CNC)
compound
concentric
cross feed
cutter bit
cutting speed
dead center
Distributive Numerical
  Control (DNC)
dog
drill
end mill
facing

head stock
high speed steel
horizontal milling
  machine
independent chuck
Jacobs chuck
knurl
lathe
machine tool
Numerical Control
  (NC)
parallel
pedestal grinder
planner
program
revolutions per minute
robot
shaper
spindle
surface grinder
tailstock
taper
threads
tool holder
turning
universal chuck
vertical milling machine

ACTIVITIES

1. Carefully clean and lubricate a lathe in your school shop. Use the lubrication chart furnished with the lathe to be certain that you do not overlook critical points.
2. After securing permission from your instructor, operate the various handwheels and levers on the lathe. THE POWER MUST BE OFF. Learn what each does. Do not force any movement.
3. Sharpen a cutter bit.
4. Prepare a check list and keep a record of the various operations you perform on the lathe. Prepare a similar chart for other machines that you operate.
5. Prepare a series of posters on the safe operation of machine tools.

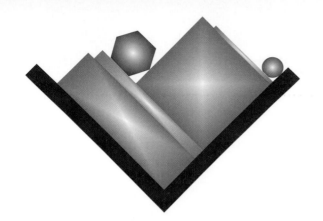

Unit 18
Entrepreneurship Activity

After completing the activity in this unit, you will be able to describe the place of the entrepreneur in the American economy. In addition, you will have gained first-hand experience in the business world. This includes experience in organizing, financing, and operating your own small manufacturing business. You will do many of the things large business does, on a smaller scale. You will obtain experience that will make you a better employee when you are ready for your first full-time job.

AN INTRODUCTION TO PRODUCTION

It has been said the average person will work in at least seven different occupations during their lifetime. It is also said that the technologies that will be involved in four of them have not yet been invented or discovered.

These new technologies will be developed and put into production by **entrepreneurs.** They are the men and women (either individually or as a group) who originate new business ventures and create jobs. They are willing to risk capital (money) and time to establish a new enterprise and make a profit if the business is a success. Entrepreneurs and the businesses they establish are thought by many to be the backbone of the American economy.

Entrepreneurs enjoy the challenge of starting a new business. However, after the venture is started and operating, many of them turn day-to-day operations over to men and women with managerial skills.

You will have the experience of being an entrepreneur if your class decides to manufacture a product by the means described in this unit.

ESTABLISHING A BUSINESS ENTERPRISE

The business activity in this unit may be changed to meet local conditions and restrictions.

The procedure to follow in a typical setup includes:

1. Deciding on a product.
2. Selecting a name for your new company.
3. Electing company officials.
4. Determining approximate amount of operating capital (money) needed to finance the business.
5. Selling shares of stock to raise operating capital.
6. Determining the manufacturing steps required, and sequence of operations.
7. Providing jigs and fixtures needed for mass production.
8. Developing sales plan; promoting product sales.
9. Dissolving the business and reimbursing stockholders.

DECIDE ON PRODUCT

At your first meeting (with your instructor serving as advisor) decide on the product to be manufactured. Select one of the tested products covered later in this unit.

SELECT COMPANY NAME

The name selected should be appropriate, businesslike, and brief.

This is a class activity in which all students are expected to participate. Therefore, each classmember should come up with a good name for the new company that is being organized.

It is suggested that each class member write one or more names on a slip of paper. The slips should be collected, proposed names discussed, and a vote taken. The vote will determine which company name is to be used.

Note: Do not include Incorporated, or Corporation after the company name. Each of these is a legal term which can be used only in cases where a state charter has been obtained.

COMPANY OFFICIALS

If your company is to be a success, it must be run in an efficient, businesslike manner. This means you will need capable company officials and capable workers. It is suggested that you elect:

1. General Manager.
2. Office Manager.
3. Purchasing Agent.
4. Sales Manager.
5. Safety Director.

Duties of the company officials are described in the following paragraphs. Each office should be discussed before holding an election.

DUTIES OF GENERAL MANAGER

Your General Manager will be expected to:

1. Exercise general supervision over entire activity.
2. Train workers, assign workers to jobs.
3. Check on, and be responsible for, product quality and manufacturing efficiency.
4. Approve bills to be paid by Office Manager. Approve purchase orders before making purchases.
5. Prepare and submit reports as required by your Advisor (instructor).

6. Cooperate with your Advisor and other Company officials.

DUTIES OF OFFICE MANAGER

The Office Manager should:

1. Maintain attendance records. Check absentees for valid excuses.
2. Keep company's financial records (8 1/2 x 11 loose leaf notebook suggested).
3. Keep a record of all money received on RECEIPTS page of record book, Fig. 18-1.
4. All cash receipts should be deposited in a local bank and a checking account established.
5. Pay all bills (invoices previously approved by your General Manager) by writing checks. Make a complete record in your checkbook. Show dates and amount of money deposited in checking account, and checks written. Show the balance. This is the amount in bank, after each check is written. See Fig. 18-2. Be sure

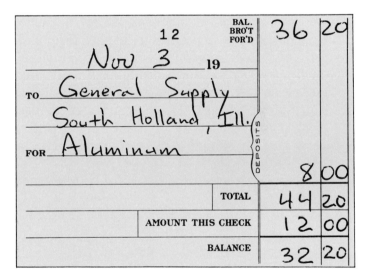

Fig. 18-2. This business activity record is a checkbook record of bills paid.

A-1 PRODUCTS, 123 W. Taft, South Holland, Ill. 60473

RECEIPTS

Date	Item	Quantity	Sold To	Sold By	Amount

Fig. 18-1. Business activity records for cash receipts.

your figures are correct. This is the only record your company has of the cash on hand.

6. Keep stock sales and ownership records.
7. Cooperate with your Advisor and other company officials.

DUTIES OF PURCHASING AGENT

The Purchasing Agent should:

1. Check with Advisor to obtain materials and supplies needed to manufacture product selected.
2. Arrange to have stock certificates, purchase orders, and other business forms printed.
3. Have purchase orders approved by General Manager before issuing order.
4. Obtain materials required from school stock or purchase from outside source.
5. Cooperate with Advisor and other company officials.

DUTIES OF SALES MANAGER

The Sales Manager should:

1. Help decide how product is to be packaged, and provide instructions on product use, to include on the package. This is done after a product sample is available.
2. Plan sales program; decide on price to charge

Fig. 18-3. Identification badge for company official.

for product, where and how product is to be sold. Note: All students participating in this activity will be expected to sell the company's products (also stock).
3. Cooperate with Advisor and other company officials.

DUTIES OF SAFETY DIRECTOR

The Safety Director should:

1. Enforce safety rules.
2. Stop all horseplay and call attention to undesirable conduct.
3. Be alert to hazards.
4. Take steps to eliminate possible accident causing problems.
5. See to it that equipment and tools are in good operating condition.
6. Make sure machine guards are in place and students are using eye protection and protective clothing as specified by the Advisor.
7. Immediately report all injuries, even minor ones, to the Advisor.
8. Cooperate with Advisor and other company officials.

All company officials should be provided with identification badges, Fig. 18-3.

RAISING CAPITAL

Operating your new business requires money (capital). This capital will be used to pay bills until cash is available from the sale of your product. Many corporations raise money by selling stock to the public. It is sugggested you do the same.

Getting stock certificates printed and sold, and keeping the necessary records, is the duty of the Office Manager. See Fig. 18-4.

A-1 PRODUCTS, 123 W. Taft, South Holland, Ill. 60473

STOCK RECORD

Date	Certif. No.	Sold To	Sold By	Amount

Fig. 18-4. Stock ownership record.

STOCK CERTIFICATE

A–1 PRODUCTS
South Holland, Ill.

One Share

Par Value 50¢

Certificate Number

_____

Redeemable Within
One Year After Issue

Date Issued

_____

This Certifies That (please print)

_____

First Name   Initial   Last Name

_____

Street Address   City   State   Zip

Is the Owner of One Share, Par Value 50¢, of the stock of A–1 Products.

_____   _____

Stockholder's Signature   for A–1 Products

Stockholder by signature, okays operation of Company by Student Officials elected
by Student Stockholders.

Fig. 18-5. Sample stock certificate form.

An example might help to illustrate this task. Let us say you need $40.00. This may be raised by selling 80 shares of stock at 50 cents each. See Fig. 18-5 for a suggested stock certificate form.

Stock certificates might be run off on a duplicating machine on two different colors of stock. Make 100 copies of each color. Yellow and white are suggested. Number the certificates 1 to 100.

Each student should purchase a share of stock at 50 cents. Sell stock on your own time. It is a good idea to make your first sale to yourself.

Use a ball point pen and carbon paper to fill out the stock certificate form. Put the white copy on top. On the share you sell to yourself sign where the stockholder's signature is required. Also sign as a representative of your company. Keep the white copy. Turn in the yellow copy (carbon) and the 50 cents to your Office Manager. It is his or her job to keep an accurate record of stock sales. The Office Manager must also see to it that the cash turned in is deposited in the bank. See Fig. 18-6.

It is suggested you limit stock sales to one share per customer. The stock should be divided so each

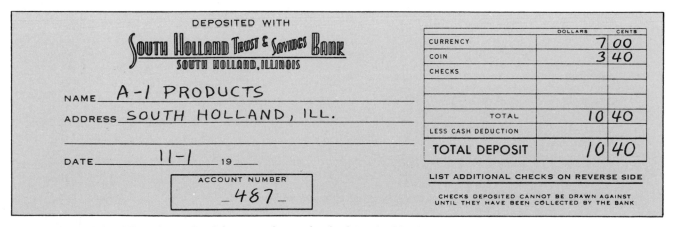

Fig. 18-6. All cash received from product sales is deposited in the bank and bills are paid by check.

student has roughly the same number of shares to sell. When contacting those persons who might purchase stock, (local businessworkers, parents, neighbors, friends) be businesslike. Briefly describe your company and its operations. Be enthusiastic. Give your prospect the idea that he or she is buying a share of a "going business," and is NOT throwing money away. Tell him or her you cannot ensure results in advance, but that your company expects to make profits. Also assure your prospect that when the business is closed, you will redeem the stock at full value (50 cents per share) and pay a small dividend.

After a share of stock has been sold, fill out the stock certificate as just described. Please print. Ask your purchaser to read the certificate to make sure everything is understood and is agreeable. Then ask purchaser to sign. Hand the top (white) copy to your purchaser.

SETTING A PRICE AND SELLING A PRODUCT

In setting a selling price for your product, determine the total cost of making each item. Then add about 25% to this to cover the cost of material wasted, and to pay a dividend to stockholders when the business is closed.

Each participant in the project is expected to help sell the company's product.

If the product selected for manufacture is small, such as the two items described in this unit, make the products in advance. Then you can provide "on the spot" delivery.

Keep an current record of all sales. This includes quantity sold, name of customer, and amount col-

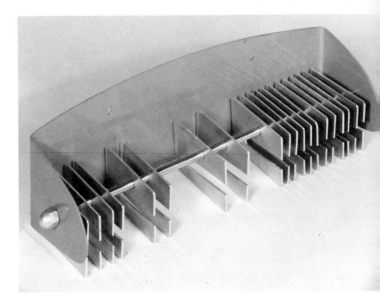

Fig. 18-8. Tie rack suitable for mass production.

lected. Turn these records and any money collected over to your Office Manager for handling.

At the end of the mass production project, the company you formed should be liquidated (closed). Money in the bank (after all bills are paid) should be divided in an equal manner among stockholders.

Include with each of the checks a memo informing each stockholder that the business is being closed. Fig. 18-7 shows a sample memo.

MASS PRODUCTION IN THE METALS LAB

Two mass production projects with proven student interest are the tie rack and the contemporary lamp.

(today's date)

TO: A-1 stockholders

FROM: A-1 company officials

SUBJECT: A-1 Products liquidation

A-1 Products, the business experience activity you helped finance by buying a share of stock, is being liquidated.

The check, in the amount of *(insert dollar amount of check)* is enclosed. This is intended to redeem the stock you purchased. The stock certificate you have does not need to be returned to us.

Thank you very much for your support of our project.

Fig. 18-7. Sample memo to send to stockholders telling of A-1 Products liquidation.

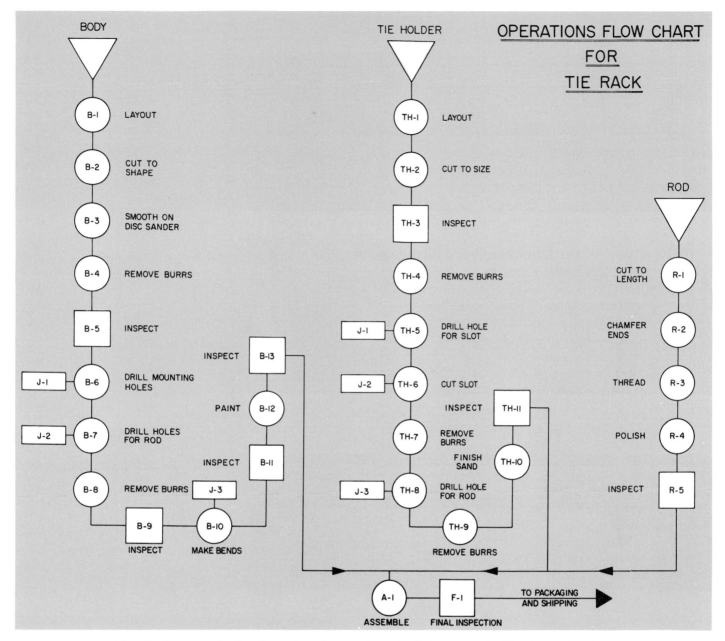

Fig. 18-9. A production flow chart for the tie rack. Only one method is shown for the manufacture of the tie holder.

IMPORTANT: No production project should be attempted until product samples are made. This will assure that the project can be made with the equipment available.

MASS-PRODUCING A TIE RACK

The TIE RACK, Fig. 18-8, provides unique production problems. But it is simple enough to be produced in a short time.

The PRODUCTION FLOW CHART, Fig. 18-9, shows a suggested sequence to follow in production. It includes such steps as manufacturing, in-specting, assembling, and finishing operations. The facilities and equipment located in your shop may require some changes.

Tie rack plans are shown in Fig. 18-10. A tie rack of the size indicated will hold about 20 ties. A smaller rack may be made up by using shorter body pieces.

BODY

The body (main part) of the tie rack is made from .080 in. (2.3 mm) half-hard aluminum sheet. Make a pattern on .030 aluminum and use it to

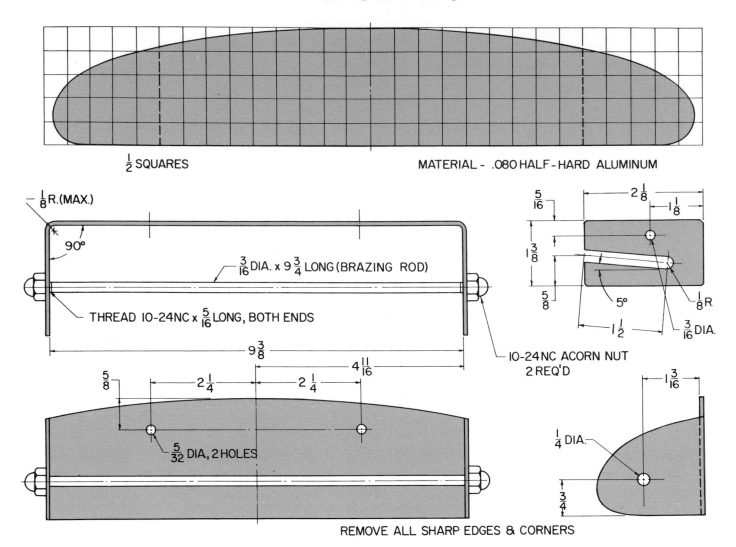

$\frac{1}{2}$ SQUARES

MATERIAL - .080 HALF-HARD ALUMINUM

$\frac{1}{8}$ R.(MAX.)

90°

$\frac{3}{16}$ DIA. x $9\frac{3}{4}$ LONG (BRAZING ROD)

THREAD 10-24NC x $\frac{5}{16}$ LONG, BOTH ENDS

$9\frac{3}{8}$

$4\frac{11}{16}$

$2\frac{1}{8}$

$\frac{5}{16}$

$1\frac{1}{8}$

$1\frac{3}{8}$

$\frac{5}{8}$

5°

$1\frac{1}{2}$

$\frac{1}{8}$ R.

$\frac{3}{16}$ DIA.

10-24NC ACORN NUT
2 REQ'D

$\frac{5}{8}$

$2\frac{1}{4}$

$2\frac{1}{4}$

$\frac{5}{32}$ DIA, 2 HOLES

$1\frac{3}{16}$

$\frac{1}{4}$ DIA.

$\frac{3}{4}$

REMOVE ALL SHARP EDGES & CORNERS

Fig. 18-10. Tie rack plans.

Fig. 18-11. Make a metal template and use it to transfer the outline to the aluminum sheet.

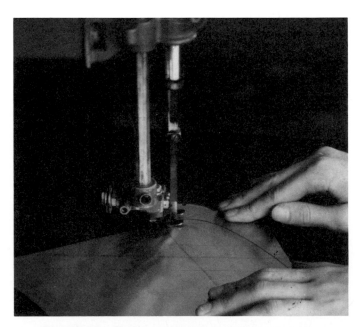

Fig. 18-12. Cutting outlined shape on a jig saw.

Fig. 18-13. Using disk sander to smooth edges.

Fig. 18-15. Jig shown in Fig. 18-14 may also be used to position work for drilling the mounting holes.

trace the body outline on the metal, Fig. 18-11. Use a soft lead pencil or fine-point marker. Cut the metal to shape on a jigsaw, Fig. 18-12. Use a fine-toothed (24T-32T) blade or a bandsaw, if available.

Rough edges of the metal are smoothed on a disk sander, Fig. 18-13. Use a fine (3/0-4/0) abrasive disc. Or, rough edges can be smoothed by hand. Be very careful. Make certain all burrs are removed.

The 1/4 in. (6.35 mm) holes for the 3/16 in. (1.01 mm) rod are drilled using a simple jig to hold

the part in position, Fig. 18-14. The same jig can be used to position the work for drilling the mounting holes, Fig. 18-15. Remove all burrs.

The 90 deg. bends are formed using a shop-made bending jig. Fig. 18-16 shows a bend being

Fig. 18-14. Jig used to position metal for drilling holes that will hold rod.

Fig. 18-16. Making bends with a shop-made bending jig.

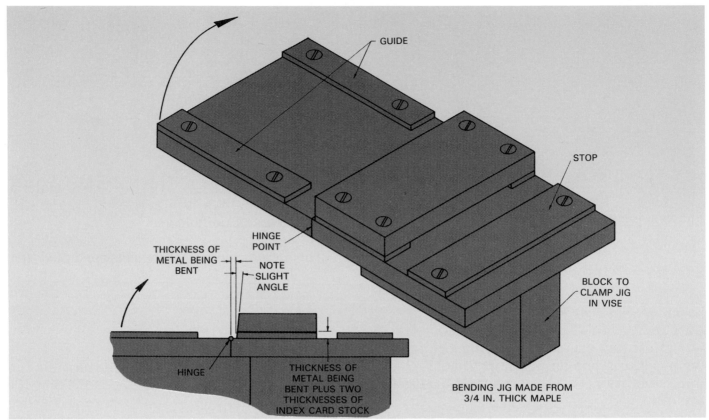

Fig. 18-17. Drawing of jig shown in Fig. 18-16.

made with the jig. Fig. 18-17 shows a working drawing of a shop-made jig. There are two guides on the bending jig. Therefore, it can be used to make both of the required bends.

After carefully inspecting the tie rack body, it is ready to be cleaned and painted, Fig. 18-18. Spray paints are easier to clean-up afterwards than are conventional paints. Store the completed bodies until needed for final assembly.

TIE HOLDERS

There are several ways the holders (for individual ties) can be made, Fig. 18-19. They may

Fig. 18-18. Painting tie rack body with spray paint.

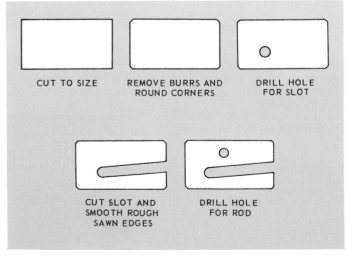

Fig. 18-19. Follow these steps when making individual tie holders.

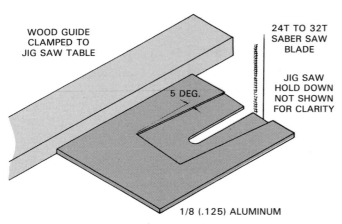

Fig. 18-20. Fixture and setup for cutting slots in individual tie holders.

be cut to individual size on squaring shears, on a jigsaw or bandsaw, Fig. 18-20. Or, the slots may be cut quickly using a small milling machine. After removing burrs, round corners to remove the sharp edges. Then, carefully sand until smooth.

Inspect each part. Has each part been properly cleaned and smoothed?

When the slot in the holder is cut on a jigsaw or bandsaw, drill a hole at the end of the slot. Use a drill jig to position the part to drill the hole. It

is not necessary to lay out each holder separately, if the fixture shown in Fig. 18-20 is used.

It is also possible to manufacture the holders in strip form. Using squaring shears, cut a strip of aluminum sheet as wide as the tie holder is long. The length of the strip will be set by the size of the squaring shears. The strip can also be cut on a jigsaw or bandsaw.

Lay out tie holders on the strip. Allow material for the cut, if a saw is to be used to cut the strip into individual tie holders.

The holes at the base of the slots are drilled first. Drill a hole in the first tie holder on the strip. Then place the hole over the aligning dowel on the drill jig, Fig. 18-21. After CAREFULLY aligning the jig, drill a hole in the second holder. Do not forget to clamp the jig firmly to the drill press. Drill the second hole. Remaining holes are positioned for drilling by moving the strip over so the last drilled hole is fitted over the aligning pin. Drill the required holes and remove the burrs. This may be done by using a countersink mounted in the drill press.

The slots in the tie holder are cut using the fixture shown in Fig. 18-22. A guide must be clamped to the jigsaw table. Make first cut in all holders,

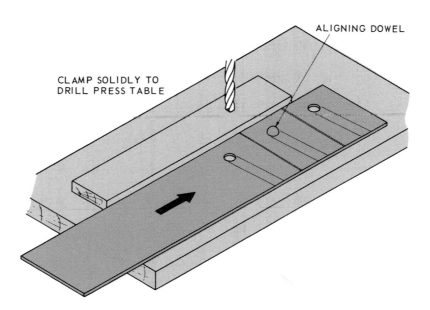

Fig. 18-21. Drill jig used to drill hole at bottom of slot, when tie holders are made in strip form.

201

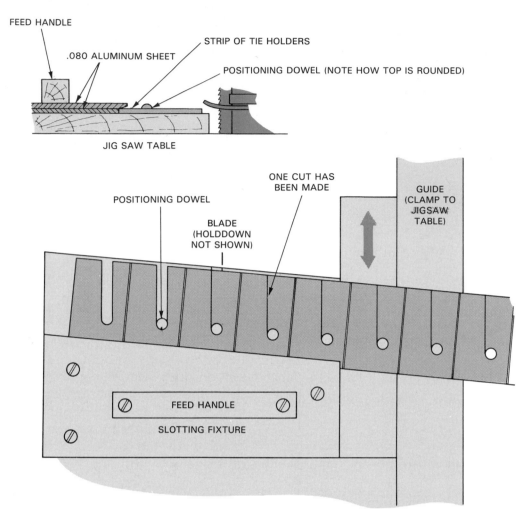

FEED HANDLE

.080 ALUMINUM SHEET

STRIP OF TIE HOLDERS

POSITIONING DOWEL (NOTE HOW TOP IS ROUNDED)

JIG SAW TABLE

POSITIONING DOWEL

BLADE
(HOLDDOWN
NOT SHOWN)

ONE CUT HAS
BEEN MADE

GUIDE
(CLAMP TO
JIGSAW
TABLE)

FEED HANDLE

SLOTTING FIXTURE

Fig. 18-22. Fixture used to position strip of tie holders so slots can be cut.

then adjust position of guide for second cut. The fixture can be mounted on the miter gauge if the slots are to be cut on a bandsaw.

The same fixture can be used if a jigsaw or bandsaw is used to cut the individual holders from the strip.

Remove all burrs and rough edges with a file.

The hole that will mount the tie holder on the rod can now be drilled. Use the jig shown in Fig. 18-23 to position the piece. In drilling it is important that the parts be fitted into the jig as shown.

After burrs are removed, sand, clean, and inspect the holder. Store until needed for assembly.

Fig. 18-23. This jig positions holder for drilling mounting hole. It is important that the part be fitted in the jig as shown.

ROD

The rod is made from 3/16 in. dia. (4.7 mm) brazing rod. However, 3/16 in. dia. cold finished steel rod may be substituted.

Cut the material to length. Using a grinder, round each end slightly. This will make threading easier.

Thread both ends. Remove any burrs, polish, and inspect.

FINAL ASSEMBLY

Parts ready for assembly are shown in Fig. 15-24. The assembly line can be operated as outlined.

1. Attach one acornnut to rod.
2. Slide rod through one end of body.
3. Fit holders to rod.
4. Slide rod thorugh other end of body.
5. Attach second acorn nut to rod.
6. Make a final inspection.

MASS-PRODUCING A TABLE LAMP

Mass-producing the lamp, shown in Fig. 18-25, involves a number of metalworking procedures. For construction details on the lamp, see Fig. 18-26, on pages 204 and 205.

Fig. 18-25. Mass-produced table lamp.

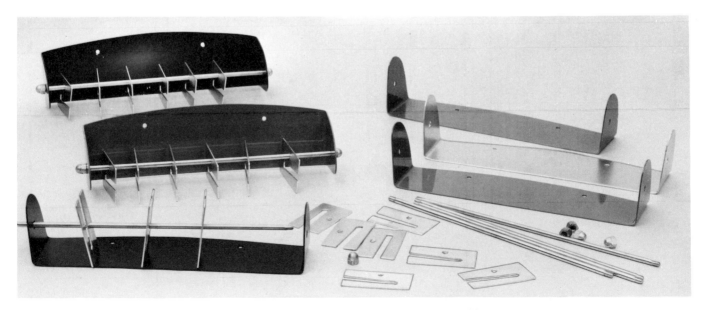

Fig. 18-24. These parts are ready for assembly.

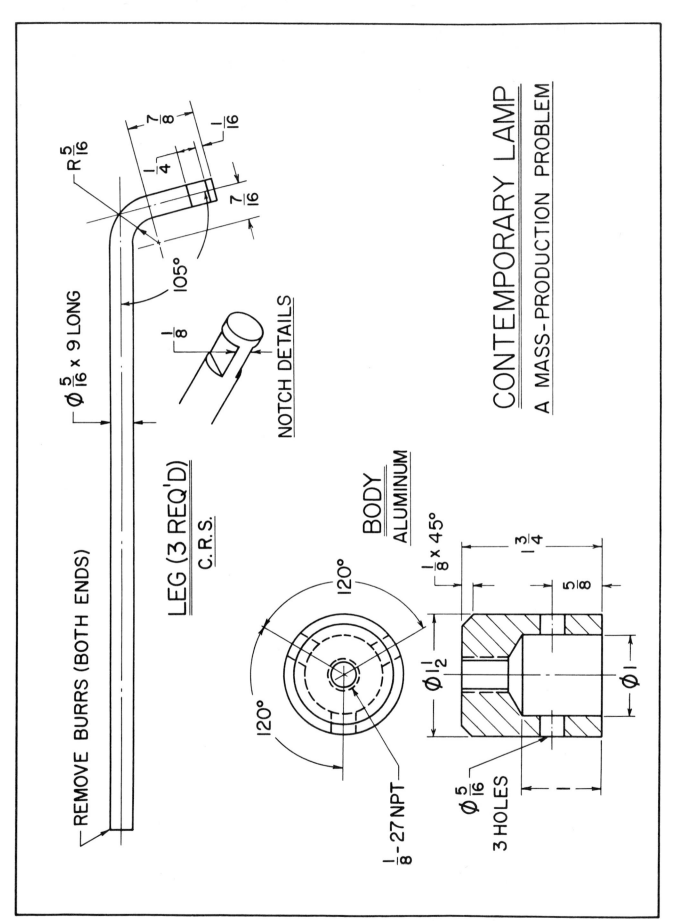

R $\frac{5}{16}$

$\frac{7}{8}$

$\frac{1}{16}$

$\frac{1}{4}$

$\frac{7}{16}$

105°

Ø $\frac{5}{16}$ × 9 LONG

$\frac{1}{8}$

NOTCH DETAILS

REMOVE BURRS (BOTH ENDS)

LEG (3 REQ'D)
C.R.S.

CONTEMPORARY LAMP
A MASS-PRODUCTION PROBLEM

BODY
ALUMINUM

120°

120°

120°

$\frac{1}{8}$ - 27 NPT

$\frac{1}{8}$ × 45°

1 $\frac{3}{4}$

$\frac{5}{8}$

Ø 1 $\frac{1}{2}$

Ø 1

Ø $\frac{5}{16}$
3 HOLES

Fig. 18-26. Plans for the lamp shown in Fig. 18-25.

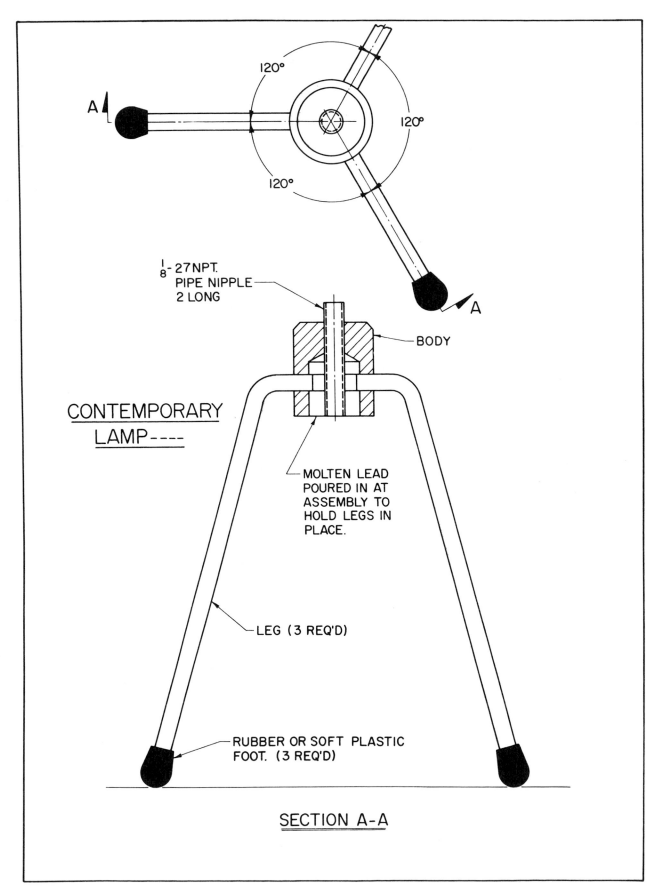

120°

120°

120°

A

A

$\frac{1}{8}$-27 N.P.T.
PIPE NIPPLE
2 LONG

BODY

CONTEMPORARY
LAMP----

MOLTEN LEAD
POURED IN AT
ASSEMBLY TO
HOLD LEGS IN
PLACE.

LEG (3 REQ'D)

RUBBER OR SOFT PLASTIC
FOOT. (3 REQ'D)

SECTION A-A

Fig. 18-26. Plans for lamp shown in Fig. 18-25. (continued)

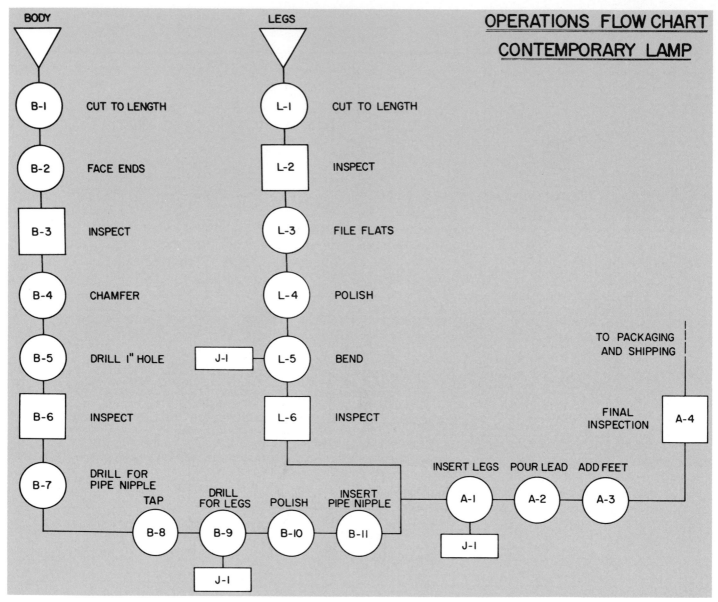

Fig. 18-27. Flow chart shows manufacturing sequences of various parts.

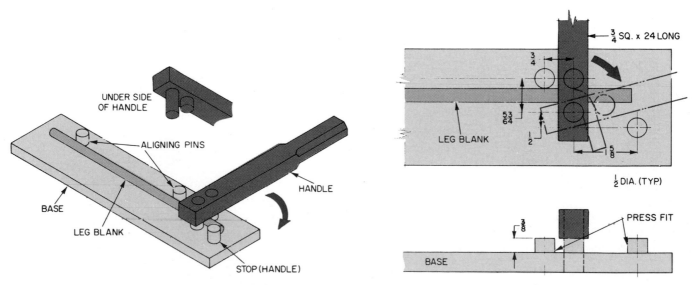

Fig. 18-28. Bending tool used to shape legs.

An OPERATIONS FLOW CHART is given in Fig. 18-27. At each step, students assigned the task of preparing production equipment should ask if there is any way to improve production or if there is any way to improve the product. Any approved changes should be noted and incorporated into the plan.

A bending tool used to form the legs is shown in Fig. 18-28. A drill jig used to locate the holes for the legs is shown in Fig. 18-29.

Figs. 18-30 and 18-31 show gauges designed to make the length of the legs and the angle at which the legs are bent the same.

To attach the legs to the body, molten metal is used. It is poured into the body cavity of the base. This is done after positioning the pipe nipple and legs, Fig. 18-32. Special training for handling of molten metals is a "must."

Another solution is to pack plastic steel around the positioned legs. The rubber or plastic feet, lamp fixtures, wire, and plug must be purchased commercially. Choose a shade of appropriate size and shape as the final step.

PRODUCTION PROBLEM IDEAS

You have seen how two student-designed production problems have been accomplished. The following projects are offered as ideas. Your class may use them or research and develop a project of your own design. In either case, you will have the opportunity to solve a production problem in your school metals lab.

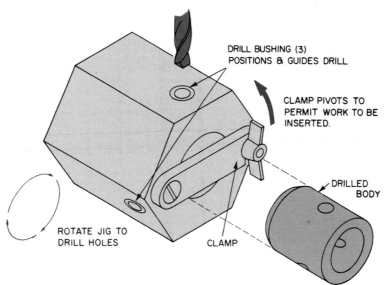

Fig. 18-29. Drill jig used to hold lamp body while being drilled.

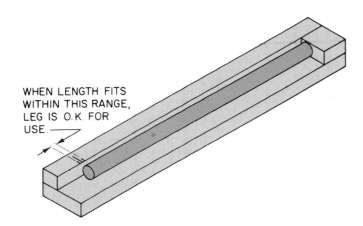

Fig. 18-30. Gauge used to check length of legs.

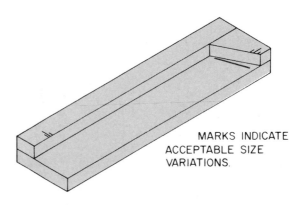

Fig. 18-31. Gauge used to check angle of bend in leg.

207

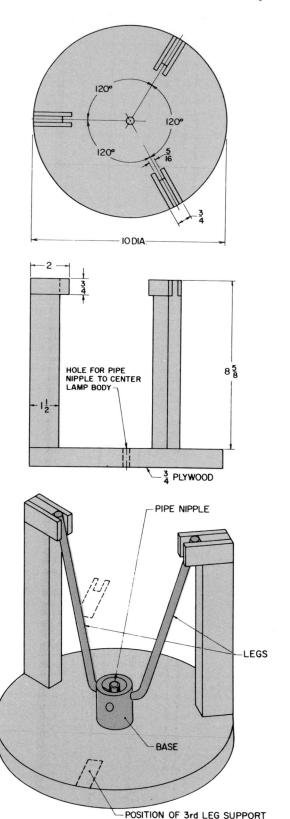

Fig. 18-32. Fixture holds legs while molten lead is poured.

PRODUCTION PROBLEM IDEA I: TRIVET

A trivet is a device used to prevent hot dishes, pans, etc., from damaging the surface of a table or a countertop. The trivet is fabricated from nine pieces of metal. The tooling needed is shown in Fig. 18-33.

PRODUCTION PROBLEM IDEA II: GARDEN TROWEL

The garden trowel is inexpensive to produce. Although it has only two pieces, a number of operations are required to produce them.

When manufacturing the trowel, all of the information needed is shown in Fig. 18-34 (top), except that the length of the 1/8 in. x 1/2 in. hot rolled steel handle is not furnished. Can you calculate the length of the metal?

The trowel can be painted flat black or a bright color.

PRODUCTION PROBLEM IDEA III: LETTER HOLDER

The letter holder, Fig. 18-34 (bottom), offers many design variations. The accent strip can be initials or any simple geometric form. Assemble by soldering.

After the letter holder has been assembled and cleaned and polished, it should be given a thin coat of clear lacquer to prevent the brass from tarnishing.

SAFETY NOTE: WEAR APPROPRIATE SAFETY EQUIPMENT AND HAVE AMPLE VENTILATION WHEN SPRAYING LACQUER OR PAINT.

REVIEW OF ACTIVITY

1. What problems were encountered in mass-producing your project? How could they have been avoided?
2. How do you think that the project could have been improved?
3. What was the most interesting part of the activity? Why was it interesting?
4. What was the most challenging part of the activity? Why was it challenging?

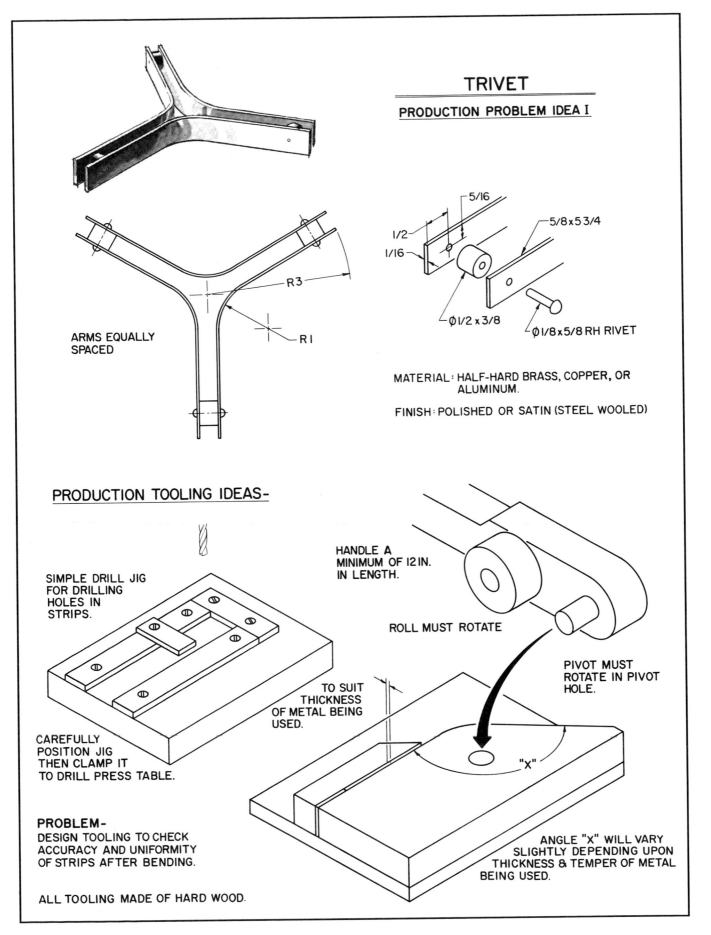

TRIVET

PRODUCTION PROBLEM IDEA I

ARMS EQUALLY
SPACED

R3

RI

5/16

1/2

1/16

5/8 x 5 3/4

Ø1/2 x 3/8

Ø1/8 x 5/8 RH RIVET

MATERIAL: HALF-HARD BRASS, COPPER, OR
ALUMINUM.

FINISH: POLISHED OR SATIN (STEEL WOOLED)

PRODUCTION TOOLING IDEAS-

SIMPLE DRILL JIG
FOR DRILLING
HOLES IN
STRIPS.

CAREFULLY
POSITION JIG
THEN CLAMP IT
TO DRILL PRESS TABLE.

PROBLEM-
DESIGN TOOLING TO CHECK
ACCURACY AND UNIFORMITY
OF STRIPS AFTER BENDING.

ALL TOOLING MADE OF HARD WOOD.

HANDLE A
MINIMUM OF 12 IN.
IN LENGTH.

ROLL MUST ROTATE

PIVOT MUST
ROTATE IN PIVOT
HOLE.

TO SUIT
THICKNESS
OF METAL BEING
USED.

"X"

ANGLE "X" WILL VARY
SLIGHTLY DEPENDING UPON
THICKNESS & TEMPER OF METAL
BEING USED.

Fig. 18-33. Trivet project.

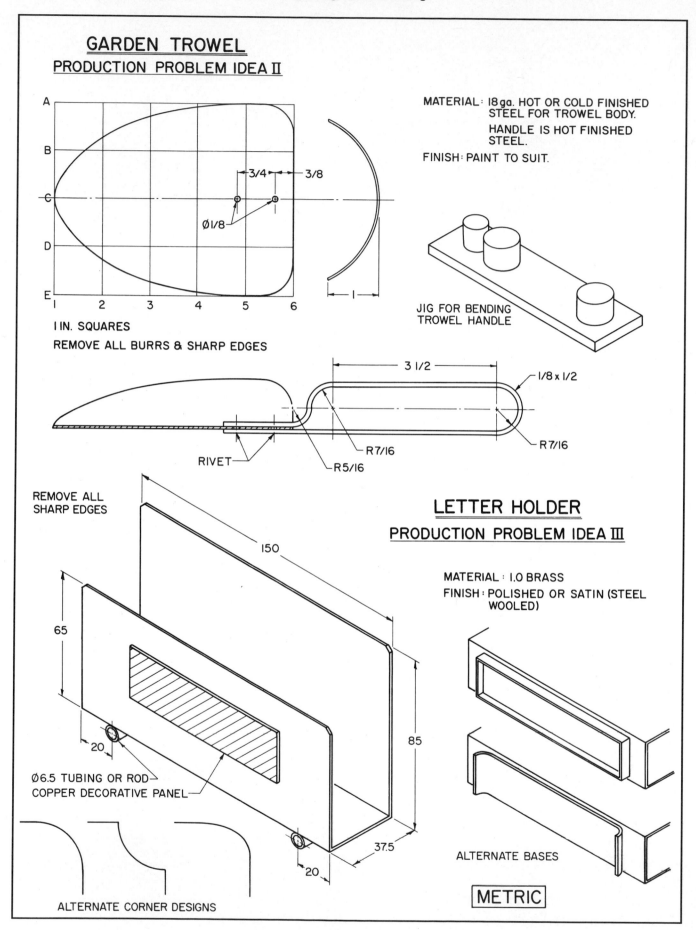

GARDEN TROWEL
PRODUCTION PROBLEM IDEA II

A
B
C
D
E

3/4 | 3/8

Ø1/8

1

1 IN. SQUARES

REMOVE ALL BURRS & SHARP EDGES

MATERIAL: 18 ga. HOT OR COLD FINISHED
STEEL FOR TROWEL BODY.

HANDLE IS HOT FINISHED
STEEL.

FINISH: PAINT TO SUIT.

JIG FOR BENDING
TROWEL HANDLE

3 1/2

1/8 x 1/2

R7/16

R7/16

RIVET

R5/16

REMOVE ALL
SHARP EDGES

150

65

20

Ø6.5 TUBING OR ROD
COPPER DECORATIVE PANEL

85

37.5

20

ALTERNATE CORNER DESIGNS

LETTER HOLDER
PRODUCTION PROBLEM IDEA III

MATERIAL: 1.0 BRASS
FINISH: POLISHED OR SATIN (STEEL
WOOLED)

ALTERNATE BASES

METRIC

Fig. 18-34. Garden trowel and letter holder projects.

TEST YOUR KNOWLEDGE, Unit 18

Please do not write in the text. Place your answers on a separate sheet of paper.

1. A company name should be appropriate, businesslike, and _____.
2. Which of the following does NOT need to be elected when electing company officials?
   a. Art director.
   b. General manager.
   c. Office manager.
   d. Safety director.
3. Which company official is expected to conduct general supervision over the entire activity?
4. Who is expected to maintain attendance records?
5. Many companies raise money by selling _____.
6. What is the formula for determining product selling price?

TECHNICAL TERMS TO LEARN

business activity
employee
entrepreneur
financing
flow chart
General Manager
liquidate
loss
manufacturing
mass production
Office Manager
product

profit
prospect
Purchasing Agent
receipts
records
Safety Director
sales
Sales Manager
stock
stockholder
supervision

ACTIVITIES

1. Visit a factory that mass-produces a product. Then, visit a shop that produces objects on a limited basis.
2. Discuss the financing of a small business with a banker. Report to your class.
3. Discuss the steps followed to advertise a product with an advertising professional. How would you advertise the tie rack described in this unit?

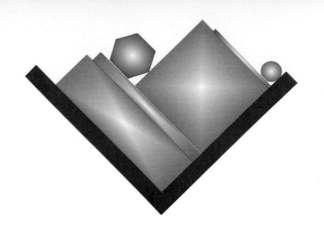

Unit 19
Projects

Designing and constructing metalworking projects will teach you to think and to plan.

Your projects should be items you are interested in making. The projects, when completed, should compare favorably, in quality, with items available commercially.

NAPKIN RINGS

Material: Aluminum.

Finish: Polished or satin finish.
Remove burrs, sharp edges.

The designs shown are but a few of the many geometric forms that can be used for this project. Use your imagination and design your own.

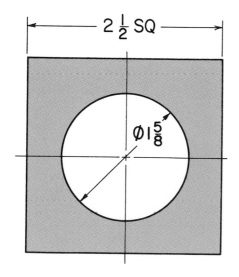

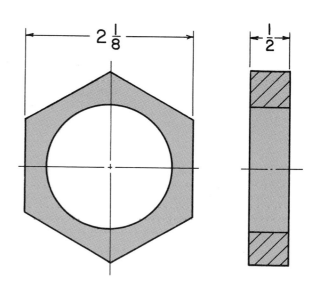

SCREWDRIVER

1. Material: Handle—Aluminum.
   Shank—Drill Rod.

2. Forge blade to shape.

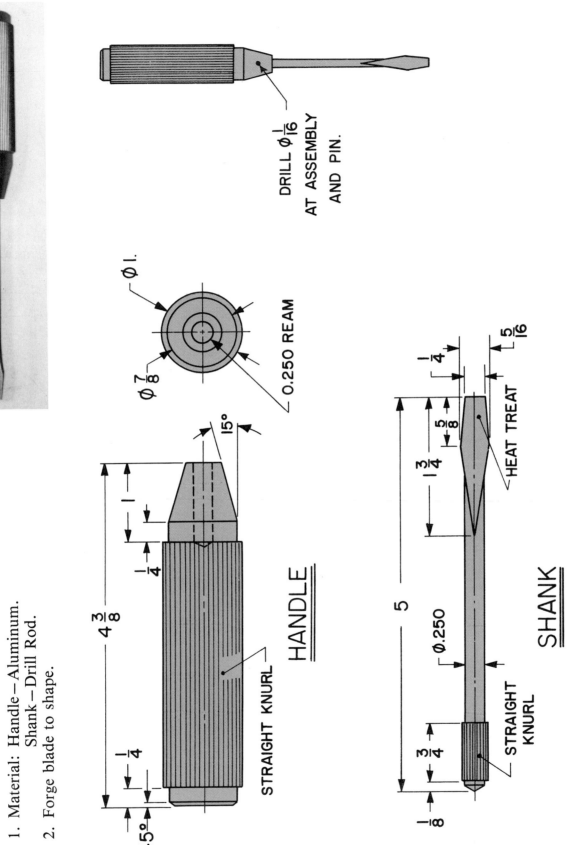

DRILL $\phi\frac{1}{16}$
AT ASSEMBLY
AND PIN.

$\phi 1$.

$\phi\frac{7}{8}$

0.250 REAM

15°

1

$\frac{1}{4}$

$4\frac{3}{8}$

$\frac{1}{4}$

$\frac{1}{16} \times 45°$

STRAIGHT KNURL

HANDLE

$\frac{1}{4}$

$\frac{5}{16}$

$\frac{5}{8}$

$1\frac{3}{4}$

HEAT TREAT

5

$\phi.250$

$\frac{3}{4}$

STRAIGHT
KNURL

$\frac{1}{8}$

SHANK

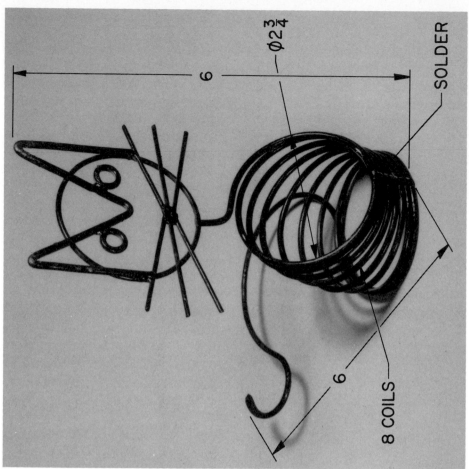

1. Use 1/8 in. (3.2 mm) diameter wire to form the figure. This may be copper wire, brazing rod, or steel coat hangers.
2. Solder all joints.
3. Finish by painting flat black.

WIRY CAT

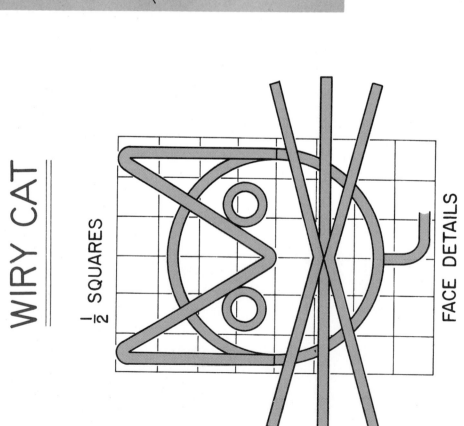

$\frac{1}{2}$ SQUARES

FACE DETAILS

SMALL PLANTER

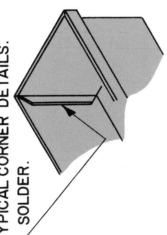

TYPICAL CORNER DETAILS. SOLDER.

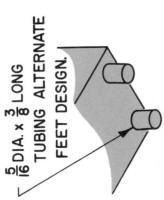

$\frac{5}{16}$ DIA. x $\frac{3}{8}$ LONG TUBING ALTERNATE FEET DESIGN.

1. Material: 18 ga. copper or brass.
2. Finish: Planish. Buff all exterior surfaces. Apply satin finish to interior surfaces. Protect polished surfaces with clear lacquer spray.

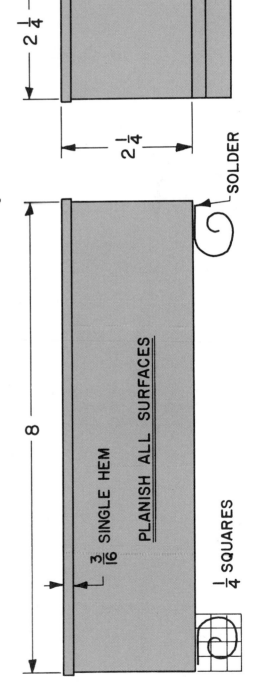

$2\frac{1}{4}$

$2\frac{1}{4}$

SOLDER

8

$\frac{3}{16}$ SINGLE HEM

PLANISH ALL SURFACES

$\frac{1}{4}$ SQUARES

COLD CHISEL

1. Material: 7/16 hexagonal or octagonal tool steel.
2. Heat and forge to shape.
3. File smooth and grind chamfer on chisel head.
4. Heat treat and sharpen.

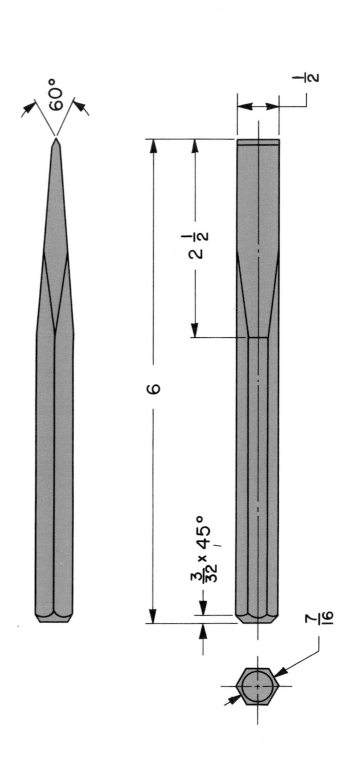

60°

$\frac{1}{2}$

$2\frac{1}{2}$

6

$\frac{3}{32} \times 45°$

$\frac{7}{16}$

PENCIL HOLDERS

1. Material: Aluminum.
2. Finish: Polished or as machined.
3. Attach felt to bottom.
4. Remove all burrs and sharp edges.

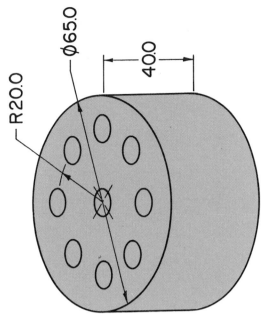

R20.0

Ø65.0

40.0

6-8 HOLES EQUALLY SPACED

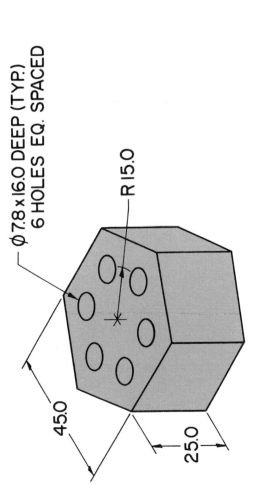

Ø7.8 × 16.0 DEEP (TYP.)
6 HOLES EQ. SPACED

R 15.0

45.0

25.0

ALL DIMENSIONS IN mm

TIC-TAC-TOE GAME

Material: Aluminum and as noted.
Finish: As machined.

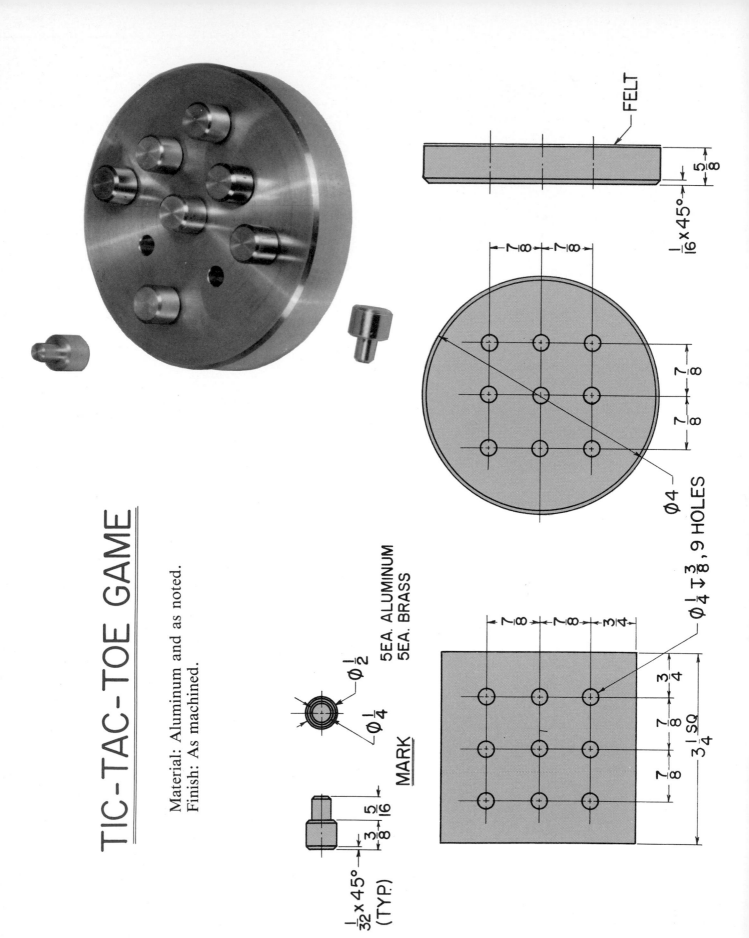

FELT

$\frac{5}{8}$

$\frac{1}{16} \times 45°$

$\frac{7}{8}$ $\frac{7}{8}$

$\frac{7}{8}$ $\frac{7}{8}$

Ø4

$Ø\frac{1}{4} \downarrow \frac{3}{8}$, 9 HOLES

$\frac{7}{8}$ $\frac{7}{8}$ $\frac{3}{4}$

$\frac{3}{4}$

$3\frac{1}{4}$ SQ

$\frac{7}{8}$

$\frac{7}{8}$

$Ø\frac{1}{2}$

$Ø\frac{1}{4}$

5 EA. ALUMINUM
5 EA. BRASS

MARK

$\frac{5}{16}$

$\frac{3}{8}$

$\frac{1}{32} \times 45°$
(TYP.)

218

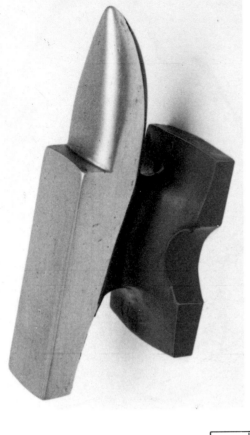

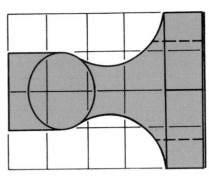

$\frac{1}{2}$ SQUARES

CAST ANVIL

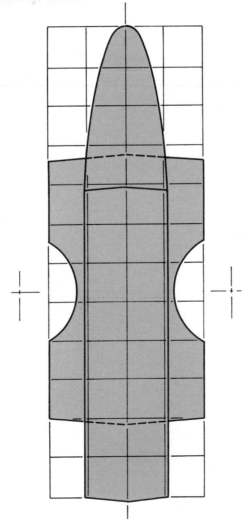

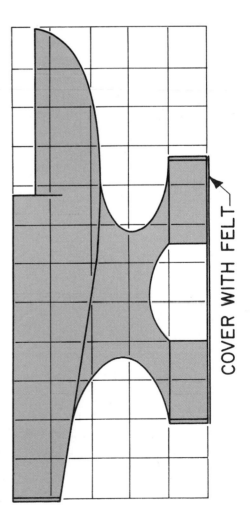

COVER WITH FELT

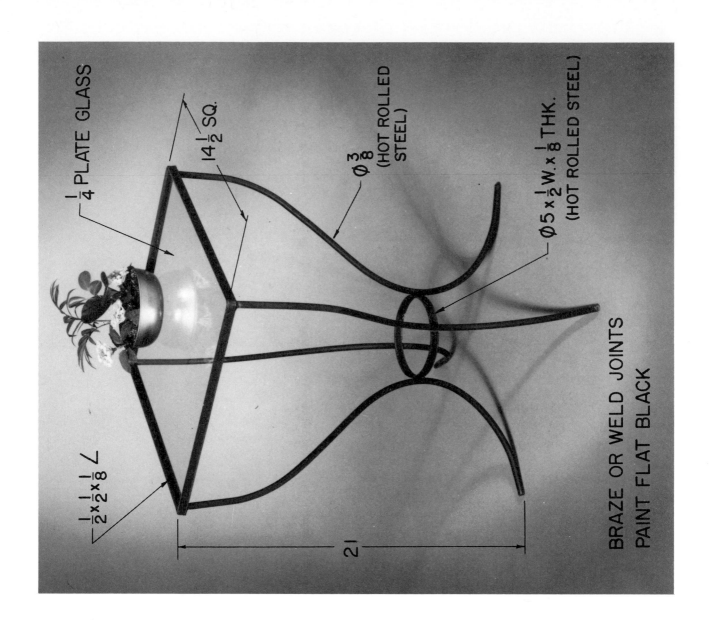

¼ PLATE GLASS

14½ SQ.

Ø⅜ (HOT ROLLED STEEL)

Ø5 x ½ W. x ⅛ THK. (HOT ROLLED STEEL)

½ x ½ x ⅛ ∠

21

BRAZE OR WELD JOINTS
PAINT FLAT BLACK

PATIO TABLE

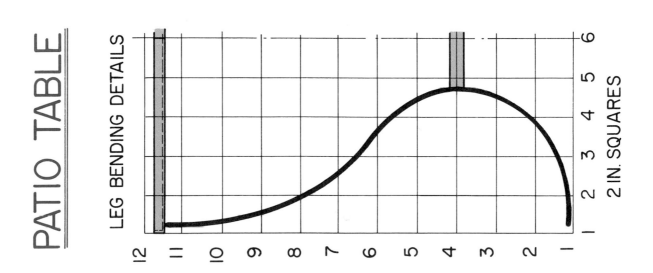

LEG BENDING DETAILS

2 IN. SQUARES

MAIL BOX DESIGN PROBLEM

Design and develop patterns, then manufacture a mail box that will meet postal regulations. Get ideas and basic dimensions from mail order catalogs and home planning magazines. Use 26 ga. sheet metal in either a plain or embossed pattern. It may be made from copper, brass, or aluminum. Larger mail boxes can be made from galvanized sheet steel. Join seams by spot welding or soldering.

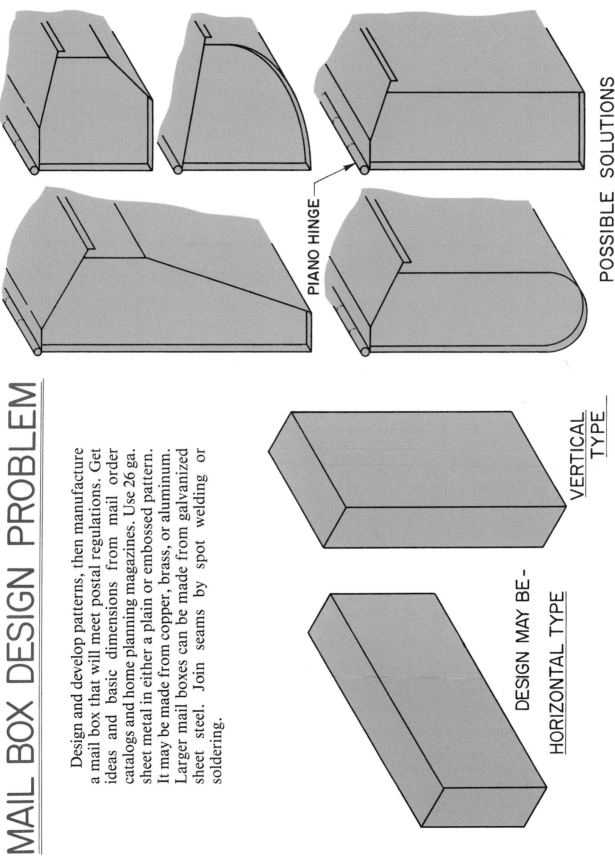

PIANO HINGE

POSSIBLE SOLUTIONS

VERTICAL TYPE

DESIGN MAY BE -

HORIZONTAL TYPE

MACHINED PAPER WEIGHT

Material: Aluminum, brass, or cold finished steel.
Finish: As machined or polished.

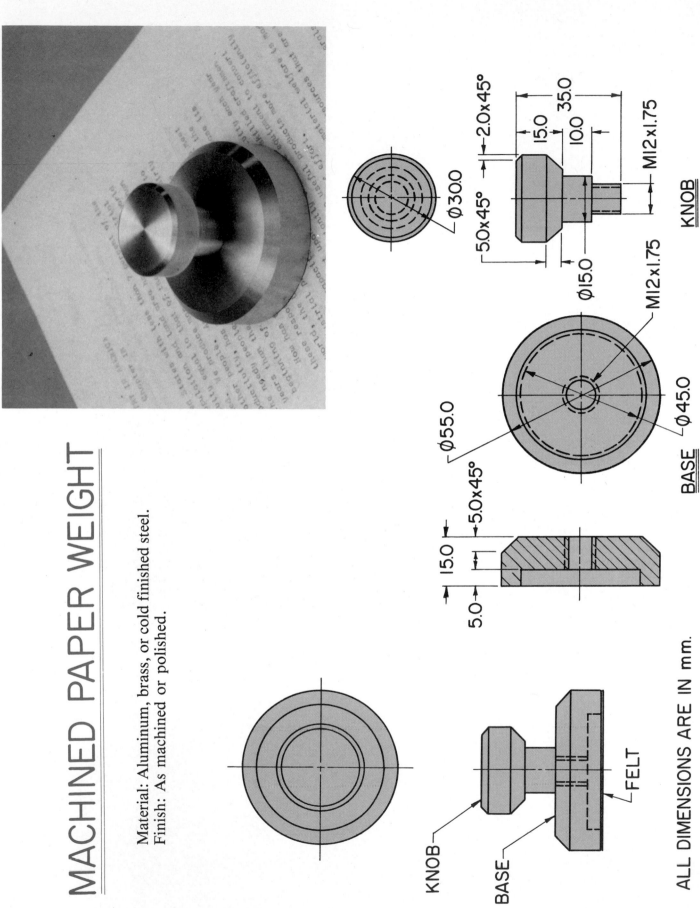

KNOB

BASE

MI2x1.75

Ø30.0

Ø15.0

Ø55.0

Ø45.0

35.0

15.0

10.0

2.0x45°

5.0x45°

5.0x45°

15.0

5.0

KNOB

BASE

FELT

ALL DIMENSIONS ARE IN mm.

222

HANDYMAN'S SPECIAL

1. Cold finished steel may be used. However, precision ground tool and die steel is preferred.
2. Saw and file or machine the head to shape. It is recommended that the handle slot be milled.
3. Shape the handle to the size shown. Drill holes for the handle. Fit the handle to the head and drill mounting holes.
4. Heat treat the hammer head, screwdriver, and bottle opener end of the handle. Harden and temper if tool and die steel is used. Case harden if cold finished steel is used.
5. Polish.
6. Mount the wooden handles.

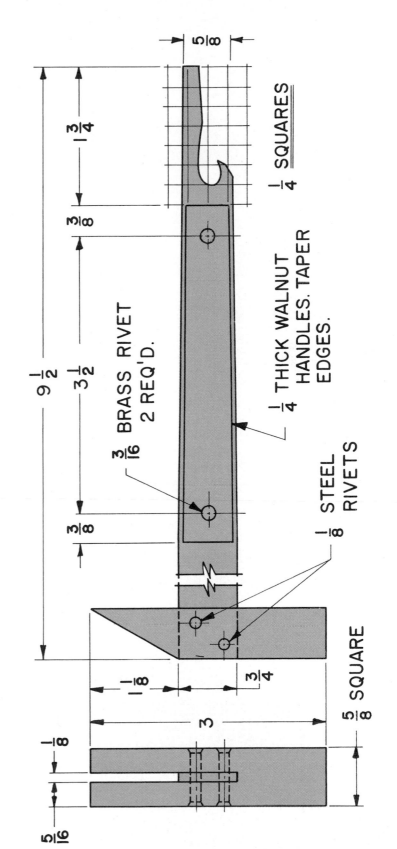

$9\frac{1}{2}$

$1\frac{3}{4}$

$\frac{3}{8}$

$3\frac{1}{2}$

$\frac{3}{8}$

$\frac{3}{8}$

$\frac{5}{8}$

$\frac{1}{4}$ SQUARES

$\frac{3}{16}$ BRASS RIVET 2 REQ'D.

$\frac{1}{4}$ THICK WALNUT HANDLES. TAPER EDGES.

$\frac{1}{8}$ STEEL RIVETS

$1\frac{1}{8}$

$\frac{3}{4}$

3

$\frac{5}{8}$ SQUARE

$\frac{1}{8}$

$\frac{5}{16}$

1. Material: 0.025 aluminum. The roof should be made of 0.025 embossed aluminum.
2. Use round head rivets and sheet metal screws for assembly.
3. Paint the outside surfaces light green. The roof may be left natural.
4. Design a suitable mount.
5. Remove all sharp edges and burrs.

DESIGN PROBLEM

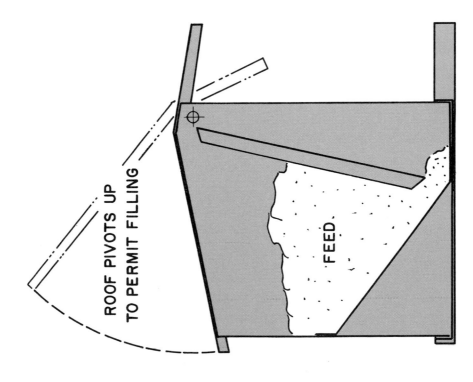

ROOF PIVOTS UP
TO PERMIT FILLING

FEED.

TYPICAL CROSS-SECTION OF BIRD
FEEDER SHOWN

224

TRIVET

Trivets are used to prevent hot pans and pots from scorching or damaging tabletops.

1. Material: Brass or hot finished steel. Walnut handle.
2. Finish: Brass—Polished. Steel—Painted flat black.
3. Remove all sharp edges.
4. Smooth all brazed joints.

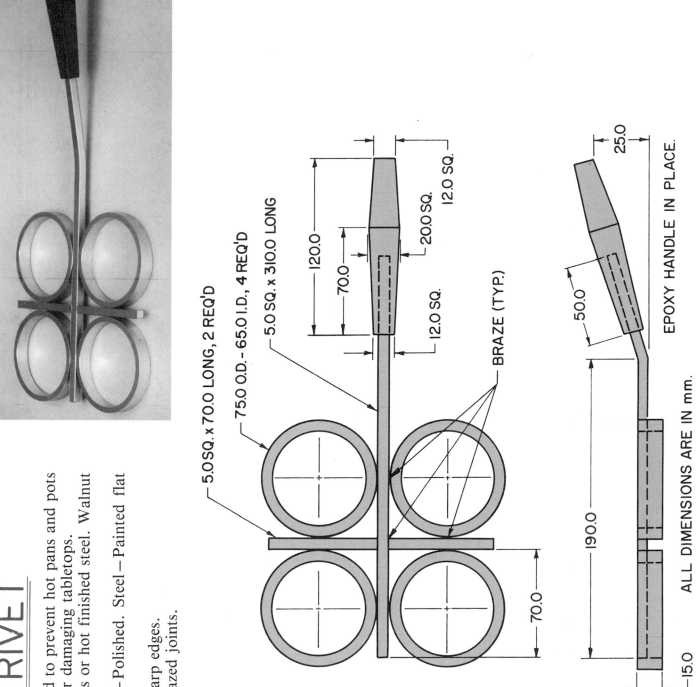

5.0 SQ. x 70.0 LONG, 2 REQ'D

75.0 O.D. - 65.0 I.D., 4 REQ'D

5.0 SQ. x 310.0 LONG

120.0

70.0

20.0 SQ.

12.0 SQ.

12.0 SQ.

BRAZE (TYP.)

70.0

25.0

50.0

190.0

15.0

EPOXY HANDLE IN PLACE.

ALL DIMENSIONS ARE IN mm.

225

DESIGN PROBLEM

Material: As noted on photo.
Finish: Paint metal figure flat black.

This is a project that can be used as a gift for a friend who is a golf enthusiast.

Using a little "think power" you can come up with a variation of the spike man that can be used to depict another sport or occupation.

Be careful to remove all burrs and rough edges before applying the finish.

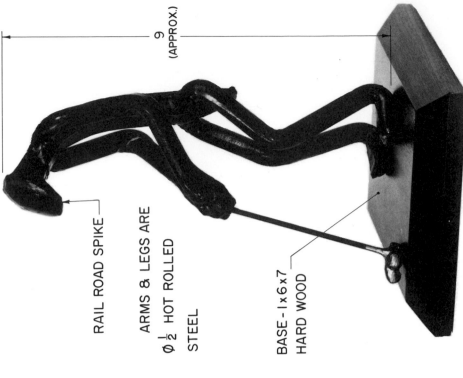

9
(APPROX.)

RAIL ROAD SPIKE

ARMS & LEGS ARE
$\phi \frac{1}{2}$ HOT ROLLED
STEEL

BASE - 1×6×7
HARD WOOD

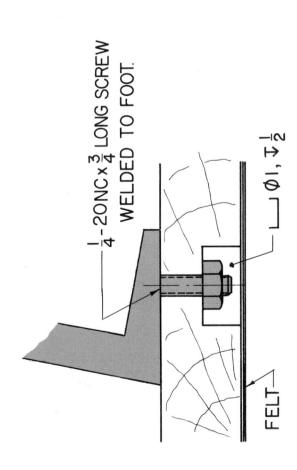

$\frac{1}{4}$-20NC × $\frac{3}{4}$ LONG SCREW
WELDED TO FOOT.

$\phi 1, \mathbf{\top} \frac{1}{2}$

FELT

CENTER PUNCH

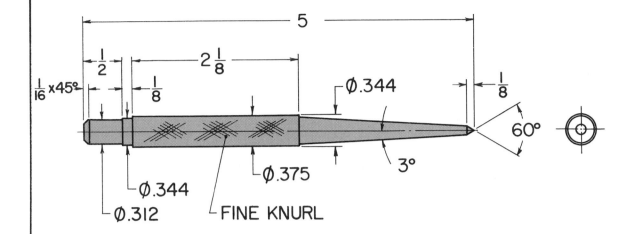

$\frac{1}{16} \times 45°$

$\frac{1}{2}$

$2\frac{1}{8}$

$\frac{1}{8}$

5

Ø.344

$\frac{1}{8}$

60°

3°

Ø.375

Ø.344

Ø.312

FINE KNURL

MACHINING SEQUENCE:

1. CUT A PIECE OF DRILL ROD $5\frac{1}{2}$ LONG.
   A. FACE BOTH ENDS.
   B. CENTER DRILL ONE END.

2. MACHINE CENTER DRILLED END AS FOLLOWS:

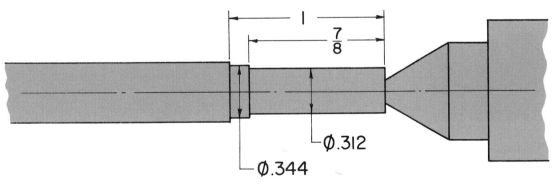

1

$\frac{7}{8}$

Ø.312

Ø.344

3. KNURL.

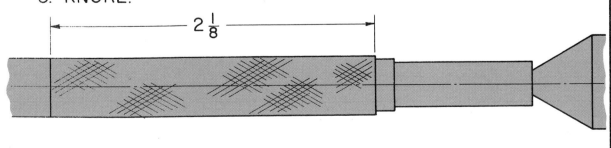

$2\frac{1}{8}$

4. REVERSE WORK IN CHUCK.

5. MACHINE TO DIAMETER.

2 $\frac{1}{4}$

Ø.344

6. SET COMPOUND TO CUT 3° TAPER.

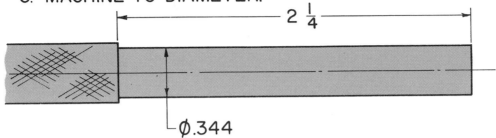

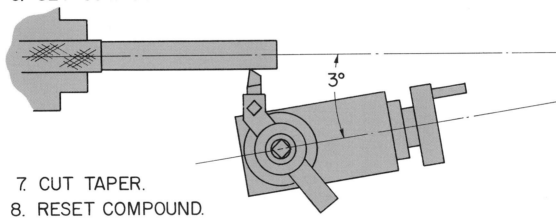

3°

7. CUT TAPER.

8. RESET COMPOUND.

9. MACHINE POINT.

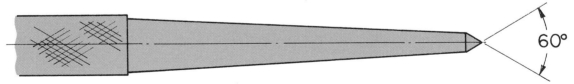

60°

10. REVERSE IN CHUCK AND FINISH MACHINE HEAD TO SIZE.

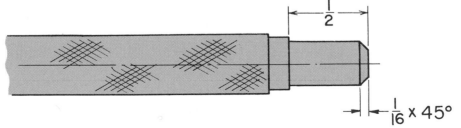

$\frac{1}{2}$

$\frac{1}{16}$ X 45°

11. HEAT TREAT COMPLETED PUNCH.

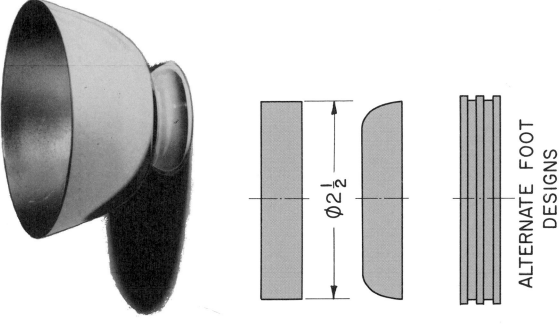

RAISED BOWL

1. Material: 14 ga. copper, brass, pewter, or aluminum.
2. Finish: Planish all surfaces and buff all exterior surfaces. Apply a satin finish to the interior of the bowl.
3. Solder base to bowl body.

$\phi 2\frac{1}{2}$

ALTERNATE FOOT DESIGNS

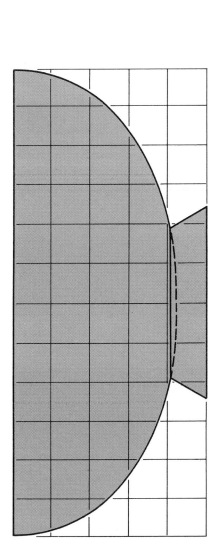

1 IN. OR $\frac{1}{2}$ IN. SQUARES

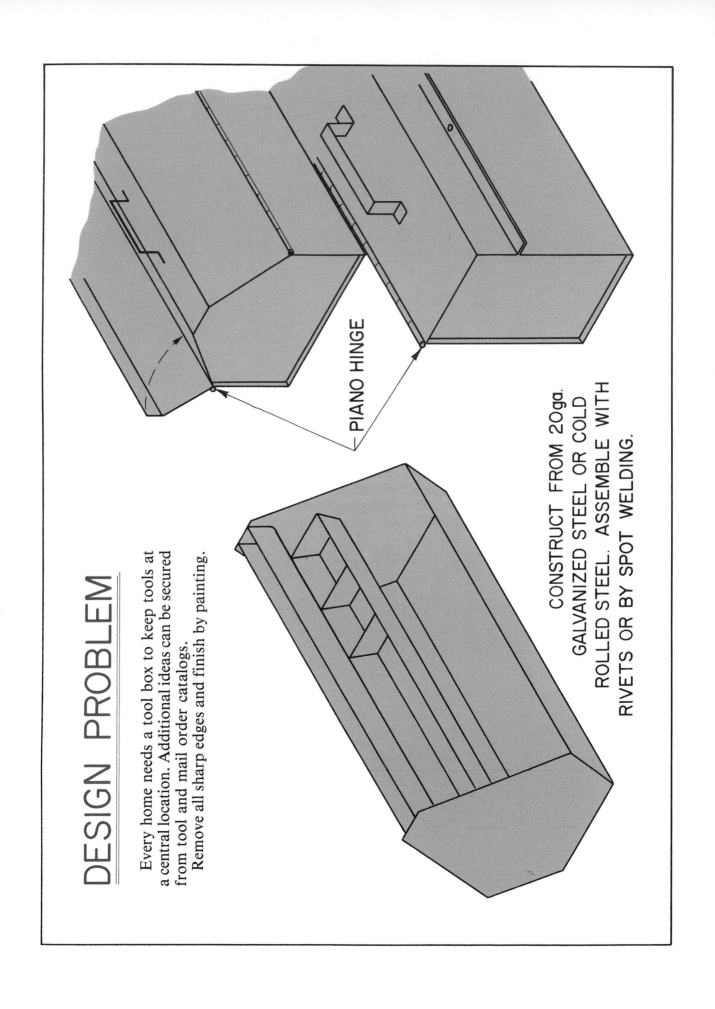

DESIGN PROBLEM

Every home needs a tool box to keep tools at a central location. Additional ideas can be secured from tool and mail order catalogs. Remove all sharp edges and finish by painting.

PIANO HINGE

CONSTRUCT FROM 20ga. GALVANIZED STEEL OR COLD ROLLED STEEL. ASSEMBLE WITH RIVETS OR BY SPOT WELDING.

CANDLE HOLDER

Material: Cast aluminum.
Finish: Polished or satin finish.

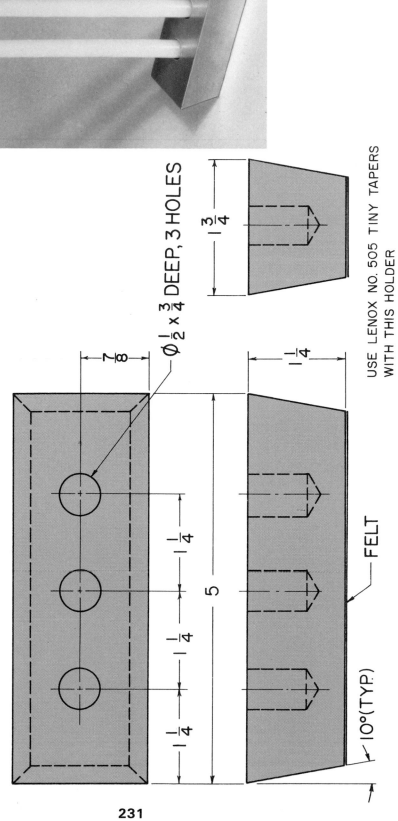

$\varnothing \frac{1}{2} \times \frac{3}{4}$ DEEP, 3 HOLES

$\frac{7}{8}$

$1\frac{1}{4}$

$1\frac{1}{4}$

5

$1\frac{1}{4}$

$1\frac{3}{4}$

$1\frac{1}{4}$

USE LENOX NO. 505 TINY TAPERS
WITH THIS HOLDER

FELT

10°(TYP.)

NAPKIN RINGS

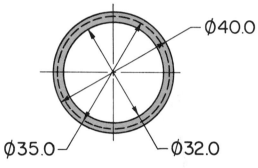

Ø40.0

Ø35.0   Ø32.0

1. Material: Aluminum or brass.
2. Finish: As machined, polished, or satin finish. Brass may be silver plated.
3. Remove all sharp edges and burrs.

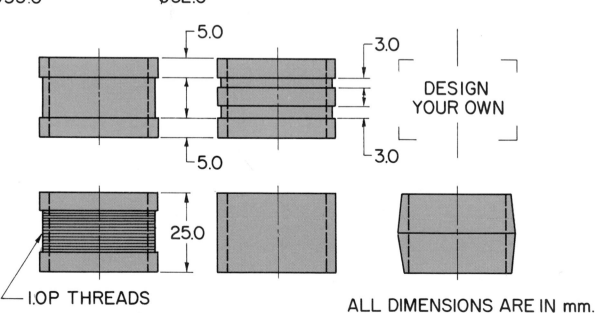

5.0

5.0

3.0

3.0

DESIGN
YOUR OWN

25.0

1.0P THREADS

ALL DIMENSIONS ARE IN mm.

CHOW TIME CHIME

Material: Hot finished mild steel.
Finish: Paint flat black.

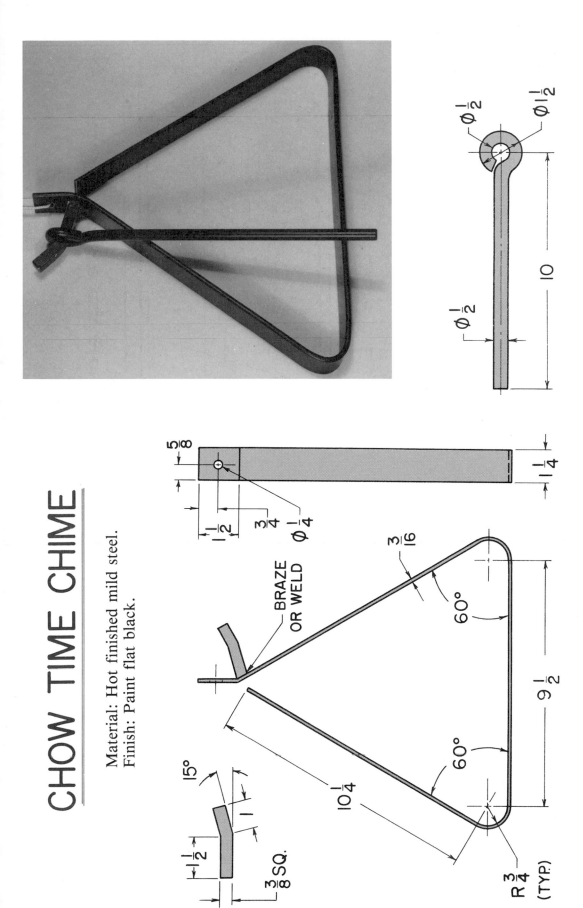

BRAZE OR WELD

60°

60°

$9\frac{1}{2}$

$10\frac{1}{4}$

$R\frac{3}{4}$ (TYP.)

$\frac{3}{16}$

$\frac{5}{8}$

$1\frac{1}{4}$

$1\frac{1}{2}$

$\frac{3}{4}$

$\varnothing\frac{1}{4}$

15°

$1\frac{1}{2}$

1

$\frac{3}{8}$ SQ.

$\varnothing\frac{1}{2}$

$\varnothing 1\frac{1}{2}$

$\varnothing\frac{1}{2}$

10

METAL STRIP SCULPTURE

This is a project for your imagination. The project illustrated is presented only as a suggestion. Metal sculptures may be made of brass, copper, or tin plate. Assemble with solder.

Finish by polishing, painting, or electroplating. Remove all burrs and sharp edges and wash away all flux used in soldering.

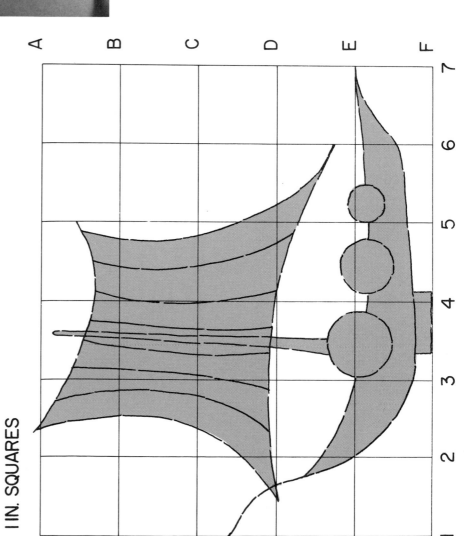

1 IN. SQUARES

234

DRAWER PULL

Material: Aluminum.

Finish: Buff to a high polish or use fine abrasive paper for a satin finish.

This project provides a real challenge for the student who likes to operate the metal lathe. It is recommended that the student practice on hardwood before using the more expensive aluminum.

A full size template will make the job easier.

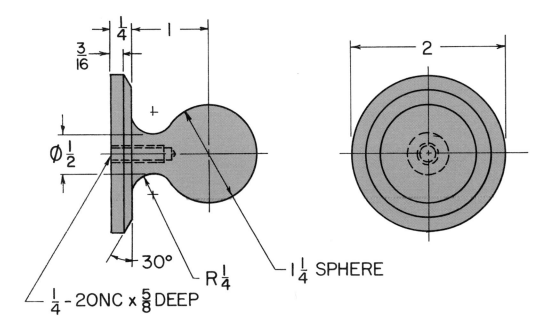

$\frac{1}{4}$ - 20NC x $\frac{5}{8}$ DEEP

30°

R $\frac{1}{4}$

1 $\frac{1}{4}$ SPHERE

$\frac{3}{16}$

$\frac{1}{4}$

1

Ø $\frac{1}{2}$

2

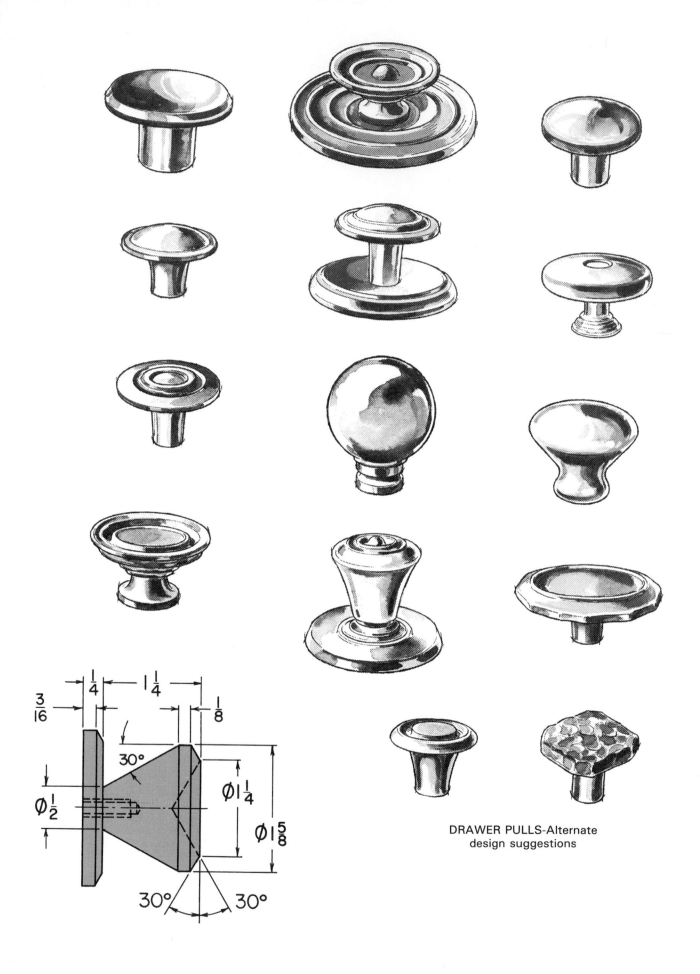

$\frac{3}{16}$ $\frac{1}{4}$ $1\frac{1}{4}$ $\frac{1}{8}$

30°

$\emptyset\frac{1}{2}$ $\emptyset 1\frac{1}{4}$

$\emptyset 1\frac{5}{8}$

30° 30°

DRAWER PULLS-Alternate
design suggestions

236

DESIGN PROBLEM

Material: Horseshoe nails and scrap metal.
Finish: Spray paint silver or gold.

These novel figures make fine gifts. Think of the many types of work and play figures you can design — ball players, golfers, swimmers, musicians, etc.

The horseshoe nails are held together by brazing, soldering, or with epoxy adhesives. Be sure the work is clean before joining the parts and painting them.

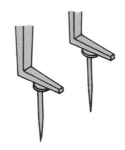

Braze or solder small nails
to feet.

$5\frac{1}{2}$

$\frac{1}{2}$ x 2 x 2 HARD WOOD

MODERN GAVEL

Material: Aluminum.
Finish: As machined.

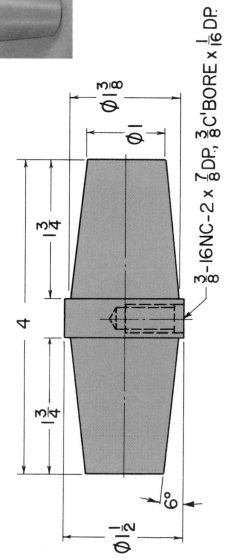

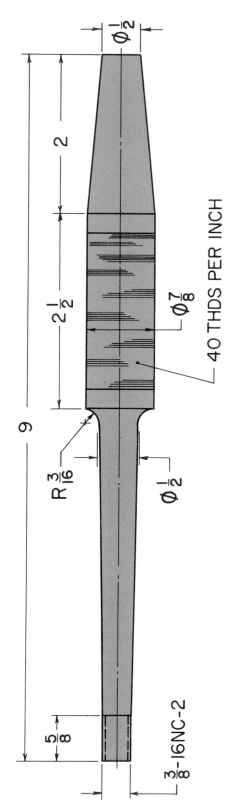

$\phi 1\frac{3}{8}$

$\phi 1$

$1\frac{3}{4}$

4

$1\frac{3}{4}$

$\frac{3}{8}$-16NC-2 x $\frac{7}{8}$ DP., $\frac{3}{8}$ C'BORE x $\frac{1}{16}$ DP.

$6°$

$\phi 1\frac{1}{2}$

$\phi\frac{1}{2}$

2

$2\frac{1}{2}$

$\phi\frac{7}{8}$

40 THDS PER INCH

9

$R\frac{3}{16}$

$\phi\frac{1}{2}$

$\frac{5}{8}$

$\frac{3}{8}$-16NC-2

238

WIRE WALL PLAQUE

1. Use 1/8 in. diameter wire to form the figures. It may be copper wire, brazing rod, or steel coat hangers.
2. Solder all joints.
3. Finish by painting flat black.

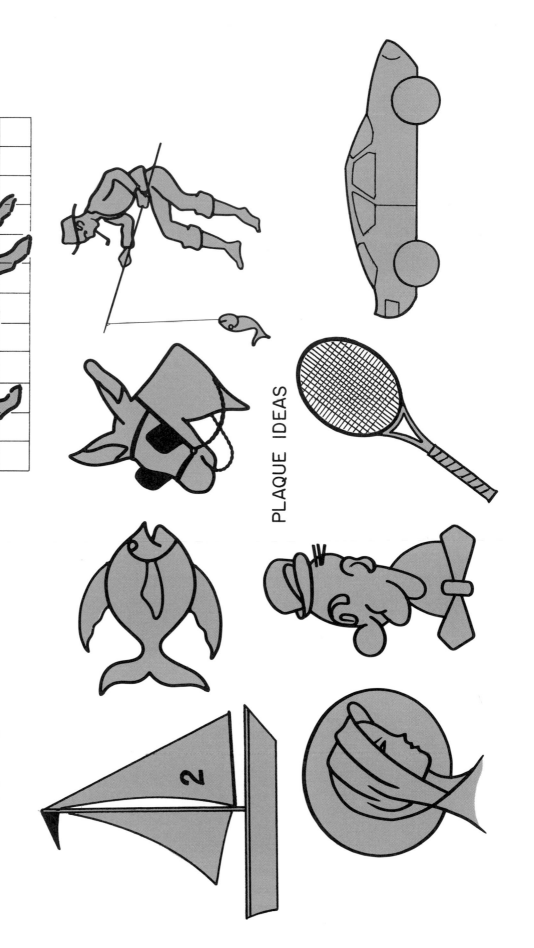

☐ = 1 SQUARES

PLAQUE IDEAS

BOOK ENDS

The book ends shown were made from 4 x 4 wide flange structural steel. Your book ends can be made from one of the many different shapes and sizes of structural steel and aluminum extrusions available.

Finish by removing all sharp edges and painting with colors of your choice.

$1\frac{1}{4}$

6-32NC x $\frac{1}{2}$ LG. F.H. MACH. SCW. 2 REQ'D

BOOK ENDS CUT FROM 4 x 4 WIDE FLANGE STRUCTURAL STEEL. OTHER SHAPES MAY BE USED.

BRACELET

Material: 16 or 18 gauge brass, aluminum, sterling silver, or nickel silver.
Finish: Buff to a high polish.

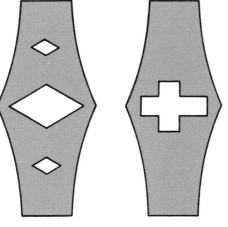
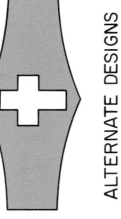

ALTERNATE DESIGNS

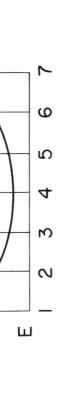

$\frac{3}{4}$ SQUARES

$\frac{1}{4} \times \frac{1}{2}$ SQUARES

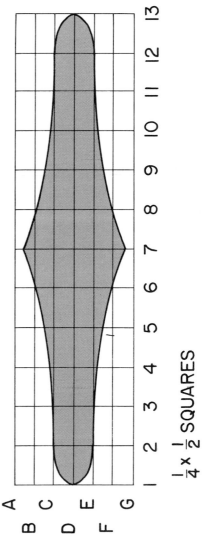

PUNCH SET

1. Material: 3/8 dia. drill rod.
2. Heat treat after machining.
3. A pin punch set can be made by making one each of these sizes: 1/16, 1/8, 3/16, and 1/4.

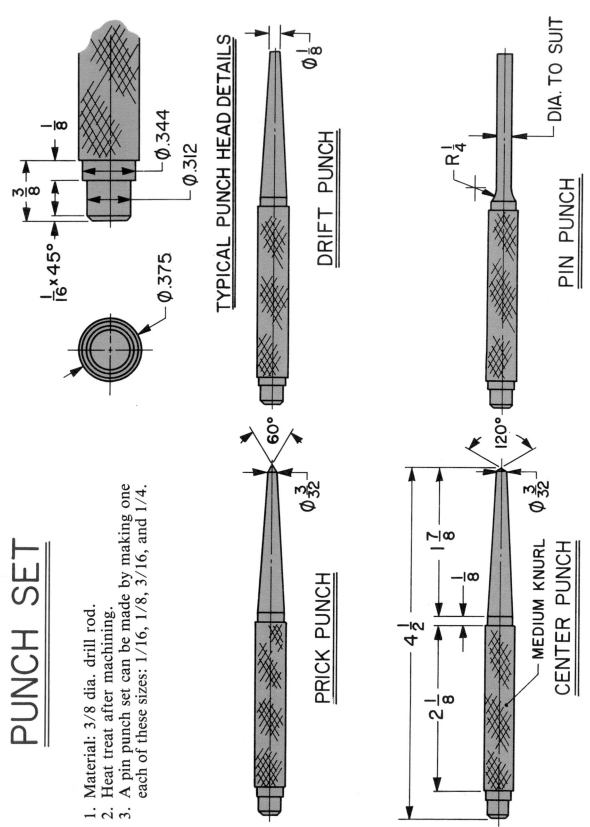

TYPICAL PUNCH HEAD DETAILS

DRIFT PUNCH

PIN PUNCH

PRICK PUNCH

CENTER PUNCH

MEDIUM KNURL

CAST PLAQUE

1. Material: Aluminum or brass.
2. Plaques may be designed to serve as awards by mounting the casting on backing boards.

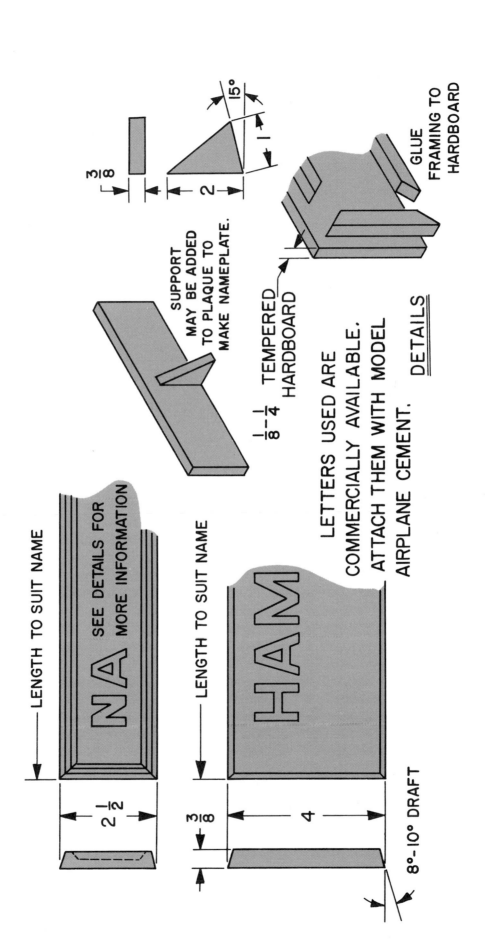

SUPPORT MAY BE ADDED TO PLAQUE TO MAKE NAMEPLATE.

$1\frac{1}{8}$ TEMPERED HARDBOARD

LETTERS USED ARE COMMERCIALLY AVAILABLE. ATTACH THEM WITH MODEL AIRPLANE CEMENT.

DETAILS

GLUE FRAMING TO HARDBOARD

LENGTH TO SUIT NAME

NA SEE DETAILS FOR MORE INFORMATION

LENGTH TO SUIT NAME

HAM

$2\frac{1}{2}$

$\frac{3}{8}$

4

8°-10° DRAFT

$\frac{3}{8}$

2

1

15°

DESIGN PROBLEM

Design a magazine rack using the project shown as a starting point.

This is a Spanish motif which fits in with most room settings and furniture styles.

The dimensions given are approximate but will help you design a rack of suitable size.

Weld or braze all joints. Remove burrs and rough edges before painting.

Small pieces of felt cemented on the bottoms of the feet will prevent rust spots.

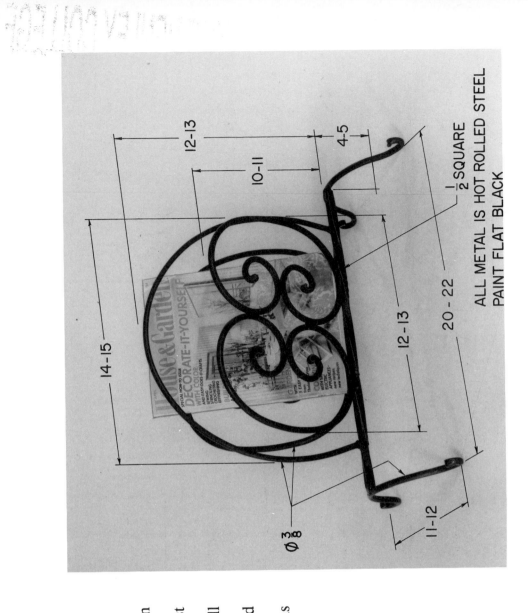

14-15

12-13

10-11

4-5

12-13

20 - 22

11-12

Ø 3/8

1/2 SQUARE

ALL METAL IS HOT ROLLED STEEL
PAINT FLAT BLACK

METAL SCULPTURE

1. Material: Figure—Aluminum or brass. Base—Walnut or cherry.
2. The project can be used as a trophy if an appropriate figure is selected. Ideas are available everywhere: a cougar, impala, road runner, fire bird, etc.
3. Use epoxy in figure and base.

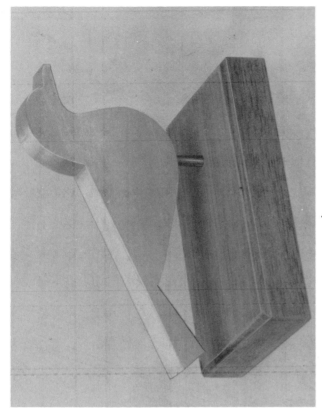

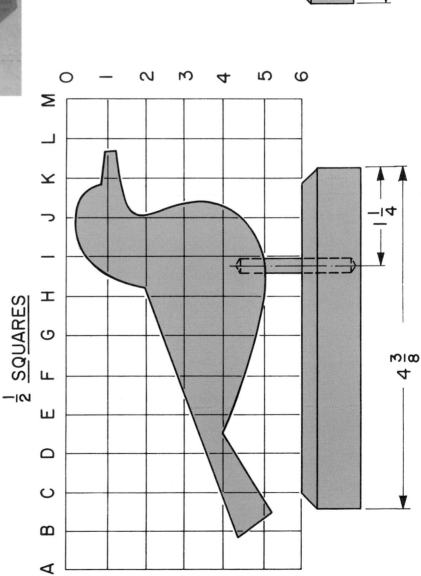

$\frac{1}{2}$ SQUARES

BOTTLE OPENER

Material: 1/8 - 3/16 hard brass or mild steel.
Finish: Polish.

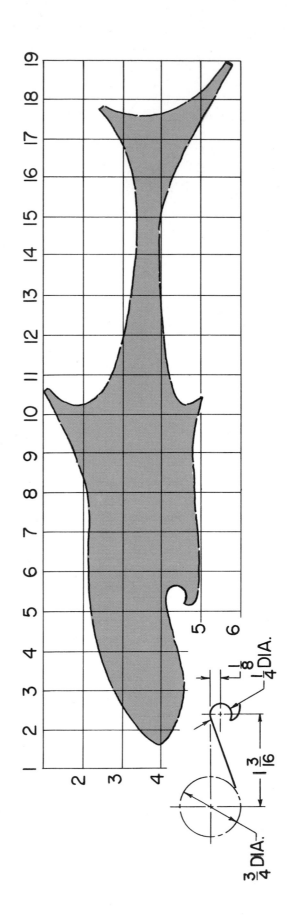

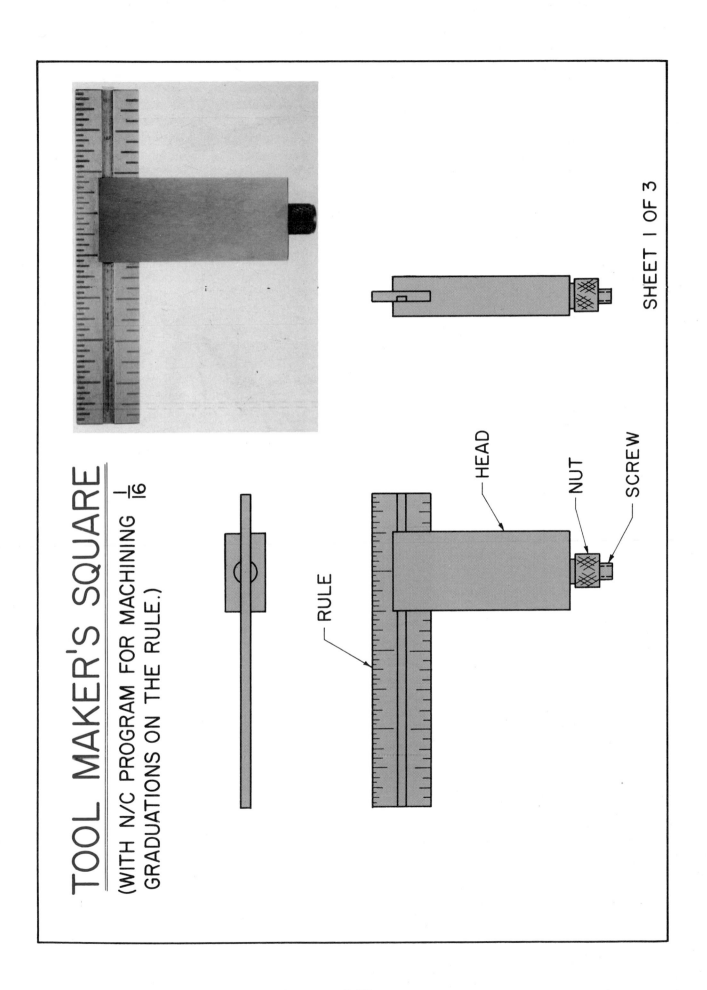

TOOL MAKER'S SQUARE

(WITH N/C PROGRAM FOR MACHINING $\frac{1}{16}$ GRADUATIONS ON THE RULE.)

RULE

HEAD

NUT

SCREW

TOOL MAKER'S SQUARE

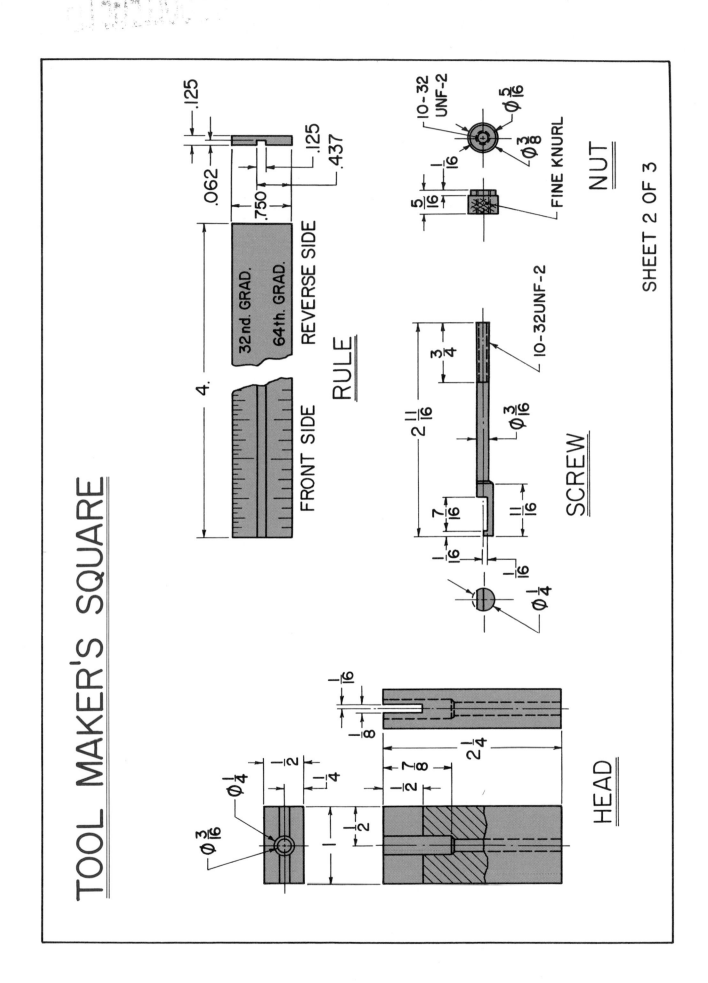

RULE

FRONT SIDE

REVERSE SIDE

32nd. GRAD.

64th. GRAD.

4.

.125

.125

.437

.750

.062

NUT

10-32 UNF-2

Ø 5/16

Ø 3/8

FINE KNURL

1/16

5/16

SCREW

10-32 UNF-2

3/4

Ø 3/16

2 11/16

7/16

11/16

1/16

1/16

Ø 1/4

HEAD

1/16

1/8

2 1/4

7/8

1/2

Ø 1/4

Ø 3/16

1/2

1/4

1

1/2

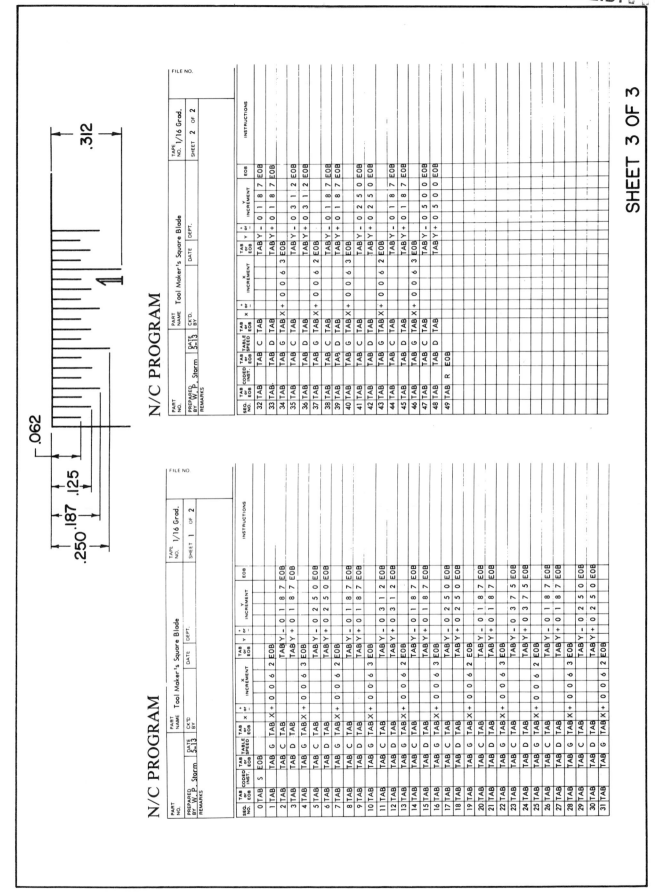

N/C PROGRAM

N/C PROGRAM

SHEET 3 OF 3

CONTEMPORARY CANDLE HOLDER

1. Material: Aluminum or brass.
2. Finish: As machined.
3. Attach felt to base.

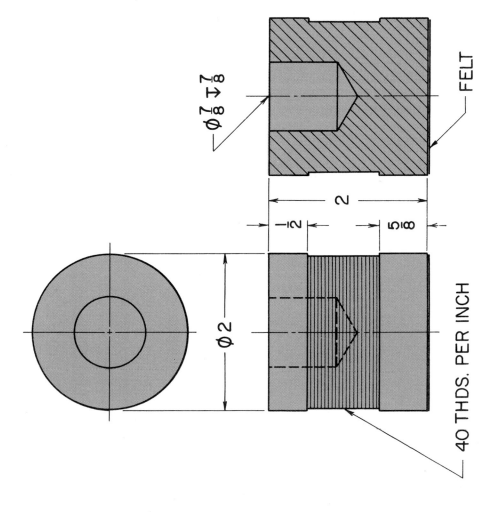

$\phi \frac{7}{8} \, \overline{\downarrow} \, \frac{7}{8}$

FELT

$\frac{1}{2}$

2

$\frac{5}{8}$

$\phi 2$

40 THDS. PER INCH

250

SCOTTY TABLE LAMP

Material: 3/8 aluminum, or a pattern can be made and the figure cast as shown in photograph. The base may be made from mahogany, walnut, or cherry.

Finish: Paint flat black or use a fine abrasive and apply a satin finish.

Design Problem: Provide a way to hold the light socket and shade. It may be made from 3/8 dia. brass tubing.

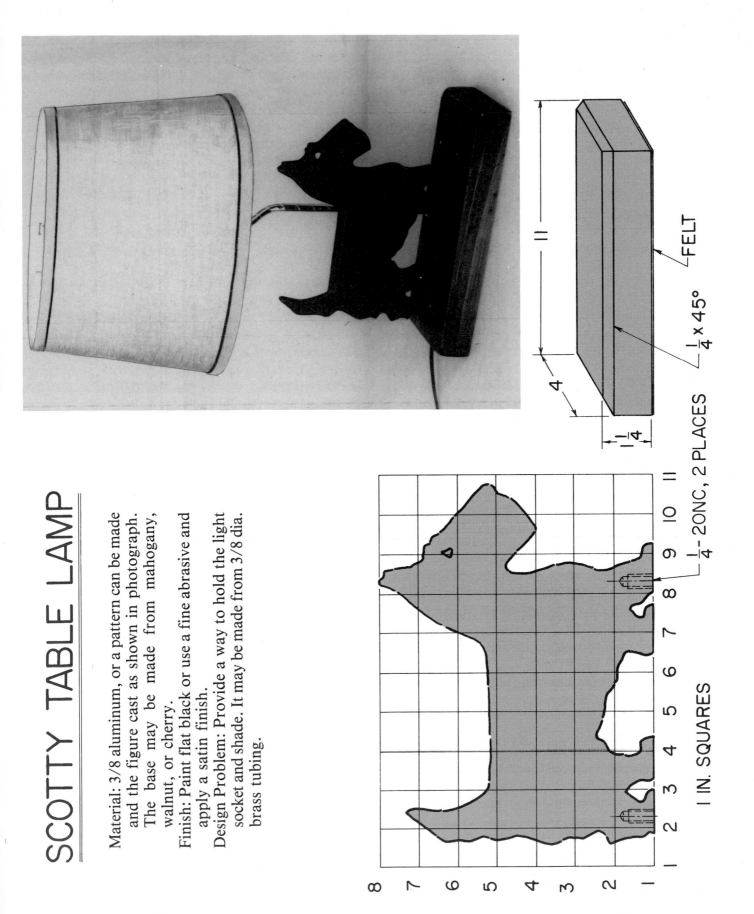

FELT

$\frac{1}{4}$ × 45°

11

4

$\frac{1}{4}$

$\frac{1}{4}$ - 20NC, 2 PLACES

1 IN. SQUARES

Glossary

A

ABRASIVE: A material that cuts a material softer than itself.

ACUTE ANGLE: An angle of less than 90 deg.

ALIGN: Adjusting to given points.

ALLOWANCE: The limits permitted for satisfactory performance of machined parts.

ALLOY: A mixture of two or more metals fused or melted together to form a new metal.

ALUMINUM OXIDE: A manufactured abrasive. It has largely replaced emery as an abrasive when large quantities of metal must be removed.

ANNEALING: The process of heating metal to a given temperature and allowing it to cool slowly to induce softness. The exact temperature and the period of time the temperature is held depends on the metal.

ANODIZING: A surface finish for aluminum that protects it against corrosion. The process also permits the surface to be dyed a variety of colors.

ANVIL: A block of iron or steel upon which metal is forged.

ASSEMBLY: A unit fitted together from manufactured parts.

ASSEMBLY DRAWING: A drawing that shows the machinist how to assemble an object. The component parts in the drawing are usually key-numbered.

AUTOMATION: An industrial technique that substitutes mechanical labor and mechanical control for human labor and human control.

AXIS: A centerline that passes through an object about which it could rotate. Also used as a point of reference.

B

BERYLLIUM: A metal that weighs about one fourth as much as steel, yet is almost as strong. It is an "exotic metal," used in rockets and aircraft where weight is critical, and in nuclear reactors.

BEVEL: An angle that is not at right angles to another line or surface.

BLOWHOLE: A hole produced in a casting when gases are trapped during the pouring operation.

BRASS: An alloy of copper and zinc. It is bright yellow in color.

BRAZING: A process for joining metals by the fusion of nonferrous alloys that have melting temperatures above 800°F (425°C) but lower than the metals joined.

BRINELL: A term used to designate the hardness of a piece of metal.

BRITTLENESSS: Characteristics that cause a material to break easily.

BRONZE: An alloy of copper and tin. It is reddish gold in color.

BUFFING: The technique of bringing out the luster of metal. Polishing.

BURR: The sharp edge remaining on the metal after cutting or machining.

BUSHING: A bearing or a guide for a cutting tool in a fixture.

C

CARBON STEEL: Steel in which the physical and mechanical properties depend primarily on the carbon content of the metal.

CASE HARDENING: A process of surface hardening iron base metals so the surface layer or "case" of the metal is made harder than the interior or core.

CASTING: An object made by pouring molten metal into a mold.

CHAMFER: To bevel a sharp external edge.

CHASING THREADS: Cutting threads on a lathe.

CHATTER: Vibrations caused by the cutting tool springing away from the work. This produces small ridges on the machine surface.

CHIP: To cut with a chisel.

CHUCK: A device on a machine tool which holds work or cutting tools.

CLEARANCE: The distance by which one part clears another part.

CLOCKWISE: From left to right in a circular motion; the direction clock hands move.

COINING: A metalworking technique that impresses images/characters on a die and punch onto a plain metal surface.

COLOR CODING: A method used to identify steel. Each type of commonly used steel is identified by a different color.

COLOR TEMPER: Using the color range steel passes through when heated to determine the proper degree of hardness.

CONCAVE SURFACE: A curved depression in the surface of an object.

CONCENTRIC: Having a common center.

CONTOUR: The outline of an object or figure; particularly a curved or irregular outline.

CONVENTIONAL: Not original. Customary or traditional.

CONVEX SURFACE: Rounded surface raised on an object.

COOLANT: A fluid or gas used to cool the cutting edge of a tool to prevent it from burning up during the machining operation.

COPPER: A reddish brown base metal.

CORE: A body of sand or other material that is formed to the desired shape and placed in a mold to produce a cavity or opening in a casting.

COUNTERBORE: Enlarging a hole to a given depth and diameter.

COUNTERCLOCKWISE: From right to left in a circular motion. The opposite direction clock hands move.

COUNTERSINK: Chamfering a hole to receive a flat-head screw.

CUTTING FLUID: A liquid used to cool and lubricate a cutting tool and to remove chips.

D

DEAD CENTER: A stationary or non-rotating center.

DIE: A tool used to cut external threads. Also, a tool used to impart a desired shape to a piece of metal.

DIE CASTING: A method of casting metal under pressure by injecting it into the metal dies of a die casting machine.

DIVIDING HEAD: An attachment for machine tools that is used to accurately space holes, slots, gear teeth, and flutes on round metal stock.

DRAFT: The clearance on a pattern that allows easy withdrawal of the pattern from the mold.

DRIFT: A tapered piece of flat steel used to separate tapered shank tools (like drills) from sleeves, sockets, and machine spindles.

DRILLING: Cutting round holes using a cutting tool with a sharpened point.

DRILL ROD: A carbon steel rod accurately and smoothly ground to size. Available in a large range of sizes.

DROP FORGING: Shaping metal by heating and hammering into impressions in dies.

E

ECCENTRIC: Not on a common center. A device that converts rotary motion into a reciprocating (back and forth) motion.

ECM: Abbreviation for the metal removal process, Electro Chemical Machining.

EDM: Abbreviation for the metal removal process, Electrical Discharge Machining.

EMERY: A natural (not manufactured) abrasive for grinding and polishing.

EMERY CLOTH: Cloth with emery abrasive cemented to its surface. Used to clean and polish metal.

ENTREPRENEUR: Person who originates a new business venture and creates jobs. Entrepreneurs take risks to achieve success.

EXTRUDE: To force metal through a die to produce a desired shape.

F

FACE: To make a flat surface by machining.

FACEPLATE: A circular plate that fits on to the headstock spindle and drives or carries work to be machined.

FERROUS: Denotes a family of metals in which iron is the major ingredient.

FILLET: The curved surface that connects two intersecting surfaces that form an angle.

FIXTURE: A device used to hold metal while it is being machined.

FLASK: A wooden or metal form consisting of a cope (the top portion) and a drag (the bottom portion). It is used to hold the sand that forms the mold in metal casting.

FLUTE: A groove machined in a cutting tool to help in the removal of chips. Permits coolant to reach the cutting point of the tool.

FLUX: Chemicals used in soldering, brazing, and welding to prevent oxidation and promote better fusion of the metals.

FORGE: To form metal with heat and/or pressure.

G

GALVANIZE: To coat steel with zinc.

GATE: The point where molten metal enters the mold cavity.

GAUGE: A tool used for checking the size of metal parts.

GEARS: Toothed wheels that are employed to transmit rotary motion from one shaft to another shaft without slippage.

GREEN SAND: Foundry sand moistened with water.

H

HARDENING: The process whereby certain iron-base alloys are heated and quenched (cooled) to produce a hardness superior to that of the untreated metal.

HEAT TREATMENT: The careful application of a combination of heating and cooling cycles to a metal or an alloy in the solid state to bring about certain desirable conditions such as hardness and toughness.

HERF: Abbreviation for the metal forming process, High Energy Rate Forming.

I

I.D.: Abbreviation for inside diameter.

INDEPENDENT CHUCK: A chuck in which each jaw can be moved independently of the other jaws.

INSPECTION: The measuring and checking of finished parts to determine whether they have been made according to specifications.

INTERCHANGEABLE: Refers to a part that has been made to specific dimensions and tolerances and is capable of being fitted in a mechanism in place of a similarly made part.

J

JIG: A device that holds work in position, and positions and guides the cutting tool.

K

KEY: A small piece of metal fitted in a shaft and hub to prevent a gear or pulley from rotating on a shaft.

KEYWAY: The slot or recess cut in a shaft that holds the key.

KNURLING: Operation that presses grooves into the surface of cylindrical work while it rotates in the lathe.

L

LAPPING: Technique of producing a smooth, accurate surface to a bearing or other mating part by using fine abrasives.

LATHE DOG: A device for clamping work so that is can be machined between centers.

LAY OUT: To locate and scribe points for machining and forming operations.

LIVE CENTER: A rotating center.

M

MACHINABILITY: The characteristic of a material that describes the ease or difficulty of machining it.

MACHINIST: A person who is skilled in the use of machine tools and is capable of making complex machine setups.

MAJOR DIAMETER: The largest diameter of a thread.

MANDREL: A cylindrical piece of steel used to support work for machining operations.

MILL: To remove metal with a rotating cutter on a milling machine.

MILLING MACHINE: A machine that removes metal by means of a rotary cutter.

MINOR DIAMETER: The smallest diameter of a screw thread. Also known as the "root diameter."

N

NC: Abbreviation for the National Coarse series of screw threads.

NF: Abbreviation for the National Fine series of screw threads.

NONFERROUS: Metals containing no iron.

O

O.D.: Abbreviation for outside diameter.

OFF CENTER: Eccentric, not accurate.

P

PEENING: Using the peen (rounded) end of a hammer to decorate the surface of metal.

PEWTER: An alloy of tin (91 percent), copper (1 1/2 percent), and antimony (7 1/2 percent). Modern pewter does not contain lead.

PICKLING: Chemical treatment to remove scale from metal.

PIN PUNCH: A tool made of carbon steel with a long cylindrical end that is used to remove pins and rivets. It can also be employed to punch holes in sheet metal.

PITCH: The distance from a point on one thread to a similar point on the next thread.

PLANISH: To finish or smooth the surface of sheet metal by hammering it lightly with a hammer having a mirror smooth face.

POLISH: To produce a smooth or glossy surface by friction.

PYROMETER: A device for measuring high temperatures. Temperatures are determined by measuring the electric current generated in a thermocouple (two dissimilar metals welded together) as it heats up.

Q

QUENCHING: Rapid cooling by contact with fluids or gases.

R

REAMER: A cutting tool used to finish a drilled hole to exact size.

RELIEF: An undercut or offset surface which provides clearance.

RISER: A reservoir of molten metal provided to compensate for the contraction of cast metals as they solidify.

ROCKWELL HARDNESS: A method of measuring the hardness of a piece of material using a Rockwell hardness testing machine.

ROOT DIAMETER: The smallest or "minor diameter" of a screw thread.

RUNNER: Channel through which molten metal flows from the sprue to the casting and risers.

S

SAE: Abbreviation for the Society of Automotive Engineers.

SAFE EDGE: Edge of a file which has no teeth cut in it.

SANDBLAST: To clean surfaces of castings by using sand blown at high pressure.

SCALE: Oxidation caused on metal surfaces by heating them.

SCRIBE: To draw a line with a scriber or other sharp pointed tool.

SCROLL: A curved section widely used for decorative purposes.

SETUP: The term used to describe the positioning of the work, cutting tools, and machine attachments on a machine tool.

SHEAR: To cut steel metal between two blades.

SHIMS: Pieces of sheet metal, available in various thicknesses, used between mating parts to provide proper clearances.

SILICON CARBIDE: The hardest and sharpest of the manufactured abrasives.

SLUG: Small piece of metal used as spacing material.

SOLDERING: A method of joining metals by means of a nonferrous filler metal, without fusion of the base metals. Normally carried out at temperatures lower than 800°F (425°C).

SPOTFACE: To machine a round spot on a rough surface, usually around a drilled hole. To provide seat for screw or bolt head.

SPRUE HOLE: The opening in a mold into which the molten metal is poured.

STANDARD: An accepted base for a uniform system of measurement and quality.

STRAIGHTEDGE: A precision tool used to check the accuracy of flat surfaces.

SURFACE PLATE: A plate of cast iron, cast steel, or granite that has one or more surfaces finished to a smooth, flat surface. It is used as a base for layout measurements and inspection.

T

TAP: Tool used to cut internal threads.

TAP DRILL: The drill used to make a hole prior to tapping.

TAPPING: The operation that produces internal threads with a tap. It may be done by hand or machine. Tapping also refers to the operation of removing molten metal from a furnace.

TEMPLATE: A pattern or guide.

THREAD: The act of cutting a screw thread.

TIN: A soft, shiny metal. It is nontoxic and when used as plating provides excellent protection against corrosion.

TOLERANCE: The permissible deviation from a basic dimension.

TOOL CRIB: A room or area in a machine shop where tools and supplies are stored and dispensed as needed.

TOOLROOM: The area or department where tools, jigs, fixtures, and the like are manufactured.

TRAIN: A series of meshed gears.

TRUE: On center.

TURN: To machine on a lathe.

U

UNIVERSAL CHUCK: A chuck on which all jaws move simultaneously at a uniform rate to automatically center round or hexagonal stock.

V

VENTS: Narrow openings in molds that permit gases generated during pouring to escape.

W

WHEEL DRESSER: A device utilized to true the face of a grinding wheel.

WORKING DRAWING: A drawing that gives the machinist information on how to make and assemble a mechanism.

WROUGHT IRON: Iron with most of the carbon removed. It is tough, easy to bend and weld.

X

X-RAY: An inspection technique used to find flaws in manufactured parts. The part is not damaged by the inspection process.

ACKNOWLEDGMENTS

While it would be a most pleasant task, it would be impossible for one person to develop the material included in this text by visiting the various industries represented and observing, studying, and taking the photos first-hand.

My sincere thanks to those who helped in the gathering of the necessary material, information, and photographs. Their cooperation was most appreciated.

John R. Walker

Special thanks to Mr. Gerry Greenway and the Swift Saw and Tool Supply Company in Hazel Crest, Illinois for loan of the tools for the cover design and Dr. William L. Schotta of Millersville University, Millersville, Pennsylvania for photographs used throughout the book.

Tables

FRACTIONAL INCHES
INTO DECIMALS AND MILLIMETERS

INCH	DECIMAL INCH	MILLIMETER	INCH	DECIMAL INCH	MILLIMETER
1/64	0.0156	0.3967	33/64	0.5162	13.0968
1/32	0.0312	0.7937	17/32	0.5312	13.4937
3/64	0.0468	1.1906	35/64	0.5468	13.8906
1/16	0.0625	1.5875	9/16	0.5625	14.2875
5/64	0.0781	1.9843	37/64	0.5781	14.6843
3/32	0.0937	2.3812	19/32	0.5937	15.0812
7/64	0.1093	2.7781	39/64	0.6093	15.4781
1/8	0.125	3.175	5/8	0.625	15.875
9/64	0.1406	3.5718	41/64	0.6406	16.2718
5/32	0.1562	3.9687	21/32	0.6562	16.6687
11/64	0.1718	4.3656	43/64	0.6718	17.0656
3/16	0.1875	4.7625	11/16	0.6875	17.4625
13/64	0.2031	5.1593	45/64	0.7031	17.8593
7/32	0.2187	5.5562	23/32	0.7187	18.2562
15/64	0.2343	5.9531	47/64	0.7343	18.6531
1/4	0.25	6.5	3/4	0.75	19.05
17/64	0.2656	6.7468	49/64	0.7656	19.4468
9/32	0.2812	7.1437	25/32	0.7812	19.8437
19/64	0.2968	7.5406	51/64	0.7968	20.2406
5/16	0.3125	7.9375	13/16	0.8125	20.6375
21/64	0.3281	8.3343	53/64	0.8281	21.0343
11/32	0.3437	8.7312	27/32	0.8437	21.4312
23/64	0.3593	9.1281	55/64	0.8593	21.8281
3/8	0.375	9.525	7/8	0.875	22.225
25/64	0.3906	9.9218	57/64	0.8906	22.6218
13/32	0.4062	10.3187	29/32	0.9062	23.0187
27/64	0.4218	10.7156	59/64	0.9218	23.4156
7/16	0.4375	11.1125	15/16	0.9375	23.8125
29/64	0.4531	11.5093	61/64	0.9531	24.2093
15/32	0.4687	11.9062	31/32	0.9687	24.6062
31/64	0.4843	12.3031	63/64	0.9843	25.0031
1/2	0.50	12.7	1	1.0000	25.4

MEASUREMENT SYSTEMS

ENGLISH SYSTEM	METRIC SYSTEM
MEASURES OF TIME 60 sec. = 1 min. 60 min. = 1 hr. 24 hr. = 1 day 365 dy. = 1 common yr. 366 dy. = 1 leap yr. **DRY MEASURES** 2 pt. = 1 qt. 8 qt. = 1 pk. 4 pk. = 1 bu. 2150.42 cu. in. = 1 bu. **MEASURES OF LENGTH** 12 in. = 1 ft. 3 ft. = 1 yd. 5 1/2 yd. = 1 rod 320 rods = 1 mile 5,280 ft. = 1 mile 1,760 yd. = 1 mile 6,080 ft. = 1 knot **LIQUID MEASURES** 16 fluid oz. = 1 pt. 2 pt. = 1 qt. 32 fl. oz. = 1 qt. 4 qt. = 1 gal. 31 1/2 gal. = 1 bbl. 231 cu. in. = 1 gal. 7 1/2 gal. = 1 cu. ft. **MEASURES OF AREA** 144 sq. in. = 1 sq. ft. 9 sq. ft. = 1 sq. yd. 30 1/4 sq. yd. = 1 sq. rod 160 sq. rods = 1 acre 640 acres = 1 sq. mile **MEASURES OF WEIGHT** (Avoirdupois) 7,000 grains (gr.) = 1 lb. 16 oz. = 1 lb. 100 lb. = 1 cwt. 2,000 lb. = 1 short ton 2,240 lb. = 1 long ton **MEASURES OF VOLUME** 1,728 cu. in. = 1 cu. ft. 27 cu. ft. = 1 cu. yd. 128 cu. ft. = 1 cord	The basic unit of the metric system is the meter (m). The meter is exactly 39.37 in. long. This is 3.37 in. longer than the English yard. Units that are multiples or fractional parts of the meter are designated as such by prefixes to the word "meter". For example: 1 millimeter (mm.) = 0.001 meter or 1/1000 meter 1 centimeter (cm.) = 0.01 meter or 1/100 meter 1 decimeter (dm.) = 0.1 meter or 1/10 meter 1 meter (m.) 1 decameter (dkm.) = 10 meters 1 hectometer (hm.) = 100 meters 1 kilometer (km.) = 1000 meters These prefixes may be applied to any unit of length, weight, volume, etc. The meter is adopted as the basic unit of length, the gram for mass, and the liter for volume. In the metric system, area is measured in square kilometers (sq. km. or $km.^2$), square centimeters (sq. cm. or $cm.^2$), etc. Volume is commonly measured in cubic centimeters, etc. One liter (1) is equal to 1,000 cubic centimeters. The metric measurements in most common use are shown in the following tables: **MEASURES OF LENGTH** 10 millimeters = 1 centimeter 10 centimeters = 1 decimeter 10 decimeters = 1 meter 1000 meters = 1 kilometer **MEASURES OF WEIGHT** 100 milligrams = 1 gram 1000 grams = 1 kilogram 1000 kilograms = 1 metric ton **MEASURES OF VOLUME** 1000 cubic centimeters = 1 liter 100 liters = 1 hectoliter

CONVERSION TABLES

TO REDUCE	MULTIPLY BY	TO REDUCE	MULTIPLY BY
LENGTH			
miles to km.	1.61	km. to miles	0.62
miles to m.	1609.35	m. to miles	0.00062
yd. to m.	0.9144	m. to yd.	1.0936
in. to cm.	2.54	cm. to in.	0.3937
in. to mm.	25.4	mm. to in.	0.03937
VOLUME			
cu. in. to cc. or ml.	16.387	cc. to cu. in.	0.061
cu. in. to l.	0.0164	l to cu. in.	61.024
gal. to l.	3.785	l to gal.	0.264
WEIGHT			
lb. to kg.	0.4536	kg. to lb.	2.2
oz. to gm.	28.35	gm. to oz.	0.0353
gr. to gm.	0.0648	gm. to gr.	15.432

DECIMAL EQUIVALENTS NUMBER SIZE DRILLS

NO.	SIZE OF DRILL IN INCHES	NO.	SIZE OF DRILL IN INCHES	NO.	SIZE OF DRILL IN INCHES	NO.	SIZE OF DRILL IN INCHES
1	.2280	21	.1590	41	.0960	61	.0390
2	.2210	22	.1570	42	.0935	62	.0380
3	.2130	23	.1540	43	.0860	63	.0370
4	.2090	24	.1520	44	.0820	64	.0360
5	.2055	25	.1495	45	.0810	65	.0350
6	.2040	26	.1470	46	.0810	66	.0330
7	.2010	27	.1440	47	.0785	67	.0320
8	.1990	28	.1405	48	.0760	68	.0310
9	.1960	29	.1360	49	.0730	69	.0292
10	.1935	30	.1285	50	.0700	70	.0280
11	.1910	31	.1200	51	.0670	71	.0260
12	.1890	32	.1160	52	.0635	72	.0250
13	.1850	33	.1130	53	.0595	73	.0240
14	.1820	34	.1110	54	.0550	74	.0225
15	.1800	35	.1100	55	.0520	75	.0210
16	.1770	36	.1065	56	.0465	76	.0200
17	.1730	37	.1040	57	.0430	77	.0180
18	.1695	38	.1015	58	.0420	78	.0160
19	.1660	39	.0995	59	.0410	79	.0145
20	.1610	40	.0980	60	.0400	80	.0135

NATIONAL COARSE AND NATIONAL FINE THREADS AND TAP DRILLS

SIZE	THREADS PER INCH	MAJOR DIA.	MINOR DIA.	PITCH DIA.	TAP DRILL 75 PERCENT THREAD	DECIMAL EQUIVALENT	CLEARANCE DRILL	DECIMAL EQUIVALENT
2	56	.0860	.0628	.0744	50	.0700	42	.0935
	64	.0860	.0657	.0759	50	.0700	42	.0935
3	48	.099	.0719	.0855	47	.0785	36	.1065
	56	.099	.0758	.0874	45	.0820	36	.1065
4	40	.112	.0795	.0958	43	.0890	31	.1200
	48	.112	.0849	.0985	42	.0935	31	.1200
6	32	.138	.0974	.1177	36	.1065	26	.1470
	40	.138	.1055	.1218	33	.1130	26	.1470
8	32	.164	.1234	.1437	29	.1360	17	.1730
	36	.164	.1279	.1460	29	.1360	17	.1730
10	24	.190	.1359	.1629	25	.1495	8	.1990
	32	.190	.1494	.1697	21	.1590	8	.1990
12	24	.216	.1619	.1889	16	.1770	1	.2280
	28	.216	.1696	.1928	14	.1820	2	.2210
1/4	20	.250	.1850	.2175	7	.2010	G	.2610
	28	.250	.2036	.2268	3	.2130	G	.2610
5/16	18	.3125	.2403	.2764	F	.2570	21/64	.3281
	24	.3125	.2584	.2854	I	.2720	21/64	.3281
3/8	16	.3750	.2938	.3344	5/16	.3125	25/64	.3906
	24	.3750	.3209	.3479	Q	.3320	25/64	.3906
7/16	14	.4375	.3447	.3911	U	.3680	15/32	.4687
	20	.4375	.3725	.4050	25/64	.3906	29/64	.4531
1/2	13	.5000	.4001	.4500	27/64	.4219	17/32	.5312
	20	.5000	.4350	.4675	29/64	.4531	33/64	.5156
9/16	12	.5625	.4542	.5084	31/64	.4844	19/32	.5937
	18	.5625	.4903	.5264	33/64	.5156	37/64	.5781
5/8	11	.6250	.5069	.5660	17/32	.5312	21/32	.6562
	18	.6250	.5528	.5889	37/64	.5781	41/64	.6406
3/4	10	.7500	.6201	.6850	21/32	.6562	25/32	.7812
	16	.7500	.6688	.7094	11/16	.6875	49/64	.7656
7/8	9	.8750	.7307	.8028	49/64	.7656	29/32	.9062
	14	.8750	.7822	.8286	13/16	.8125	57/64	.8906
1	8	1.0000	.8376	.9188	7/8	.8750	1-1/32	1.0312
	14	1.0000	.9072	.9536	15/16	.9375	1-1/64	1.0156
1-1/8	7	1.1250	.9394	1.0322	63/64	.9844	1-5/32	1.1562
	12	1.1250	1.0167	1.0709	1-3/64	1.0469	1-5/32	1.1562
1-1/4	7	1.2500	1.0644	1.1572	1-7/64	1.1094	1-9/32	1.2812
	12	1.2500	1.1417	1.1959	1-11/64	1.1719	1-9/32	1.2812
1-1/2	6	1.5000	1.2835	1.3917	1-11/32	1.3437	1-17/32	1.5312
	12	1.5000	1.3917	1.4459	1-27/64	1.4219	1-17/32	1.5312

LETTER SIZE DRILLS

A	0.234	J	0.277	S	0.348
B	0.238	K	0.281	T	0.358
C	0.242	L	0.290	U	0.368
D	0.246	M	0.295	V	0.377
E	0.250	N	0.302	W	0.386
F	0.257	O	0.316	X	0.397
G	0.261	P	0.323	Y	0.404
H	0.266	Q	0.332	Z	0.413
I	0.272	R	0.339		

Tables

SCREW THREAD ELEMENTS FOR UNIFIED AND NATIONAL
FORM OF THREAD

THREADS PER INCH (n)	PITCH (p) $p = \frac{1}{n}$	SINGLE HEIGHT	DOUBLE HEIGHT	83 1/3 PERCENT DOUBLE HEIGHT	BASIC WIDTH OF CREST AND ROOT FLAT $\frac{p}{8}$	CONSTANT FOR BEST SIZE WIRE ALSO SINGLE HEIGHT OF 60 DEG. V-THREAD	DIAMETER OF BEST SIZE WIRE
		SUBTRACT FROM BASIC MAJOR DIAMETER TO GET BASIC PITCH DIAMETER	SUBTRACT FROM BASIC MAJOR DIAMETER TO GET BASIC MINOR DIAMETER	SUBTRACT FROM BASIC MAJOR DIAMETER TO GET MINOR DIAMETER OF RING GAGE			
3	.333333	.216506	.43301	.36084	.0417	.28868	.19245
3 1/4	.307692	.199852	.39970	.33309	.0385	.26647	.17765
3 1/2	.285714	.185577	.37115	.30929	.0357	.24744	.16496
4	.250000	.162379	.32476	.27063	.0312	.21651	.14434
4 1/2	.222222	.144337	.28867	.24056	.0278	.19245	.12830
5	.200000	.129903	.25981	.21650	.0250	.17321	.11547
5 1/2	.181818	.118093	.23619	.19682	.0227	.15746	.10497
6	.166666	.108253	.21651	.18042	.0208	.14434	.09623
7	.142857	.092788	.18558	.15465	.0179	.12372	.08248
8	.125000	.081189	.16238	.13531	.0156	.10825	.07217
9	.111111	.072168	.14434	.12028	.0139	.09623	.06415
10	.100000	.064952	.12990	.10825	.0125	08660	.05774
11	.090909	.059046	.11809	.09841	.0114	.07873	.05249
11 1/2	.086956	.056480	.11296	.09413	.0109	.07531	.05020
12	.083333	.054127	.10826	.09021	.0104	.07217	.04811
13	.076923	.049963	.09993	.08327	.0096	.06662	.04441
14	.071428	.046394	.09279	.07732	.0089	.06186	.04124
16	.062500	.040595	.08119	.06766	.0078	.05413	.03608
18	.055555	.036086	.07217	.06014	.0069	.04811	.03208
20	.050000	.032475	.06495	.05412	.0062	.04330	.02887
22	.045454	.029523	.05905	.04920	.0057	.03936	.02624
24	.041666	.027063	.05413	.04510	.0052	.03608	.02406
27	.037037	.024056	.04811	.04009	.0046	.03208	.02138
28	.035714	.023197	.04639	.03866	.0045	.03093	.02062
30	.033333	.021651	.04330	.03608	.0042	.02887	.01925
32	.031250	.020297	.04059	.03383	.0039	.02706	.01804
36	.027777	.018042	.03608	.03007	.0035	.02406	.01604
40	.025000	.016237	.03247	.02706	.0031	.02165	.01443
44	.022727	.014761	.02952	.02460	.0028	.01968	.01312
48	.020833	.013531	.02706	.02255	.0026	.01804	.01203
50	.020000	.012990	.02598	.02165	.0025	.01732	.01155
56	.017857	.011598	.02320	.01933	.0022	.01546	.01031
60	.016666	.010825	.02165	.01804	.0021	.01443	.00962
64	.015625	.010148	.02030	.01691	.0020	.01353	.00902
72	.013888	.009021	.01804	.01503	.0017	.01203	.00802
80	.012500	.008118	.01624	.01353	.0016	.01083	.00722
90	.011111	.007217	.01443	.01202	.0014	.00962	.00642
96	.010417	.006766	.01353	.01127	.0013	.00902	.00601
100	.010000	.006495	.01299	.01082	.0012	.00866	.00577
120	.008333	.005413	.01083	.00902	.0010	.00722	.00481

Using the Best Size Wires, the measurement over three wires minus the Constant for Best Size Wire equals the Pitch Diameter.

Exploring Metalworking

MACHINE SCREW AND CAP SCREW HEADS

FILLISTER HEAD

SIZE	A	B	C	D
#8	.260	.141	.042	.060
#10	.302	.164	.048	.072
1/4	3/8	.205	.064	.087
5/16	7/16	.242	.077	.102
3/8	9/16	.300	.086	.125
1/2	3/4	.394	.102	.168
5/8	7/8	.500	.128	.215
3/4	1	.590	.144	.258
1	1 5/16	.774	.182	.352

FLAT HEAD

SIZE	A	B	C	D
#8	.320	.092	.043	.037
#10	.372	.107	.048	.044
1/4	1/2	.146	.064	.063
5/16	5/8	.183	.072	.078
3/8	3/4	.220	.081	.095
1/2	7/8	.220	.102	.090
5/8	1 1/8	.293	.128	.125
3/4	1 3/8	.366	.144	.153

ROUND HEAD

SIZE	A	B	C	D
#8	.297	.113	.044	.067
#10	.346	.130	.048	.073
1/4	7/16	.1831	.064	.107
5/16	9/16	.236	.072	.150
3/8	5/8	.262	.081	.160
1/2	13/16	.340	.102	.200
5/8	1	.422	.128	.255
3/4	1 1/4	.526	.144	.320

HEXAGON HEAD

SIZE	A	B	C
1/4	.494	.170	7/16
5/16	.564	.215	1/2
3/8	.635	.246	9/16
1/2	.846	.333	3/4
5/8	1.058	.411	15/16
3/4	1.270	.490	1 1/8
7/8	1.482	.566	1 5/16
1	1.693	.640	1 1/2

SOCKET HEAD

SIZE	A	B	C
#8	.265	.164	1/8
#10	5/16	.190	5/32
1/4	3/8	1/4	3/16
5/16	7/16	5/16	7/32
3/8	9/16	3/8	5/16
7/16	5/8	7/16	5/16
1/2	3/4	1/2	3/8
5/8	7/8	5/8	1/2
3/4	1	3/4	9/16
7/8	1 1/8	7/8	9/16
1	1 5/16	1	5/8

PHYSICAL PROPERTIES OF METALS

METAL	SYMBOL	SPECIFIC GRAVITY	SPECIFIC HEAT	MELTING POINT*		LBS. PER CUBIC INCH
				DEG. C	DEG. F.	
Aluminum (Cast)	Al	2.56	.2185	658	1217	.0924
Aluminum (Rolled).	Al	2.71	–	–	–	.0978
Antimony	Sb	6.71	.051	630	1166	.2424
Bismuth	Bi	9.80	.031	271	520	.3540
Boron.	B	2.30	.3091	2300	4172	.0831
Brass.	–	8.51	.094	–	–	.3075
Cadmium.	Cd	8.60	.057	321	610	.3107
Calcium	Ca	1.57	.170	810	1490	.0567
Carbon.	C	2.22	.165	–	–	.0802
Chromium	Cr	6.80	.120	1510	2750	.2457
Cobalt	Co	8.50	.110	1490	2714	.3071
Copper.	Cu	8.89	.094	1083	1982	.3212
Columbium . . .	Cb	8.57	–	1950	3542	.3096
Gold	Au	19.32	.032	1063	1945	.6979
Iridium.	Ir	22.42	.033	2300	4170	.8099
Iron.	Fe	7.86	.110	1520	2768	.2634
Iron (Cast) . . .	Fe	7.218	.1298	1375	2507	.2605
Iron (Wrought) .	Fe	7.70	.1138	1500–1600	2732–2912	.2779
Lead	Pb	11.37	.031	327	621	.4108
Lithium	Li	.057	.941	186	367	.0213
Magnesium . . .	Mg	1.74	.250	651	1204	.0629
Manganese . . .	Mn	8.00	.120	1225	2237	.2890
Mercury	Hg	13.59	.032	38.7	37.7	.4909
Molybdenum. . .	Mo	10.2	.0647	2620	4748	.368
Monel Metal. . .	–	8.87	.127	1360	2480	.320
Nickel	Ni	8.80	.130	1452	2646	.319
Phosphorus . . .	P	1.82	.177	43	111.4	.0657
Platinum.	Pt	21.50	.033	1755	3191	.7767
Potassium. . . .	K	0.87	.170	62	144	.0314
Selenium.	Se	4.81	.084	220	428	.174
Silicon.	Si	2.40	.1762	1427	2600	.087
Silver.	Ag	10.53	.056	961	1761	.3805
Sodium.	Na	0.97	.290	97	207	.0350
Steel	–	7.858	.1175	1330–1378	2372–2532	.2839
Strontium	Sr	2.54	.074	–	–	.0918
Sulphur.	S	2.07	.175	115	235.4	.075
Tantalum	Ta	10.80	–	2850	5160	.3902
Tin	Sn	7.29	.056	232	450	.2634
Titanium.	Ti	5.3	.130	1900	3450	.1915
Tungsten	W	19.10	.033	3000	5432	.6900
Uranium	U	18.70	–	–	–	.6755
Vanadium	V	5.50	–	1730	3146	.1987
Zinc	Zn	7.19	.094	419	786	.2598

* Circular of the Bureau of Standards No. 35, Department of Commerce and Labor.

CONVERSION TABLE ENGLISH TO METRIC

WHEN YOU KNOW	MULTIPLY BY: VERY ACCURATE	MULTIPLY BY: APPROXIMATE	TO FIND
	* = Exact		
LENGTH			
inches	*25.4		millimeters
inches	*2.54		centimeters
feet	*0.3048		meters
feet	*30.48		centimeters
yards	*0.9144	0.9	meters
miles	1.609344	1.6	kilometers
WEIGHT			
grains	15.43236	15.4	grams
ounces	28.349523125	28.0	grams
ounces	*0.0283495523125	.028	kilograms
pounds	*0.45359237	0.45	kilograms
short ton	*0.90718474	0.9	tonnes
VOLUME			
teaspoons		5.0	milliliters
tablespoons		15.0	milliliters
fluid ounces	29.57353	30.0	milliliters
cups		0.24	liters
pints	*0.473176473	0.47	liters
quarts	*0.946352946	0.95	liters
gallons	*3.785411784	3.8	liters
cubic inches	*0.016387064	0.02	liters
cubic feet	*0.028316846592	0.03	cubic meters
cubic yards	*0.764554857984	0.76	cubic meters
AREA			
square inches	*6.4516	6.5	square centimeters
square feet	*0.09290304	0.09	square meters
square yards	*0.83612736	0.8	square meters
square miles		2.6	square kilometers
acres	*0.4046564224	0.4	hectares
TEMPERATURE			
Fahrenheit	*5/9 (after subtracting 32)		Celsius

CONVERSION TABLE METRIC TO ENGLISH

WHEN YOU KNOW	MULTIPLY BY: VERY ACCURATE	MULTIPLY BY: APPROXIMATE	TO FIND
	* = Exact		
LENGTH			
millimeters	0.0393701	0.04	inches
centimeters	0.3937008	0.4	inches
meters	3.280840	3.3	feet
meters	1.093613	1.1	yards
kilometers	0.621371	0.6	miles
WEIGHT			
grains	0.00228571	0.0023	ounces
grams	0.03527396	0.035	ounces
kilograms	2.204623	2.2	pounds
tonnes	1.1023113	1.1	short tons
VOLUME			
milliliters	0.06667	0.2	teaspoons
milliliters	0.03381402	0.067	tablespoons
milliliters		0.03	fluid ounces
liters	61.02374	61.024	cubic inches
liters	2.113376	2.1	pints
liters	1.056688	1.06	quarts
liters	0.26417205	0.26	gallons
cubic meters	0.03531467	0.035	cubic feet
cubic meters	61023.74	61023.7	cubic inches
cubic meters	35.31467	35.0	cubic feet
cubic meters	1.3079506	1.3	cubic yards
cubic meters	264.17205	264.0	gallons
AREA			
square centimeters	0.1550003	0.16	square inches
square centimeters	0.00107639	0.001	square feet
square meters	10.76391	10.8	square feet
square meters	1.195990	1.2	square yards
square kilometers		0.4	square miles
hectares	2.471054	2.5	acres
TEMPERATURE			
Celsius	*9/5 (then add 32)		Fahrenheit

Tables

METALS WE USE

	SHAPES	LENGTH	HOW MEASURED	*HOW PURCHASED	OTHER
	Sheet less than 1/4 in. thick	to 144 in.	Thickness x width widths to 72 in.	Weight, foot, or piece	Available in coils of much longer lengths
	Plate more than 1/4 in. thick	to 20 ft.	Thickness x width	Weight, foot, or piece	
	Band	to 20 ft.	Thickness x width	Weight or piece	Mild steel with oxide coating
	Rod	12 to 20 ft.	Diameter	Weight, foot, or piece	Hot-rolled steel to 20 ft. length; cold-finished steel to 12 ft. length; steel drill rod 36 in.
	Square	12 to 20 ft.	Width	Weight, foot, or piece	
	Flats	Hot rolled 20-22 ft. Cold finished	Thickness x width	Weight, foot, or piece	
	Hexagon	12 to 20 ft.	Distance across flats	Weight, foot, or piece	
	Octagon	12 to 20 ft.	Distance across flats	Weight, foot, or piece	
	Angle	Lengths to 40 ft.	Leg length x leg length x thickness of legs	Weight, foot, or piece	
	Expanded sheet	to 96 in.	Gauge number (U.S. Standard)	36 x 96 in. and size of openings	Metal is pierced and expanded (stretched) to diamond shape; also available rolled to thickness after it has been expanded
	Perforated Sheet	to 96 in.	Gauge number (U.S. Standard)	30 x 36 in. 36 x 48 in. 36 x 96 in.	Design is cut in sheet; many designs available.

* Charge made for cutting to other than standard lengths.

Index

How to Decorate

Bounty
BOOKS

First published in Great Britain in 2006
by Hamlyn, a division of
Octopus Publishing Group Limited

This edition published in 2007
by Bounty Books, a division of
Octopus Publishing Group Limited
2–4 Heron Quays, London E14 4JP

ISBN-13: 978-0-753715-26-0
ISBN-10: 0-753715-26-0

A CIP catalogue record for this book is
available from the British Library

Printed and bound in China

Contents

Introduction

Whether you have a whole new home or just one room that requires redesigning and updating, consider what needs doing in terms of practical repair and decoration. You'll need to think, too, about design style, colours, patterns and textures; ideas for floors, walls and fabrics; and about everything from essential furniture to those all-important finishing touches. But where do you start?

Bright ideas and basic skills

Decorating can be a daunting task, but once you've completed the boring yet necessary preparatory work such as filling cracks, stripping or sanding, it can also be great fun. It's exciting and satisfying to see your design scheme come to life as you choose a style of décor you'd like to live with, paint or paper walls, lay your chosen tiles and flooring and employ colour and pattern in your home – designing, decorating and furnishing it to suit your lifestyle and personality.

Today's ranges of paint, wallpaper, fabric, tiles and flooring options are immense, and many of the materials and equipment that go into making and maintaining a comfortable and attractive home are more widely accessible and easier to use than ever before. But how do you choose the right colours, patterns and finishes for the effects you want to create? Is there a special technique for papering around a doorway? Do you paint the walls or the ceiling first? How do you calculate how many tiles to buy? What colour works best with another? How can you create the illusion of spaciousness? How do you make your own curtains? How can you give your bedroom a contemporary look, or a classic look?

Focusing on bright ideas and basic skills, this comprehensive book provides the solutions to these and many more decorating and design issues. You'll

Many do-it-yourself stores offer a paint-mixing service, which adds hundreds more possible shades to the already huge selection of ready-mixed paint colours.

find it a useful source of practical advice and inspiration, which you can read at leisure or follow project by project. There are step-by-step instructions for repairing and preparing surfaces prior to decorating; for painting, papering and tiling; for sanding and laying floors; and for making various soft furnishings from bed linen to cushions. It includes information on the tools you'll need for specific projects, suggestions for choosing materials and guidance on estimating quantities, plus plenty of ideas on decorating styles, on colour as well as on treatments for floors and walls. Lastly, the chapters on the main rooms in any home – kitchen, living room, bathroom, bedrooms and children's rooms – give you detailed advice and imaginative suggestions on how to tackle each of these rooms.

Plan ahead

Planning goes hand in hand with interior design and decorating. If you can live in your home for a while before imposing any major changes, you will gain a better understanding of what is required. Plan the space and furniture arrangements so that the rooms relate to their proposed purpose and function, and so that people will be able to move around comfortably. Take the time to discover where you need hardwearing and easy-to-clean surfaces, where you can opt for more fragile, delicate materials and where thick, sound-absorbing textures might make an appreciable difference. Seeing how the light levels in different rooms change with the seasons can help you assess which rooms would benefit from visually warming up in winter or cooling down in summer, or from shiny, light-reflecting surfaces.

Use this book to help you think about the style and ambience you would like to create. There are so many possibilities, but in a smaller property it is better to aim for continuity and use one style throughout. You may find it helpful to make up a mood board (see page 106), especially if you have existing items to incorporate in your design scheme.

Decorating involves more than just painting or papering. You have to consider elements like space, light and texture, too, so live with a room for a while before you decide how to style it.

The colours, materials and accessories you choose for each room you decorate are what define your personal style and give your home character.

If you do decide to do a lot of decorating, floor-laying and other do-it-yourself jobs, always plan things efficiently. Make sure that you have sufficient materials to complete the project (this includes the right tools, adhesives and sealants, for example, as well as paint or fabric) and, especially, allow enough time. Trying to finish painting a ceiling at midnight, or papering an awkward wall with a heavily patterned design after a busy day at work is a recipe for disaster. Working in good light, at a pace that suits you, will give a much more satisfactory, professional-looking result.

Above all, however, enjoy decorating and restyling your home. Discover the satisfaction to be had from planning your decorating project, seeing it through and delighting in the look of your newly decorated room(s). The sooner you start, the sooner you'll be finished – so get decorating!

Paint
basics

Painting tool kit

The way you apply paint is largely a matter of personal choice. Brushes in a variety of sizes are essential, and paint rollers or paint pads are also useful. You will also need a few vital specialist items in your painting tool kit.

Paintbrushes

For a good finish, choose brushes made with genuine bristle or good-quality synthetic fibres – as a rule, the more expensive brushes really do give the best results. Cheaper brushes are ideal, however, for outside work, such as applying preservative to wooden fencing or painting masonry.

Good-quality brushes that are well cared for (see box, opposite) improve with use. Loose bristles are shed and the tips become nicely rounded. Start a new brush on primer and undercoat, then use it for fine finishing as it ages.

Useful brush sizes include 12 mm, 18 mm and 25 mm (½ in, ¾ in and 1 in) brushes for painting edges and windows, and 10 cm, 12.5 cm, 15 cm and 20 cm (4 in, 5 in, 6 in and 8 in) brushes for painting walls. A radiator brush has an extra-long metal handle that allows you to paint behind radiators.

Rollers

An alternative to brushes, rollers are an easy and quicker way to apply paint to large, flat areas without leaving defined brushstrokes. Available in various widths and fabrics, they are best suited to applying water-based paints, which can be easily washed off the roller. When using solid emulsion paint, lift the roller direct from the container, but use a paint tray with liquid paint. Use a roller extension handle or tape the roller to a broom handle to paint ceilings or high walls. The different types of roller sleeve include:

Foam Easy to clean, but doesn't give the finest finish, and will tear if used on rough surfaces.

Mohair A hard roller with a very close pile, which gives a fine finish to smooth surfaces. It is not suitable for textured surfaces.

Sheepskin Expensive, but hardwearing and good for use on rough surfaces.

Shaggy pile Its deep, floppy pile makes it suitable for textured surfaces or for applying textured paint.

Radiator roller A small, deep-pile roller with a very long handle for reaching behind radiators and getting to other awkward spots.

Texturing roller A specialized roller for use with textured paint to produce a rag-rolled or other textured effect.

Paint roller with sheepskin sleeve

Radiator brush

Assorted-sized paintbrushes

Radiator roller

Paint pads

Light and easy to use, paint pads come in squares or rectangles and consist of fine mohair pile stuck to a layer of foam, bonded to a metal or plastic handle. Sizes range from 25 mm to 20 cm (1 to 8 in); some have a hollow handle to take an extension handle for painting ceilings or other out-of-reach areas without using stepladders. Paint pads are suitable for smooth or textured surfaces, but not rough finishes. A pad does give a very fine finish when gloss-painting flush doors.

Clean pads immediately after use. Note that proprietary cleaners can attack the adhesive holding the mohair to the foam.

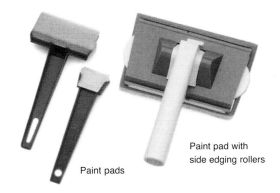

Paint pads

Paint pad with side edging rollers

Other painting tools

If you are using liquid paint, it's useful to have a paint kettle into which to decant some. This makes carrying paint – especially up ladders – much easier, since not all paint tins have built-in handles. Paint kettles are also useful because, should the paint become contaminated in any way, then only the paint in the kettle is affected.

Other useful accessories include:

- Triangular or combination shavehooks, flexible scrapers and a Skarsten scraper for removing old paint
- Masking tape for protecting surfaces not intended to be painted
- Metal or plastic paint shield to restrict paint to the area being painted
- Tack cloth to pick up dust
- Protective sheets, old newspapers and clean, soft, lint-free rags
- Paint stirrers
- Brush cleaner suited to the type of paint

Paint kettle

Combination shavehook

Triangular shavehook

Cleaning and caring for paintbrushes

Remove excess paint from brushes, then clean those that have been used with water-based paint in warm, soapy water, and rinse well. Clean brushes covered in solvent-based paint using white spirit or a proprietary brush cleaner, then with hot soapy water, and again rinse well.

To clean roller sleeves and paint pads, remove the excess paint on old newspaper, then clean as for brushes, using water or white spirit as appropriate. Allow to dry.

After cleaning, shake vigorously to remove excess water. Then, while still damp, slip an elastic band over the brush tip to hold the bristles firmly together. As long as the elastic band is not too tight, it will ensure that the bristles keep in good shape, with no stray whiskers. Hang the brush up to dry, bristles facing downwards. When completely dry, place the brush in a sealed polythene bag to keep free from dust, and store so that the bristles remain flat.

Choosing paint

Paint is usually the most economical choice for decorating. There is a vast range of paints, formulated to meet all the many different requirements of the home decorator. The difficulty comes in deciding which colour and finish you want from the thousands available.

Types of paint

Traditionally, painting bare wood or metal involves a three-step application of primer, one or two coats of undercoat and a top coat, with a light sanding down between coats. Walls and ceilings are much simpler, however, requiring only one to three coats of emulsion paint, although a primer on bare plaster or an undercoat before a change from a darker to a lighter shade may be necessary first. Developments in paint technology mean that many modern paints now combine two or more of these steps, making life much easier. Paints have also become cleaner to use and more environmentally friendly. There are more water-based versions of paints than ever before, and some paints are even described as 'organic'. Free of solvents or other harmful chemicals, these paints are biodegradable and allow walls to breathe. Your choice of paint will depend to a large extent on the type and condition of the surface you are painting and on your personal preference.

Paints can generally be divided into one of two categories – water- or solvent-based paint. The label on the paint tin will tell you which is which. It will also list its recommended uses and, most importantly, covering power. Generally speaking, you can apply solvent-based paint over existing water-based paint provided it is sound, but you cannot use a water-based paint over a solvent-based one.

Water-based paint Quick drying and without a strong solvent smell, water-based paints include emulsion, distemper and water-based gloss and eggshell. The brushes and equipment may be cleaned with warm soapy water.

Solvent-based paint Solvent-based paint, which includes oil- and alkyd-based paint, is made from a mixture of oils and resins. It is a slower-drying paint and has a harder finish. Examples include traditional eggshell and gloss paint, and enamel. You can thin solvent-based paint that has thickened with white spirit, unless the manufacturer's instructions dictate otherwise. The brushes and equipment need cleaning with a solvent such as white spirit or turpentine.

Primer

Primer seals a surface to prevent subsequent layers of paint sinking in and disappearing. There are primers created specifically for bare wood, metal and bare plaster, while universal primers are designed to suit all three surfaces. Other types of primer among the many available include aluminium primer, used for specific surfaces that need a high level of protection, and stabilizing primer, which is needed to seal the powdery or flaky surface of walls painted with old paints like limewash or distemper (see opposite).

Where possible, it is a good idea to work with the same brand of primer, undercoat and top coat, since they are formulated to work together.

Undercoat

Undercoat obliterates the previous colour and gives body to the next coating. Note that some gloss paints are self-undercoating and will say so on the label.

Personalizing a room by using a number of different colours is fun, but avoid colours that clash – you will quickly tire of the combination.

Emulsion

Emulsion is mostly used for walls and ceilings, although some modern high-performance emulsions can also be used on wood and metal. Because it is water-based, emulsion is quick-drying, easy to use (and to rinse out of brushes) and reasonably odourless. It comes in traditional liquid form and a non-drip jelly-like form, while emulsions designed for ceilings tend to be solid or semi-solid and come in a tray, ready for use with a roller. Emulsion that is too thick can be thinned with a little water.

The modern range of emulsions includes metallic effects and paints with subtle textures like suede, which are ideal for feature walls, but the widest choice of colours is found in vinyl matt and vinyl silk emulsions. Vinyl matt emulsions have an attractive, sometimes chalky look, which shows up scuffs and finger marks; vinyl silk emulsions have a light-reflective quality that will magnify any surface imperfections, but are wipeable. Emulsions known as vinyl satin or soft sheen are designed for use in kitchens and bathrooms. They often contain fungicide to deter mould growth and stand up better to washing down and condensation. There are also extra-tough, scrubbable paints in bright colours that are designed for children's rooms.

At the other end of the scale, there are several ranges of non-vinyl emulsions that produce the soft, almost chalky finish of old period paints. These are particularly good for old houses, but their painted surfaces are more difficult to clean.

One-coat emulsion paint has extra covering power and will save you time, but perhaps not money, and the choice of colours is not always as great.

Emulsion paint specially formulated for kitchens is moisture- and stain-resistant. It is designed to withstand steamy hot atmospheres and be easily wiped clean if splattered with food.

Traditional paints

The original precursors to emulsion, traditional paints tend to be water-based and are particularly suitable for use in historic houses. Such paints include limewash, which is made from slaked lime, and distemper. Made with natural resins, ordinary distemper has a flat, powdery or chalky finish, which allows the plaster beneath to breathe. Generally speaking, distemper cannot be washed, although some strengthened versions are wipeable.

Textured paint

This special paint contains fine aggregate, which gives it a thicker, rougher texture. It can be used on walls and ceilings to hide surface imperfections such as small cracks or joins in plasterboard. Apply the paint thickly with a textured roller or use an ordinary roller and, before the paint starts to dry, use a rubber-bristled stippling brush to produce a variety of different effects. It is a permanent form of decoration and difficult to remove.

Gloss and eggshell paint

These provide a more durable surface than emulsion and are mostly used on wood and metal. Traditionally solvent-based, these paints are now also available in water-based versions, some of which are quick-drying.

Paint is an inexpensive way to transform a room, but do choose the appropriate finish of paint. For example, shinier finishes are more likely to show up any surface imperfections.

The infinite range of colours available can make choosing paint difficult. Try and use inexpensive tester pots first to envisage how a colour will look in a certain room.

Liquid gloss (usually solvent-based), when properly applied, will produce a high-sheen, perfectly smooth and very hard surface, but is more prone to drips and runs; the jelly-like non-drip version (both solvent- and water-based) gives nearly as good a finish. Alternatives to gloss are flat matt, often chosen for an authentic period look for wood, and various mid- to low-shine finishes, typically called eggshell, satin or silk. These have a less 'hard' look than gloss and, although not as hardwearing, can be washed down to remove sticky finger marks.

When painting bare wood, you will need to use a primer first. If painting over another colour, a self-undercoating paint should cover the old colour without the need for a separate undercoat. As with emulsions, there are one-coat gloss and eggshell paints designed to save time by condensing the traditional three-stage application.

Floor paints

Floor paints provide a protective finish for wood, concrete and tiled floors, and are designed to withstand heavy wear and tear. Applied by brush or roller, they can be solvent- or water-based. The latter are a good alternative to using coloured stains (see opposite) on wooden floorboards – they give a more solid effect and are particularly effective at giving a new lease of life to unattractive floorboards or where some boards have at some point been replaced and the different age of the woods shows (see page 76).

Specific-use paints

There are a number of paints designed for specific surfaces. Radiator enamel, for example, is formulated to maintain its whiteness where ordinary solvent-based paint would crack and yellow with the heat of a radiator. Other heat-resistant paints suitable for pipes and hearths are available in a limited range of colours.

There are also various special paints designed to spruce up domestic kitchen appliances like dishwashers and refrigerators, to cover melamine and acrylic surfaces like kitchen units and to overpaint existing wall tiles. Use these according to the manufacturer's instructions – you may need to use a primer and/or key the surface to be painted.

Rust-resistant paint is an enamel-formulated paint, which inhibits the penetration of rust. No primer or undercoat is needed. It is available in a smooth or indented 'hammered' finish.

Paint your floorboards to disguise unattractive and differently aged boards. Use purpose-made floor paint for a hardwearing finish designed to withstand the movement of furniture and people.

The specialist paints now readily available allow you to completely modernize junk shop finds like this previously dark-stained hall table and make a feature wall using pearlized or metallic paints.

Decorative-effects paint

You'll find a huge range of paints and glazes in craft shops and do-it-yourself stores for achieving special effects such as sponging, rag-rolling, bagging and stencilling (see pages 28–33). Pearlized and metallic paints are another alternative for decorating primed wood, walls and metal.

Woodstains and varnishes

All articles made of wood need treating with a preservative or finish to preserve and protect the surface. The quality of the surface of the bare wood will affect your choice of finish. Painting, for example, would hide any slight surface defects, whereas any blemish in the wood is immediately accentuated when a clear finish is applied. It is therefore important that all woodwork is clean and smooth before decorating work begins.

All finishes alter the colour of wood to some extent. Woods like mahogany and walnut, for example, turn much darker even when a completely clear finish is applied. You can get a rough idea of the colour the wood will take when treated with a clear finish by dampening a small area with ordinary water. If this colour is too light for your needs, you can stain the wood before finishing.

Stains

Woodstains can be used to alter the natural appearance of wood. They are available in natural wood colours and other colours designed to harmonize with decorative schemes but that still let the grain show through. You can stain wood only to a darker colour; for a lighter shade you must apply bleach or a limewash. Do test the stain on an offcut or on an area that is normally out of sight to check that you like the effect, as it is notoriously difficult to remove stain, even immediately after it has been applied. If the wood has an open grain, but you want a smooth finish, apply grain filler to fill the pores, rubbing it across the grain. If you need to use a wood filler for cracks or holes in the wood, be aware that fillers do not take up the colour of stain as well as the wood.

Varnish stains colour and finish the wood in a single operation. Bear in mind that each extra coat of varnish stain will darken the colour of the wood and, unless brushed on very evenly, the colour will vary with the thickness of the film. When wood is stained with a penetrating dye, the colour will not vary – no matter how many coats of clear finish you later apply.

Varnishes

Available in water- and solvent-based versions, varnish provides a clear and very durable gloss, satin or matt finish. Polyurethane is a varnish with a plastic finish.

Oil finishes

Oil finishes, such as teak oil and Danish oil, are easier to apply than polyurethane and other varnishes. On new wood, you will need to apply two coats, either with a brush or a mildly abrasive pad, cleaning off excess oil with a cloth. Teak oil and Danish oil both leave the wood with a soft, lustrous finish that is truly resistant to liquids. You can also add stains to them.

Waxing

This is popular for treating newly stripped pine. Some waxes colour the wood a little – improving the bleached-out effect of pine that has been stripped, and giving the surface an 'antiqued' look. To preserve the natural look, make sure that the wax is colourless.

Always rub the wood down first with fine-grade steel wool. If you apply the wax with a soft cloth, it will produce a natural satin finish. For a higher gloss, allow the new wax to dry completely and then buff it vigorously with a soft duster or clean, soft shoe brush.

How much paint do I need?

Before you start decorating, you need to work out how much paint to buy. This will depend on the surface to be painted and how many coats of paint it will require. This will be affected by the porosity of the surface as well as its overall texture and possibly its colour.

Buying paint

Household paints are generally sold in 500 ml, 1 litre, 2½ litre and 5 litre tin sizes. A few paints, such as white emulsion and paints for exterior walls, come in larger sizes. Smaller quantities – for example, 250 ml, 100 ml and 50 ml – are available in some brands. Always buy sufficient paint and check the batch number on each tin to ensure that they come from the same mixing, as colour can vary subtly but noticeably between batches.

How much paint?

To work out the total area of wall to be painted, measure all around the room (or just one long wall and one short wall and double the calculation if the room is a simple rectangle with no recesses) and multiply by the height. This will give you a figure in square metres or yards. Subtract 1.8 sq m (2 sq yd) for each door, then measure the width and height of the windows (ignore any small ones) and subtract their area, too. Remember to multiply by the number of coats the walls will need (see opposite).

To work out the quantity of paint required for standard windows with several panes, simply multiply the width of the overall frame by the depth, and treat it as a solid area. For large picture windows, make the same calculation, but deduct 50 per cent. For metal windows, deduct 25 per cent. For flush doors, multiply

Calculating quantities

To cover a smooth, sealed surface, be guided by the coverage indicated on the tin. As a general guide, 1 litre of paint covers an area as follows:

General-purpose primer	10–12 sq m (12–14 sq yd)
Undercoat	15 sq m (18 sq yd)
Gloss/eggshell paint	15 sq m (18 sq yd)
One-coat gloss	10 sq m (12 sq yd)
Emulsion paint	10–14 sq m (12–17 sq yd)
One-coat emulsion	8 sq m (10 sq yd)

Use a paint tray and roller rather than a paintbrush to cover large areas like walls and ceilings more quickly. Don't overload the roller or you will splash paint as you work.

the height of the door by the width and add a further 10 per cent for the edges. For a panelled door, add 25 per cent. Once you know the total surface area to be covered, you can work out how much paint to buy by dividing this figure by the one given for coverage (spread rate) on your tin of paint (see box, opposite).

How many coats of paint?

The number of coats of paint you apply depends on the existing colour (if any), the type of surface you are working on (its porosity and condition) and the quality of the paint you are using. Bear in mind that you will need more paint to cover a dark colour than if you are painting a light surface with a similar or darker colour. Also, highly textured and very porous surfaces like newly plastered walls may be very 'thirsty', so double the quantity suggested on the paint tin label (or, for the latter, heavily dilute the first coat of emulsion or use specially formulated 'new-plaster emulsion').

It is always better to apply two, or even three, thin coats of paint rather than a single thick one of topcoat, which could result in a patchy-looking finish.

If, despite your calculations, you find yourself running out of paint, stop at a convenient corner, angle or other natural feature while you still have paint left. Try not to start a new tin of paint in the middle of a wall or ceiling in case of colour variation.

When calculating how much paint to buy, bear in mind the porosity of the surface being covered. New plaster, for example, takes up a lot of paint, so dilute the first coat.

To overpaint an existing colour with a lighter one, it is more economical to coat it with inexpensive white emulsion first rather than apply several coats of your final colour.

Preparing to paint

Preparing the surface to be painted is paramount for good-looking and long-lasting paintwork. Preparation may involve simple cleaning, stripping away layers of old paint or repair work such as filling cracks to ensure that the surface is sound and ready to paint.

Cleaning

When preparing painted surfaces for redecoration – whether walls, ceilings, woodwork or floors – you don't need to remove the old paintwork if it is in good condition. Simply wash the surface with a strong sugar soap solution or another degreasing agent to break any glaze and remove grease to which the new coat will not adhere. Then rub over the surface with a flexible sanding pad to improve the key (see page 14), and dust it down before applying the new paint.

Stains on walls or ceilings, due for example to water damage, will sometimes bleed through new decoration if not treated beforehand. Seal the stained area with an aluminium primer-sealer or with a proprietary aerosol stain block.

Unpainted metal should need only cleaning and priming (radiators often come ready-primed). Follow with an undercoat.

If you are able to empty the room completely of furniture before decorating, it makes the job much easier. Protect the floor with sheeting unless it will ultimately be carpeted over.

Stripping

If the existing paintwork has been badly applied or consists of so many layers that it causes windows and doors to stick, then you should remove the paint. Surfaces that have been poorly painted, or overpainted several times, can take on a treacle-like appearance, robbing architraves and mouldings of their fine detail. Plaster ceiling roses look particularly unattractive when they have been painted too often.

The main choices for stripping are using either chemicals or heat. However, dry scraping may be possible using a Skarsten scraper for convex surfaces, such as banister handrails. Use a shavehook on finer mouldings. Make sure that you protect your eyes and hands, and wear a simple face mask.

If the paint is old, buy a simple detector kit from a hardware (houseware) store to check whether it contains lead. If it does, chemical stripping is the best way to remove the paint because the alternatives could result in lead poisoning. You could inhale lead particles if you sand or scrape lead-based paint, and heating lead paint produces toxic fumes.

Think carefully before stripping back wood in order to stain and varnish it. You'll need to do a great deal of sanding to remove all the primer from the pores of the wood; if any remains, the stain will look patchy.

Rub down previously painted metal with emery paper. Remove rust with a wire brush and use a filler to fill chips and dents. Apply a coat of metal primer and follow with an undercoat.

Chemical stripping

Chemical stripping can be expensive if large areas are involved. The secret is to be patient while the chemical works – otherwise you may have to apply more coats to get down to the bare wood or metal.

Chemical strippers are traditionally solvent-based, although safer and environmentally friendly solvent-free strippers are becoming available. If working with the former, protect any exposed skin and your eyes, as some products could cause minor burns or localized irritation to the skin. In addition, avoid breathing in the vapours and keep the stripper away from vulnerable items.

Chemical strippers generally come as a liquid or gel, which you apply with a brush and scrape away when the paint bubbles and breaks up. Alternatively, there is a paste that you trowel on and allow to set before lifting it away with the point of the trowel. This works well on mouldings, since it lifts paint out of hollows. You must always neutralize a chemical stripper immediately after use using water or white spirit according to the manufacturer's instructions.

Instead of using chemical stripper yourself, you could take large, relatively portable items like doors, window shutters and unglazed sashes to specialists for stripping. This involves immersing the item in a paint-stripping solution. However, there is a risk of the wood discolouring, and the joints in woodwork may open up as any old glue or filler dissolves.

Heat stripping

Most paint softens quickly when you apply heat with a blowtorch or hot-air gun, wrinkling and bubbling up so that you can scrape it away fairly easily, but take great care not to overdo it, since the paint may ignite. Don't use newspaper on the floor to catch the hot, flaking paint because of the risk of fire, and always have a bucket of water to hand in case of emergencies. Hold the scraping tool so that the stripped paint cannot fall on to your hand and burn you. Cotton gloves provide adequate protection.

A blowtorch A blowtorch burns liquid gas either from a small container attached to the torch head or from a larger cylinder connected to the torch by a tube. The naked flame can be extremely hot, so keep it moving all the time to avoid scorching the wood or cracking glass if you are stripping a window frame.

A hot-air gun This tool resembles a powerful hair dryer, blowing a jet of air through a very hot electric element. Having no flame, it is safer than a blowtorch, but is still sufficiently hot to char wood and crack glass, if not used carefully. It usually comes with different nozzles, allowing you to adjust the airflow.

Using a heat stripper

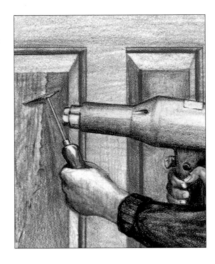

1 Hold a hot-air gun or blowtorch in one hand about 15–20 cm (6–8 in) from the paintwork and a scraper in the other and keep the tools moving together. Use a shavehook for scraping mouldings. Try not to dig into the wood or scorch it – scorch marks can bleed through new paint. If you are burning paint off window frames, keep the heat away from the glass or it will crack.

2 After you have removed the paint, rub the wood down with medium-grade glasspaper following the direction of the grain and paying particular attention to any mouldings.

Stripping radiators

If you have any radiators in your home whose paint surface is in need of attention, it's a good idea to strip the existing paint from them before repainting them, since they lose their heat efficiency if covered with too many layers of paint. To do this, use a sanding attachment for a power drill.

Stripping around door handles

You will achieve a neater finish if you remove door handles and any other door furniture before stripping paint from a door. If for some reason a metal door handle cannot be removed, use a thin piece of plywood to shield it from the direct heat of a hot-air gun or blowtorch. If you are using a chemical stripper, it may not affect the metal, but test it first on a small area that cannot be readily seen – such as the inside surface of the door handle.

Making repairs

Once you have stripped and cleaned the walls and ceiling, you need to examine them for any defects, such as cracks and gaps in the plaster. Most such problems should have become evident during the preparation stages; some others, however, may not initially be obvious. The most common of these is 'blown' plaster, which occurs when patches of plaster lift away from the underlying wall. When rapped with your knuckle, blown plaster has a distinctive hollow sound. Ideally, you should hack out this defective plaster and patch with fresh material.

Although replastering an entire room is a major job, and requires considerable skill and professional expertise, it is relatively easy to undertake minor repairs yourself.

Gaps between woodwork and walls

You commonly find cracks in the plaster where a wall abuts door and window frames, and gaps between walls and skirting boards (baseboards). They are usually caused by the movement of the woodwork with temperature and humidity changes. Clear any loose material out of these cracks or gaps with the point of a small trowel, then fill the gap with a gun-applied acrylic frame sealant or expanding foam filler. These grip better than cellulose filler and tolerate some movement. Frame sealant needs to touch both surfaces so as to seal the gap – if the gap is particularly wide or deep, bridge it with a strip of wood or polystyrene first. Expanding foam filler sinks right in and expands to fill the gap – trim off the excess when set.

Cracks between walls and ceilings

These weak spots rarely stay sealed, whatever material you use, because any slight movement of the building will reopen the cracks. An expanding foam filler is the easiest material to use, but the best solution to a very noticeable gap that keeps reappearing is to hide it with decorative coving, covering the angle where the two surfaces meet. Available in different styles, sizes and materials (such as polyurethane, gypsum or plaster), coving is available ready-coloured or you can paint it.

Cracks and holes in walls

Fill small cracks with a cellulose filler and large cracks with a plaster repair filler (see illustrations, opposite). If you are not using a ready-mixed product, make sure that you use clean water to mix the plaster and clean the mixing bucket after each mix, as any residues of set plaster will reduce the setting time of subsequent mixes and may weaken them.

If the cracks are particularly wide or deep, fill them in stages. Each layer should be no more than about 1 cm (½ in) deep. Score the surface of each layer to provide a key for the next layer, which you apply when the previous layer is set but not completely dry. The filling material is likely to absorb more paint or wallpaper paste than the original plaster, so prime it with one or two coats of universal primer or emulsion before starting to redecorate.

If you have to fill a large cavity in a wall, packing it with filler is wasteful and the filler may crack unless you apply it in a laborious series of layers. Instead, where possible, pack the hole with broken brick or stone and push cement in to fill the gaps and firm it all up – small bags of ready-mixed mortar are convenient for these minor jobs. Once you have packed the hole to within 5 cm (2 in) of the wall surface, you can apply an all-in-one do-it-yourself plaster. (Developed for do-it-yourself use, this plaster requires no undercoat plaster and can be applied up to about 5 cm (2 in) thick, directly on to brickwork or similar walls.)

For plastering large areas, it is a good idea to nail temporary guide battens to the wall. These enable you to bring all the plastered surface to the same level. When the plaster has set, remove the battens and fill the grooves level with the rest of the job.

Filling a crack

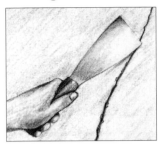

1 Start by carefully raking out any loose material from the crack, using the corner of a flexible filling knife or a narrow wallpaper scraper.

2 Using a small paintbrush or plant mister, wet the crack with water. This stops the plaster from drying out the filler too quickly, causing it to crack or fall out.

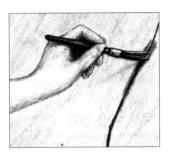

3 Now fill the crack by drawing your filling knife loaded with filler across it, at right angles to the crack. Repeat until the crack has disappeared, leaving the filler slightly proud of the surrounding plaster.

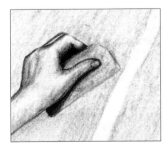

4 Leave the filler to set hard (a deep crack may need a second application), then sand it back flush with the surrounding plaster using medium-grade glasspaper.

Patching plasterboard

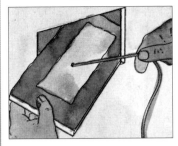

1 Cut a piece of plasterboard just larger than the hole. Tie string to a nail and feed it through a hole in the centre. Dab the edges with plaster.

2 Insert the new piece through the hole and pull the string tight while you fill the recess in front with plaster. When the plaster is set, cut the string and fill the hole.

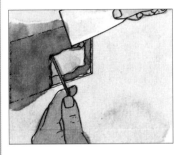

3 To repair a corner, pin a batten on one side flush with the corner and fill the opposite side with plaster. When set, move the batten and repeat on the other edge.

How to paint

Never rush a painting job – you must wait for one coat of paint to dry completely before applying another. Emulsion paint dries quickly, so you don't have to wait too long between coats, but oil-based paint requires a little more patience. Refer to the label on the paint tin for the manufacturer's instructions regarding drying and recoating time.

Handy hints

- Make sure that the paint tin is free of loose dust and dirt before opening the lid. If the label instructs you to stir the paint, use a length of wood to do so.

- Never use a brush with a rusty ferrule, as this will discolour the paint.

- Store paint in a cool, dry place free from exposure to frost.

- Paint the lid of the tin so that you can see the colour at a glance when you need to do a little touching up.

- Before you use a new brush, or one that has been stored for any length of time, manipulate the bristles by rubbing them briskly in the palm of your hand in order to loosen and remove any dust and broken bristles.

- Dip no more than one-third of the paintbrush bristles in the paint and wipe off the excess (string or thin wire tied across the top of the paint tin or kettle is useful for this purpose).

How to paint interior walls

Choose your decorating tools according to the finish you require. For a smooth surface, use a brush, a foam or mohair roller or a paint pad; for a deeper texture, or to cover a rough surface, use a brush or a shaggy pile roller (see page 10). Whatever painting tool you use for the main area of the walls, you will need a small paintbrush for the edges.

1 If painting with a brush, choose the widest brush you can comfortably hold. Begin at the top of a wall and run bands of paint downwards in vertical strips, leaving a slight gap between the bands. Then brush across the wall to blend the bands of paint, finishing with light, vertical strokes of the paintbrush.

2 Rollers and pads are easiest to use with shallow rectangular trays. Most have a built-in shelf or, for pads, a rolling edge, against which you can squeeze off the excess paint. To use a roller, work it in all directions, taking care to merge the joins and fill in any gaps, and finish off with light strokes in a single direction. It is best to work over a small area at a time. Don't roll too quickly or the paint will fly off the roller. Allow the roller to shed all its paint before reloading it. Don't paint too thick a coating; instead, apply two (or more) thin coats as you would with a brush.

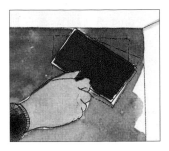

3 Use a paint pad in much the same way as a roller, applying paint in a criss-cross pattern. A pad applies only a very thin coat of emulsion, so you will have to repeat this process to give two or even three coats – especially if you have a base colour to hide. Paint pads are especially useful for running along wall edges.

How to paint woodwork

If using gloss or eggshell paint, first seal bare wood with a wood primer. When dry, rub down lightly with fine glasspaper, dust off with a lint-free rag and apply a layer of undercoat (to give body to the final coat), unless you are using a paint that combines undercoat and topcoat. Dispense with the primer if covering a sound painted surface. If painting over a different, especially a darker, colour, use a self-undercoating paint or as many coats of undercoat as necessary.

Apply the paint in parallel strips a short distance apart following the grain of the wood. Then, without reloading the brush, draw the bristles across the grain to spread the paint over the uncoated area. Finish off with light strokes following the grain again to ensure

When painting woodwork, always follow the grain of the wood. Choose a good-quality paintbrush that won't leave hairs in the paintwork and don't overload the brush with paint.

Masking paintwork

Masking around glass window or door panes saves a lot of time and effort later. Run strips of masking tape around the edges of the glass where they meet the woodwork. Leave a narrow margin to allow a thin line of paint to overlap the glass and form a seal. Using a small brush, and holding it like a pencil, apply paint in the direction of the grain. Peel away the tape as soon as the paint is touch dry.

even coverage and a smooth finish. Always brush out towards an edge to prevent the paint forming an unsightly ridge.

Paint care

Fill small, screw-top jars with any leftover paint. If all air is excluded, the paint will keep indefinitely and can be used later for touching up. Label the jars carefully, indicating paint type and where you've used it. If you leave leftover paint in the bottom of its original container, it will soon evaporate.

Paint that has been stored for some time can be affected in two ways. If there is a brown liquid floating on top, just stir it thoroughly back into the rest of the paint. If a skin has formed on the surface, cut around this with a knife and discard once you have scraped away and returned to the tin any paint on its underside. Strain the paint before reuse, either with a paint strainer or by passing it through a clean nylon stocking. Fix the stocking loosely over the tin with an elastic band and push the nylon down into the paint, then dip your brush into the uncontaminated paint. If an emulsion tin has become rusty, transfer the unaffected paint to a clean container.

Interior painting

Ideally, you should decorate in a well-ventilated area and during daylight hours to take advantage of the natural light. If you paint under artificial lighting, you may overlook gaps in the coverage. For best results, paint an interior in a strict order, from the highest to the lowest point in the room.

From start to finish

To paint any room, always start with the ceiling, since some paint will inevitably drip or be splattered on to the walls.

Handy hints

• Although you can paint a ceiling from a stepladder, it involves a lot of tiring leg work, since you have to climb up and down to move the steps. It is much simpler to make a platform using stepladders or trestles and a scaffold board (see page 39).

• When painting walls and ceilings, allow yourself enough time to complete the work. You can do one wall at a time, but you must paint the whole ceiling in one go.

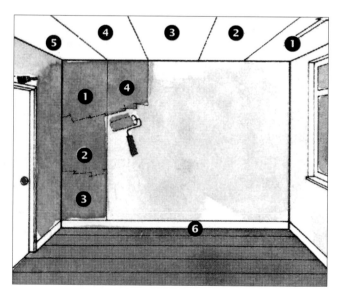

1 In an average room with a window and a door, begin painting the ceiling at the window end and work across the ceiling covering the surface in parallel bands approximately a metre or a yard wide until you reach the door end (see numbers **1–5** on the illustration above). Work an area that is comfortable to reach from your standing point, whether you are on a makeshift platform or a stepladder. If there is a ceiling rose or light fitting in the middle of the ceiling, paint neatly around this with a small paintbrush and then continue painting in parallel bands.

2 The next step, if necessary, is to apply primer and undercoat to all the woodwork – frames, door and skirting boards (baseboards) – in the room (see page 23) and allow it to dry completely, preferably overnight.

3 Next, paint the walls or any other large surfaces in the room. Start in the top left-hand corner of one wall and work with a brush, roller or paint pad, covering an area of about a square metre or yard at a time. Work your way down the wall from the ceiling to the floor in vertical bands (see numbers **1–4** on the illustration, left). Continue to paint the rest of the first wall and then the remaining walls in this order. Complete one wall in full before you begin the next. To paint around a window, paint around the frame with a small brush and then fill in the surrounding wall space with a larger brush, roller or paint pad.

4 Once this is done, give the door and window frames a topcoat. Then paint the door of the room and add any covering to the floor last. Paint the coving and skirting boards (baseboards) after the ceiling and walls are completely finished.

Painting interior doors

Remove as many fittings from the door as possible, since handles, escutcheon plates and hooks are difficult to paint around. Put old newspaper under the door to protect the floor and to prevent picking up dust on your brush. Select your painting tools according to the surface of the door. Use a 75 mm (3 in) wide brush for a flush door or flat panels of a panelled door, and smaller brushes (50 mm/2 in and 25 mm/1 in) for decorative mouldings and door edges, and for cutting in. With all doors, paint the edge opposite the hinges last so that you have something to hold on to.

Handy hints
• Smearing a little petroleum jelly over the door hinges and other fittings will protect them from being touched with paint.

• Keep checking your paintwork for runs (see page 34) so that you can brush them out before the paint dries.

Flush doors

Mentally divide up the surface area into small sections (see illustration, below), and try to work quickly to avoid tide marks or visible joins.

1 Begin at the top left-hand corner (or top right-hand corner if you are left-handed), covering to about half the width of the door (**1**). First, work using vertical brushstrokes, then brush across these with light horizontal strokes.

2 Complete the other sections (**2–6**) in the same way. It is important to work the brushstrokes consistently so that the finish of the door will be even.

3 Use a small brush to complete the edges of the door (**7**), then paint the frame of the door last (**8**).

Panelled doors

For a panelled door, the sequence for painting is a little more complicated (see illustration, below) and you'll need two or three different-sized paintbrushes.

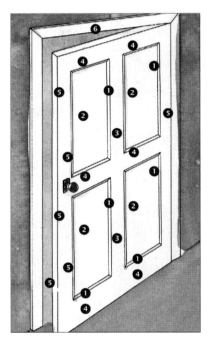

1 First, paint the mouldings around the panels (**1**) and then the panels themselves (**2**). Start with the upper panels and work from top to bottom. Begin each panel at the top and work downwards, painting vertical brushstrokes followed by light horizontal ones.

2 Next, paint the section that runs down the middle of the door and divides the panels (**3**). Continue by coating the horizontal rails, starting at the top and working down (**4**).

3 Complete the outer strips and then all the edges (**5**).

4 Lastly, paint the whole door frame (**6**).

Painting interior windows

Remove any window catches or handles if possible before painting windows. This makes it easier to paint and gives a better finish. To make a temporary window catch, fix a small nail to the underside of the bottom rail, attaching it to an old wire coathanger and hooking this into a screw hole in the window frame.

Don't be tempted to use too wide a paintbrush and do use a paint shield as you work to protect the panes, or mask the glass with tape (see page 23). Remove any dry paint from the glass carefully later, using a scalpel, razor blade or glass scraper.

see page 23

Handy hint

When painting windows, aim to paint them as early in the day as possible so that they may be dry enough to close by nightfall. If you close the windows before the paint is completely dry they are liable to stick and cause problems.

Sash windows

To paint a sash window (see illustration, below), push the rear sash down and the front one up so that at least 20 cm (8 in) of the lower rear sash is exposed.

1 Paint the bottom rail of the rear sash and as much of the exposed upright sections as possible.

2 Pull the rear sash up so that it's almost shut and paint the rest.

3 With the front sash slightly open, paint its frame.

4 When both sashes are dry, paint the surround, shut the window and paint the exposed part of the runners, but not the cords. Paint the sill last.

Casement windows

With a casement window (see illustration, below), fix the window slightly ajar.

1 First, paint the rebates (**1**) and then the horizontal and vertical crossrails (**2**).

2 Paint the horizontal top and bottom sides and edges (**3**), then the vertical sides and edges (**4**).

3 When the window is dry, paint the frame (**5**), including the edges.

4 Leave the sill (**6**) to the end, to avoid smudging; if the stay needs painting, do it last of all so that you can use it until the very end.

Painting stairwells can be tricky and requires using some form of scaffolding or platform on which to stand and a roller or paint pad fixed to an extension pole.

that your brush does not pick up dirt or fluff from below the skirting (baseboard) by moving a sheet of card or an offcut of wood along the floor as a shield as you paint.

Painting stairwells

A stairwell is one of the most difficult areas to decorate, but it is quite possible to paint this part of the home yourself, even if you are inexperienced.

The first thing you need is some form of scaffolding. The most reliable solution is to hire a staircase platform, which is designed to fit neatly on to stairs by means of adjustable legs. However, if you are not worried by heights, then you can construct your own platform using stepladders and a scaffold board (see page 39). Make frequent checks that the scaffold board is centrally placed and has not slipped out of position, particularly as you climb on and off it.

By using a roller or a paint pad with an extension pole, you can greatly increase your reach and apply paint overhead on the ceiling and high up on the stairwell walls.

Painting picture rails and covings

Before painting picture rails, make sure that they are free from dust, dirt and grease, particularly along their top surface. It is usual to work in a similar type of paint as for the walls, which is likely to be emulsion, although you may want a contrasting colour to pick out the detail of the picture rail. Using a small paintbrush, paint along the run of the rail, shielding the wall in the same way as for painting skirting boards (baseboards).

With covings it is also usual to use the same type of paint as for the walls. The paint can either match the walls or the ceiling or be in a contrasting colour, which will help to emphasize any detailing. Use a paint shield or masking tape for a neat finish.

Painting skirting boards

When the rest of the room is freshly painted, a discoloured skirting board (baseboard) will mar the overall effect. Painting the skirting (baseboard) is a quick job that is best left until after the walls are completed. Because skirtings (baseboards) are low down and narrow, work with a small brush and use a hardwearing paint that will withstand knocks. Wipe down the skirting (baseboard) with a damp cloth to remove any dust and vacuum along its bottom edge to remove any dirt or fluff in the carpet that may stick to the wet paint. Lay down newspaper to protect the floor. To protect the wall, use a piece of stiff card or a slim offcut of wood and hold it to shield the wall while you brush. If the wall has been dry for some time, then you can use low-tack masking tape – but remember to peel away the tape before the gloss or eggshell paint hardens. Work the brush following the direction of the skirting (baseboard), making horizontal strokes and working your way in one direction around the room. When you come to painting close to the floor, ensure

Handy hint

Any decorating job will be quicker if you can remove all the furniture from the room and cover the floor completely with polythene sheeting or a dust sheet.

Paint effects

Paint effects such as colourwashing, sponging or rag-rolling create depth and visual texture, and can lend interest to a bland expanse of wall. Such techniques all involve painting a base coat over a surface, then applying a glaze and manipulating it using a variety of 'tools' for a decorative effect.

Successful paint effects

Whichever paint effect you choose, the wall or ceiling to be decorated must always be properly prepared. No paint finish will disguise bad workmanship like unfilled cracks or dirty walls, so be prepared to spend some time on surface preparation. The next step is to apply a base coat or two of matt or silk emulsion to the surface to be decorated and allow to dry.

The secret of successful paint effects is to employ subtle complementary tones rather than strong contrasts and to test all the different techniques and colour combinations on odd sheets of lining paper, pieces of wood or hardboard before applying to the wall. In this way you can see beforehand the finished look and decide whether it's what you want. You may find you need to reduce the amount of pressure on the tool you are using or perhaps thin the glaze – with practice you will soon build the skills required to achieve the results you want.

Handy hints

• Keep a clean cloth to hand to remove any excess paint, drips or mistakes.

• Wait for the glaze to dry thoroughly before correcting errors or making improvements.

• If possible, complete a whole wall before finishing a painting session so that there are no visible joins in obvious places.

Materials and equipment

The following tools and materials are used for creating a wide range of broken colour effects. Other tools for patterning a glaze can be things you already have around the house such as rags, combs, sponges and even scrunched-up newspaper and plastic bags. Some specialist brushes are expensive, so look for synthetic substitutes, which can be of excellent quality. Ideally, it is best to reserve brushes either for paintwork or for varnishing, and not to mix them between the two jobs. Clean brushes scrupulously, taking care not to leave any paint or varnish near the handle end of the brush.

Acrylic paint Available in tubes in an extensive range of colours, it is water-soluble and useful for tinting emulsion paint. It becomes waterproof when dry and dries very quickly, making it convenient to use.

Flat fitch This hog's hair brush is useful for applying glazes in small or difficult areas.

Flogger A horsehair brush made with very long, coarse bristles, this is used to tap, or 'flog', a wet glaze to give a finely flecked finish.

Glaze A see-through film of colour used for creating special paint effects, available in acrylic and solvent-based versions. Acrylic glazes are easier to use, have fewer fumes and are less likely to yellow with age.

Jamb duster This is usually used to remove dust before painting a surface. It is also useful as a softening and blending brush instead of a dusting brush.

Mottler A brush made from hair, available in a range of different sizes. It is used primarily for dragging (a paint effect that gives the appearance of fabric) or to simulate woodgrain.

Polyurethane varnish A solvent-based varnish available in different finishes, it is easy to apply and provides good surface protection.

Softening brush Use just the tip of the brush to blend solvent-based glazes gently after they have been applied but are still wet.

Sponge A natural sea sponge is the best type for creating a sponged paint effect.

Universal stainer A chemical dye that dissolves in white spirit. Use it to tint emulsion and solvent-based paints and glazes.

White spirit Also called turpentine substitute. Use it to dilute solvent-based paints and varnish, and to distress solvent-based paints and glazes when they are still wet.

Painting stripes on a plain wall introduces pattern and colour to the room – simply use masking tape to define the stripes and a plumbline or spirit level to ensure straight lines.

Painting stripes

A simple way to give a plain painted surface an interesting pattern is to use masking tape to create bold stripes. First, paint the walls the colour of the lighter stripe. Decide how wide the stripes are to be (they could be uniform or alternately wide and narrow) and, when the base coat is dry, smooth a length of masking tape along one edge of the proposed stripe and a second length down the other. Press the tape down firmly with the back of a teaspoon to prevent paint seeping underneath. Apply the second stripe colour with a brush, or sponge or cloth if you wish to create a subtle ragged stripe effect. When the paint is dry, peel off the tape slowly and carefully.

Choose low-tack masking tape, specially designed for delicate painted surfaces, and remove it carefully as soon as the paint is dry. Using curved masking tape will allow you to paint uniform wavy lines neatly instead of straight ones, if preferred.

Using glazes

Paint effects are best created using a glaze, traditionally known as scumble. Although you can use thinned ordinary paint instead, a glaze is preferable because it remains workable for a long time. Since it doesn't dry quickly, you can wipe off any mistakes and start again if you are not happy with the results. You can either buy ready-made glaze or 'effects' paint in a range of translucent and pearlescent colours, or create your own glaze by mixing emulsion or eggshell paint and/or a tint or pigment with a colourless acrylic or oil glaze according to the manufacturer's instructions.

When using a glaze, the trick is to work in small sections and always keep the wet edge 'open' so that there are no obvious joins between sections. Ideally, work with someone else so that one person can apply the glaze and the other follow quickly behind, creating the special effect.

When you have completed your paint effect and it is fully dry, apply a coat of clear, matt polyurethane varnish or a water-based acrylic varnish over the top for a hard, protective finish.

How to apply a glaze

1 Apply the tinted glaze in random strokes over the eggshell base colour. Be sure to cover the walls evenly.

2 To create a dragging effect, apply the glaze, then line up a spirit level vertically on the wall and indent slightly to give a guidemark for vertical strokes.

3 With a clean brush (about 75 mm/3 in wide), drag straight down the wet glaze so that vertical strokes are left behind.

Colourwashing

Choose the same colour paint as the base coat or a colour close in tone but slightly darker and mix up a wash of emulsion and water, in equal parts. Gradually increase the water to make a thin colour that will not run down the wall in heavy droplets, but will allow the base colour to show through when it is applied on top. Alternatively, use a thinned glaze or 'effects' paint as a wash over the whole surface for a more durable finish. Once you have tested your colourwash on a sample area, continue to apply the remaining colour using either a wide decorator's brush in sweeping criss-cross strokes or a car-polishing mitt or soft cloth to achieve a cloudy effect. When this coat is dry, apply a further coat in a toning colour if you like.

How to get the look

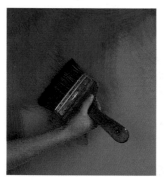

1 Apply the emulsion base coat and leave to dry. Use a large decorator's brush to apply the diluted top coat in random strokes.

2 Soften off any hard lines with a dry softening brush while the top coat is still damp.

Sponging

Sponging is one of the easiest decorative paint finishes to achieve and you can use two or more colours. Choose your first topcoat colour and make up a glaze. Dip a natural sea sponge into the glaze, taking care to wipe off any excess. Use it to sponge the wall using a dabbing, twisting movement to achieve an uneven, mottled effect. Cover the whole wall and allow to dry. If required, start at the beginning again and apply a second coat using another colour until the wall is evenly covered.

How to get the look

1 Apply the base emulsion and leave to dry. Dip a piece of natural sponge into emulsion and dab it across the surface.

2 When the first sponged layer is dry, take a new piece of clean sponge and dab on the second colour to soften the first.

Ragging

Similar in principle to sponging, ragging involves scrunching up a lint-free rag into a ball and using this to apply glaze in a contrasting shade to the base coat. Dip the rag into the glaze, remove the excess, then dab the rag straight on to the wall. Keep dipping the same side of the rag into the glaze and dabbing at the same angle with even pressure to achieve a uniform pattern. For a more random effect, frequently rearrange the paint-covered rag, dabbing with even pressure. Avoid overloading the rag with glaze.

Rag-rolling is a variation of ragging and its soft, mottled paint effect is excellent for disguising bumps or surface irregularities. It involves literally rolling a piece of rag scrunched into a sausage shape over a coat of wet glaze or paint. Either match the tone of the top colour with the background, or introduce a slightly darker tone on top.

How to get the look

1 Paint on the glaze in vertical strokes. Over large areas, join the strokes at varying levels across the wall to avoid an obvious join line.

2 While the glaze is still wet, blot the surface with a piece of hessian (burlap) to break up the glaze.

Bagging

Bagging relies on a scrunched-up plastic bag to achieve a traditional broken-paint finish. The result can be beautifully subtle, producing an effect similar to the look of crushed velvet. You can use a pale shade over a darker base coat or vice versa – experiment first with different colours.

How to get the look

1 Working on a small section at a time, use a wide brush to apply the topcoat of glaze or thinned paint in vertical strokes.

2 Now press an inside-out, scrunched-up thin plastic bag down on to the paint, immediately lifting it up and pressing down on the next spot in a dabbing motion, to produce a crinkled effect. Take care not to smudge the paint as you lift the plastic bag. Overlap the bagging slightly as you work along the wall to avoid the appearance of bands where one section finishes and another starts.

Wallpaper
basics

Choosing wallcoverings

The huge range of wallcoverings includes foil, plastic finishes, natural fibres and fabric as well as paper. The prices vary widely and some coverings are more durable and easier to hang than others. Choosing the right one for your needs can be a bit daunting unless you are familiar with the options available.

Lining paper

Lining paper is a plain, inexpensive paper designed to cover poor wall or ceiling surfaces before the application of a decorative covering of paint or an expensive wallcovering. (It is often the best way to hide thin cracks in ceilings, too.) Lining paper is hung vertically before painting or horizontally before covering with wallpaper.

Standard wallpaper

Standard wallpaper is available in a wide range of patterns and prices, including expensive hand-blocked and hand-printed wallpapers, which require more specialist handling. Standard papers are not washable or stain-resistant.

Flock wallpaper

A type of embossed paper, flock wallpaper comprises fibres of nylon, cotton, rayon or silk glued to a paper or vinyl backing, which create a raised pile with a velvety feel. Available mainly in traditional patterns, and usually in rich colours, the paper was originally developed to mimic velvet wall hangings. Flocks are expensive, and difficult to hang and to clean, although the washable vinyl versions are easier to handle.

Washable wallpaper

Easy to hang, this is ordinary wallpaper with a transparent plastic protective finish that makes it stain-resistant. It can be wiped or lightly sponged, but not saturated – too much water could weaken the wallpaper paste – and is particularly suitable for kitchens and bathrooms.

Vinyl wallcoverings

Comprising a layer of vinyl (applied as a spray, a liquid or a solid sheet) fused to a paper or fabric backing, these wallcoverings come unpasted or ready-pasted. Generally speaking, they are durable, tough and scrubbable and particularly suitable for bathrooms and kitchens. They must be hung with a paste containing a fungicide (which discourages mould), and it is important that seams do not overlap because vinyl will not stick to itself. Vinyls can be stripped quite easily.

The range of vinyls includes vinyl-coated papers, solid sheet vinyls, flock, metallic and textured vinyls.

Textured vinyls include heavy-duty 'contoured' vinyl, also called tiling-on-a-roll, which commonly features tile-effect patterns such as mosaic tiles, and blown vinyl. Blown vinyl, or 'expanded vinyl', has a relief pattern that gives a three-dimensional effect.

A vinyl wallcovering is the best choice for papering a bathroom because it is tougher than standard wallpaper and better able to withstand the moisture and hot atmosphere likely in bathrooms.

A textured wallcovering can add interest and subtle pattern to a room scheme and is useful for helping to disguise uneven walls. Some textured papers are designed to be painted over.

Good for covering imperfect surfaces, it is available in a range of designs and colours, and can also be bought uncoloured for overpainting. Blown vinyl is washable, but the surface scuffs easily.

Relief and embossed papers

These papers have a raised surface and are particularly good for covering up poor plaster. Many are designed for overpainting, usually with emulsion or eggshell, but are tough enough to take an oil-based gloss paint. Some of the papers must be handled with care so as not to flatten the raised pattern. Once hung and painted, they can be extremely difficult to strip. The main types include the following:

Woodchip paper Chips of wood and sawdust within the paper give it a raised texture like oatmeal or porridge, with different grades from fine to coarse. It is inexpensive and ideal for covering less-than-perfect surfaces.

Anaglypta This is a trade name. The heavy-duty paper can look like finely modelled plaster. There are over 100 designs, some of which relate to specific period styles. It is made from wood pulp or cotton fibres, or a vinyl version is available. Generally, the more heavily embossed the paper, the more expensive it is.

Lincrusta Invented in the late 19th century, lincrusta is made from linseed oil and fillers. It is often modelled to resemble panelling and used on the lower part of the wall, then painted or stained to resemble wood. It is difficult to hang.

Foil wallcoverings

Foils comprise either aluminium foil or metallized plastic film fused to a backing of paper or scrim and overprinted with a pattern. The shiny finish helps to reflect light, which can enhance dark areas of a room, but will show up any poor wall surfaces. Foils are hardwearing, but they can be difficult to hang, and particular care must be taken around electrical fittings.

Natural fibres

These textured wallcoverings are made of natural fibres like raffia, sisal and hemp, which have been dyed and laminated to a backing material. They are suitable for walls in poor condition as long as they are exposed to very little moisture.

Fabric wallcoverings

Silk, linen and suede are examples of the modern luxury fabric wallcoverings available. Backed with paper or latex acrylic, they are sold either in wallpaper-sized rolls or by the yard or metre. Alternatively, unbacked fabric can be stretched in place over battens. Fabric wallcoverings require expert hanging and are expensive and difficult to clean.

How much paper do I need?

Strictly speaking, the term 'wallcovering' is more accurate, but, for convenience, the word 'paper' is used here as a generic term for all types of wall and ceiling coverings. The charts here will help you determine the number of rolls of paper required, but remember to take account of roll sizes and any pattern repeats.

Measuring up and calculating

To work out how many rolls of wallpaper you need, you can't just calculate the area to be covered because you need to work in complete vertical runs or 'drops' – joins halfway down a wall are not a good idea!

First, measure your drop, from the ceiling down to the dado or skirting (baseboard). The number of drops in a roll will depend on whether you have a pattern repeat to take into account. A standard roll is 10 m (33 ft) long. A paper with no repeat would provide four drops of up to about 2.4 m (8 ft) (you will need to allow an overlap top and bottom). Over that and you will get only three drops, although the leftovers will be useful for areas over doors and below windows. Allowing for a pattern repeat will increase the wastage, but you should still get three drops unless the ceiling is very high. If papering only from dado rail to picture rail, you can expect to get five or six drops per roll.

To work out how many drops you need, simply measure all round the room. A standard roll is 53 cm (21 in) wide, so if your round-the-room measurement comes to 14 m (46 ft), that's 27 drops. At four drops per roll, you will need seven rolls; at three drops, you will need nine rolls. If in doubt, over-order, checking that you can return what you don't use.

Roll sizes

A standard roll is 10 m x 53 cm (33 ft x 21 in). The paper comes ready trimmed and is usually wrapped to keep it clean. The charts for calculating quantities (see below and opposite) are based on the standard roll size. However, some Continental European papers may be narrower than this, and American papers may be twice as wide. If selecting any of these papers, look in the manufacturer's sample book for guidance on coverage.

Repeats

Another variable to consider when calculating quantities of wallcoverings is the pattern repeat, since patterns need to be matched at every join. Sample books and the wrapping label will give the repeat

Calculating the number of rolls for walls

Measurement around room	8.5m (27ft 10in)	9.75m (32ft)	11m (36ft)	12m (39ft 4in)	13.4m (44ft)	14m (46ft)
Height from skirting (baseboard)						
2.13–2.29m (7ft–7ft 6in)	4	4	5	5	6	6
2.30–2.44m (7ft 6in–8ft)	4	4	5	5	6	6
2.45–2.59m (8ft–8ft 6in)	4	5	5	6	6	7
2.60–2.74m (8ft 6in–9ft)	4	5	5	6	6	7
2.75–2.90m (9ft–9ft 6in)	4	5	6	6	7	7
2.91–3.05m (9ft 6in–10ft)	5	5	6	7	7	8
3.06–3.20m (10ft–10ft 6in)	5	5	6	7	8	8

measurement. The larger the pattern repeat, the greater amount of wastage you will need to allow for and the more rolls you will need. For a large, bold pattern, you may also need to allow for a dominant element to 'sit' well on the wall – awkward breaks such as headless animals, for example, are less noticeable at the bottom of the wall than at the top.

Patterned paper comes in the following types:

- Free-match papers have designs that do not require pattern matching, and so involve little wastage.
- Set-match, or straight-match, patterns have motifs that repeat in a straight line across the paper. Wastage will depend on the overall size of the pattern repeat, but it's not usually too much.
- Offset, or drop-match, patterns have a repeat in diagonal lines, and so each new length of paper has to be moved either up or down to allow for this. You can minimize wastage with drop-match patterns by cutting alternate lengths from two different rolls. A drop-match pattern is a good choice if your walls are not perfectly square.

In addition, some papers have a symbol on their wrapping label indicating 'reverse alternate lengths'. This means that you need to hang each length of paper the opposite way up relative to the previous one. Some free-match wallcoverings look best when reverse hung.

Batch numbers

Each roll of paper should be stamped with a production batch number. When selecting your rolls, check that all the batch numbers are the same. If they are not, this means that the rolls have been produced in different batch runs and there may be variations in colour and shading. For this reason, it's a good idea to buy more than you need. If you run out and have to buy one or two more rolls at a later date, they may come from a different batch. If you do buy too many rolls, you can usually return any unopened ones. If you do have to finish off with rolls from a different batch, plan to use this paper in an alcove or recess, or somewhere in the room where any slight differences in colour or shade will not be immediately obvious.

Calculating the number of rolls for a ceiling

Measurement around room	Number of rolls required
9.75 m (32 ft)	2
10 m (32 ft 10 in)	2
11 m (36 ft)	2
11.5 m (37 ft 9 in)	2
12 m (39 ft 4 in)	2
12.8 m (42 ft)	3
13.4 m (44 ft)	3
14 m (46 ft)	3
14.5 m (47 ft 7 in)	3
16 m (52 ft 6 in)	4
16.5 m (54 ft 2 in)	4
17 m (55 ft 9 in)	4
17.5 m (57 ft 5 in)	4
18 m (59 ft)	5
19 m (62 ft 4 in)	5
19.5 m (64 ft)	5
20 m (65 ft 9 in)	5
20.5 m (67 ft 3 in)	6
21 m (69 ft)	6
22.5 m (73 ft 10 in)	7

16 m (52 ft 6 in)	17 m (55 ft 9 in)	18.5 m (60 ft 8 in)	19.5 m (64 ft)	20.75 m (68 ft)	22 m (72 ft 2 in)	23 m (75 ft 5 in)	
7	7	8	8	9	9	9	
7	8	8	9	9	10	10	
7	8	8	9	9	10	10	
8	8	9	9	10	11	11	
8	9	9	10	10	11	12	
9	9	10	10	11	12	12	
9	10	10	11	12	12	13	

How to hang wallpaper

Hanging wallpaper is not a difficult job if you want to do it yourself, but it is essential that you apply it to a properly prepared surface, otherwise any imperfections in the wall surface will be obvious and your efforts will be wasted.

A boldly patterned wallpaper like this results in some necessary paper wastage, as it requires very careful matching at the seams so as not to detract from the overall effect.

Order of work

If you are papering the ceiling, do this before you paper the walls. To paper the walls, start at the main window, working away from the light towards the longest unbroken wall. When you reach the corner of the room, go back and work from the other side of the window. If the room has a chimney breast or some other focal point, centre a length of paper on it and work outwards to the left and right, lapping around the edge of the chimney breast.

Before you start

All woodwork should be painted and dry, and walls and ceilings should be clean, dry and smooth. Give the walls and ceiling a final check for uneven filler or rough areas where any old wallcovering has been removed, and rub down with medium glasspaper where necessary.

If the ceiling or walls have been newly plastered, you will first have to 'size' them to prevent too much of the wallpaper paste being absorbed by the plaster and weakening the adhesion of the paper. An application of size also helps to slide your paper into place, by making the surface to be covered more slippery. Size may be bought as a paste or a powder, to be mixed with cold water. Spread newspaper around the bases of the walls to catch any splashes of size, then apply it to all the surfaces to be papered and leave it to dry – usually about an hour or two – before papering.

Collect all your tools together (see pages 38–39), set up your pasting table and place it under a window facing the light, so that you will be able to see where you have pasted. Make sure that the surface of the pasting table is dry and clean.

Wallcovering adhesives

Different wallcoverings need different types and strengths of adhesive, so check on your covering's requirements before buying the paste and make sure that you buy enough. Using the wrong type of adhesive can result in the wallcovering failing to adhere, the surface discolouring or mould growing. As a general rule, the heavier the wallpaper, the stronger the adhesive you need.

All-purpose wallpaper paste contains a fungicide and can be bought either ready-mixed (usually in a handy tub) or as a powder to mix with water. The ready-mixed paste is convenient to use and a tub usually contains enough to do about five rolls of wallpaper, but it works out very expensive if you have a lot of wallpapering to do. You can also buy tubes of wallpaper paste for very small jobs such as borders or patches. These are often described as 'strong wallpaper paste'.

Handy hints

- Allow time for the paper to absorb the paste before hanging, and give the same 'soaking' time to every length.

- If using a paste-the-wall covering, apply the adhesive using a roller to an area slightly wider than the length being pasted. You can paste the next section without smearing paste on the paper.

- If you have to make a horizontal join, do not use a straight cut – tear the paper across the drop slightly unevenly and overlap for an almost invisible join.

- Paper often bubbles when first hung, but dries flat. Prick any remaining air bubbles with a pin and smooth out. Alternatively, try shrinking bubbles back into place by blowing the surface with a hair dryer.

- To remedy curling seams, ease open the seam, dab on a little paste with a cotton bud and smooth back with a seam roller.

- When hanging flock, protect it with a piece of lining paper.

Papering walls

1 Measure the drop and cut the first piece of paper to length, adding about 12 cm (5 in) to allow for trimming to fit.

2 Cut several more lengths, matching up the pattern exactly to the neighbouring piece each time. Number them in sequence on the reverse to avoid mistakes.

3 Take a window as your starting point, not a corner. From the edge of the frame, measure the width of the paper, less 1 cm (½ in), and mark on the wall. Suspend a plumbline from ceiling height to coincide with your mark. Draw the vertical in pencil or chalk.

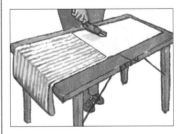

4 Lay the first length of paper face down on the pasting table, slightly overlapping the edge to avoid getting paste on the table. Weigh it down so that it does not roll up. Load the brush with paste, then apply paste evenly and liberally, working right to the edges. If the paper is longer than the table, let the pasted paper overhang while you paste the rest. Carefully fold the ends of the pasted paper back on itself to the middle, fold it again (take care not to flatten the folds) to make the length easy to move and leave to absorb the paste. This is important because the paper will continue to expand for some minutes. If it does not expand fully, you may get bubbles later on.

5 To hang, carry the first length of paper to the wall. Holding the top corners of the paper, start to unfold and press the top to the wall, sliding it to line up the edge with your vertical guideline. Smooth downwards and outwards using a smoothing or paperhanger's brush. Open the bottom fold and continue smoothing the paper down the wall. Lightly dab any bubbles flat.

6 Use the edge of the closed scissors to press the paper firmly but gently into the top edge, under the cornicing or against the ceiling. Press very lightly to avoid tearing the damp paper, then peel back the paper and cut neatly along the crease. Smooth the paper back into place. Do the same at the bottom of the wall along the skirting (baseboard) and check for air bubbles. Scissors are preferable to a knife for trimming to avoid tearing, although on heavy papers like vinyls you can use a very sharp knife and steel straight edge. Place the straight edge just above the crease and cut.

7 Paste the next length in the same way and hang edge to edge (butt joined) with the previous piece – do not overlap. Match any pattern by sliding the paper up or down. Except on embossed paper, gently run a seam roller down the seam to stick the edges down firmly.

8 Continue hanging lengths, in the order described, and mark a new vertical guideline every time you turn a corner to ensure the pattern stays truly vertical. As you progress, clean up between each length. Wipe paste from picture rails, skirtings (baseboards) and window and door frames while it is still soft. Keep any trimmed pieces of paper for patching. Wipe the pasting table clean after each length. You will need to cut slightly shorter lengths above a radiator, unless you are able to move it. Leave about 15–20 cm (6–8 in) for tucking and smoothing.

Papering a ceiling is best tackled by two people – one to smooth down the paper and one to hold the pasted paper against the ceiling using a clean broom or purpose-made roller.

Papering ceilings

When papering a ceiling, you must be able to reach it safely and in reasonable comfort, so check how you will reach the highest point of the room. A short step-stool may be sufficient, otherwise make a platform you can stand on (see page 39), so that your head is about 7.5 cm (3 in) from the ceiling. Place the platform along the line of the first length of paper to be hung.

The best starting place is parallel with the window wall. This ensures that no shadows will be thrown should any paper overlap. The simplest way to mark the position for the first length is to mark the ceiling at each end with a pencil, the width of the paper away from the wall, then snap a chalk line between the two points to leave a line of chalk on the ceiling.

1 Measure the ceiling and cut the first piece of paper to length, adding 10 cm (4 in) to allow for trimming. Paste and fold the length of paper as described for papering walls (see page 49), and leave the paper to soak.

2 Lay the folded paper over a spare roll of paper, with the edge to be stuck first uppermost. Hold the roll in one hand, grip the top edge with the other hand, turn it paste-side up and apply it to the corner where you plan to start. Slide the paper to the chalk line and smooth it on to the ceiling. Move along the platform, releasing the folds as you go. Smooth the paper to the ceiling and get someone to hold the paper in place with a broom as you move along. Continue until the whole length is in position. Run over it with your smoothing or paperhanger's brush, making sure that the edges are well stuck down all along the length.

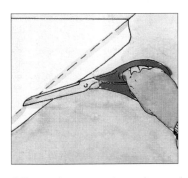

3 Press the paper into the end walls and crease it with your closed scissors as before.

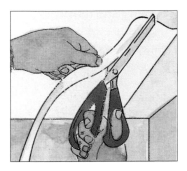

4 Trim 3 mm (1/8 in) outside the crease so that the paper just turns on to the wall. Dab the paper back into place. Hang the second length, matching it to the edge of the first, then paper the rest of the ceiling in the same way.

Ready-pasted paper

Ready-pasted paper does not stretch, so there is less chance of it bubbling through expansion. Cut and hang the paper as above, but, instead of pasting, soak each rolled length in the water trough according to the manufacturer's instructions. Pull it out slowly, top edge first, allowing excess water to drain away.

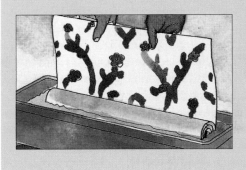

Pendant light fittings

1 To paper around a ceiling pendant, cut the paper to length, then mark on the back the position of the pendant. Allow for trimming.

2 Paste the paper and make a series of small cuts radiating out from the centre of the light position you have marked. Hang the paper normally. When you reach the light fitting, feed the pendant through the hole made by the cuts and press the paper into place around it.

3 Crease each flap with the blunt edge of the closed scissors and carefully cut off the surplus paper 3 mm (1/8 in) outside the crease marks. Press the paper into place and wipe the paste from the fitting.

Tiling
basics

Choosing tiles

Tiles come in all manner of colours, patterns, shapes, sizes and materials. Mostly hardwearing and easy to clean, tiles provide an incredibly versatile and practical covering for both floors and walls. Ceramic tiles are probably the most widely used, but there are plenty of other types of tile for use around the home.

Tile sizes

The best size of tile to use depends on the area to be covered and the desired effect. It's obviously quicker to tile an area with larger tiles and they work better for large expanses of wall, whereas smaller tiles work well in small rooms. Small tiles are easier to cut and shape around awkward areas such as built-in fixtures, recesses and window rebates. As for the effect created by different sizes of tile: smaller tiles create a busier look; rectangular tiles are good for creating interesting herringbone and other traditional brickwork patterns; while there are shapes that interlock for special effects.

The vast range of tiles available provides you with plenty of scope for mixing colours, designs and styles in any place that a durable surface is required.

Ceramic tiles

Ceramic tiles are the perfect protection against water penetration and are great for bathrooms and kitchens. They are made from a thin slab of clay decorated with a coloured glaze, and often with a pattern. There are many shapes beyond the usual squares and rectangles, including hexagonal, octagonal, curved Provençal styles and small 'drop-ins'. The most common square ceramic tiles are either 10 cm (4 in) or 15 cm (6 in). Rectangular tiles are usually 20 x 10 cm (8 x 4 in) and 20 x 15 cm (8 x 6 in).

'Universal' tiles have several glazed edges, which are useful if you wish to avoid leaving an unglazed edge exposed.

You can also buy feature panels – sets of tiles that form a large design. These can be used alone, rather like a piece of artwork, or set into a wall of plainer tiles.

Border tiles

Unless you are using 'universal' tiles, border tiles are the conventional way of finishing off half-tiling. Borders can be chosen to coordinate with the main tiling or to contrast with it and there are some highly decorative deep-profile tiles available. Check that the border tiles are compatible in width with the main tiles and that the colours and glazes work together.

Mosaic tiles

These are tiny tiles supplied on a mesh backing, or paper frontsheet, which is soaked off after the tiles have been laid, usually 30 cm (12 in) square. They may be different tones of plain colours or form a distinct pattern. The sheets are laid in a bed of adhesive and then grouted. Faux mosaics are standard tiles scored to look like mosaic.

Quarry tiles

Quarry tiles are usually used on the floor, but they make a handsome alternative to ceramic tiles on the wall. Although as equally hardwearing as ceramic tiles and requiring similar preparation and fixing, they are unglazed and have more of a country appearance – a look that is emphasized by their earthy coloration.

Cork tiles

Cork tiles are usually fixed as panels, rather than used to cover a whole wall, although cork has good insulating properties. They are ideal for pinboards or feature walls. Most come in 'natural' colours, although some may be stained or dyed. Unsealed tiles are thick and crumbly, so it is better to choose presealed. Cork can be sealed after hanging, but may need many coats.

Metallic and mirror tiles

Metallic and mirror tiles are often used for their reflective value. The metallic tiles are usually produced in a lightweight metal such as aluminium, made to simulate copper, pewter or stainless steel, and some have interesting textures and self-patterns. They are fixed by self-adhesive pads.

Mirror tiles are much heavier, and come in a wide range of sizes. Some are hung using a special adhesive, but heavier panels need to be fixed with special plates or mirror screws. Choose very high-quality tiles for a bathroom, as condensation can cause the silvering on the back of cheap mirror to perish.

You need an especially flat surface for reflective tiles or you will get a distorted reflection. If necessary, fix them to a panel of chipboard or hardboard and then attach this to the wall.

Other types of tiles

Other types of tile include those suitable for flooring (see page 74), namely vinyl, cork, linoleum, rubber and leather tiles. You can use floor tiles on walls, although

Shiny metallic tiles make a perfect splashback in a modern kitchen, but must be laid on a very flat wall to avoid distorting their reflective surface.

they are likely to be heavier and more expensive, but don't use wall tiles on floors because they are not strong enough and may be too slippery.

Where to use tiles

Consider using tiles on various horizontal and vertical surfaces:

- In the dining room, as an integral part of a food-serving surface.
- On the lower part of a hall wall.
- In recesses or alcoves use mirror tiles as a reflective background to glass shelving or to suggest a window in a dark, uninteresting area.
- In the bedroom, where mirror or metallic tiles are useful on wardrobe doors.
- On tabletops or tiled worktops use flooring-grade tiles or special worktop tiles on a base of blockboard, chipboard or plywood. Some have matching edging strips. Alternatively, use plastic trim strips designed to sandwich under the row of edge tiles as they are laid.

Tile examples

It's often difficult to judge how tiles will look when covering a large expanse. Some retailers have comprehensive catalogues showing room sets, while others have samples mounted on display boards. Avoid trendy effects or colours, which you might tire of easily, and be aware that geometric effects will work only where all walls and corners are straight and true.

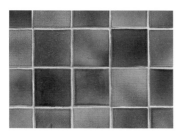

A tricolour mix of blue tiles in gloriously deep glazes of rich azure, sapphire and aquamarine. Show off these intense shades alongside natural woods or modern chromed surfaces in the kitchen. In the bathroom, lay them as a border, or be brave and cover a complete wall, creating a calming environment.

These organic blue stone floor tiles are made by mixing pure clay and water and then covering the surface with pure oxides before firing. An oxidized blue is most effective for flooring and the uneven colour surface gives depth to any room. Use in a hallway or throughout an open-plan kitchen and living area.

Mosaics look fantastic in any bathroom or kitchen – they bring a feeling of Roman grandeur to even the smallest of spaces. If you are feeling truly creative, why not make your own designs? Highlight the areas around mirrors or shower cubicles and have a go at cleverly manipulating the shape of the room.

Multi-coloured mosaic tiles are a simple yet stylish solution for all surfaces, even window sills or cupboards and shelving. Mosaics were used by the Japanese, Roman, Greek and Arabic communities to create wonderfully lavish and colourful floor and wall patterning. Today they come in easy-to-handle sheets.

Traditional elongated tiles are an attractive alternative to the more usual square ones. Reminiscent of Victorian bath houses, these pale marble colours would look great in the bathroom or kitchen. Think about continuing them as borders along kitchen work surfaces, or use all over in the bathroom.

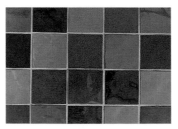

These zinc-look polished tiles are of a relatively new generation of tiles, and are available in both polished and satin finishes. Use in the bathroom and shower as a space-age alternative to coloured tiles. In the kitchen, these tiles are an ideal companion to today's stainless steel appliances and work surfaces.

How many tiles do I need?

Tiles are usually sold in boxes of 25 or 50, or in packs to cover one square metre or yard, but this varies and buying large quantities could be wasteful if you are tiling only a small area. First, you need to know how many tiles you want.

Estimating quantities

Unless you plan your tile design using a scale drawing (see below right), you need to calculate the area to be covered.

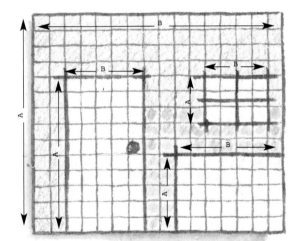

For an unbroken expanse of wall, simply multiply the height of the wall (A) by the width (B). To allow for doors, windows and radiators or other fixed heaters, multiply the height (A) by the width (B) of each feature, add the totals and subtract from the total wall area. This will give you the area to be tiled. For floor tiles, measure the width and length of the room and allow for the area taken up by cupboards, a bath or other features as above.

To determine the number of tiles required, divide the area to be tiled by the area of a single tile. Add an extra 10 per cent to allow for cutting wastage and breakages, and to keep as spares in case you need to replace a damaged tile in the future.

Designing with tiles

Instead of a vast area of plain tiles, try to be bold and create an original look:

- Hang square tiles diagonally for a more interesting, diamond-quilted look.
- Create bold stripes with two or more colours, either horizontally or vertically.
- Halved diagonally, a square tile becomes a triangle for a zigzag border.
- For a patchwork look, use tiles of the same size and thickness in a selection of different colours.

Pattern planning

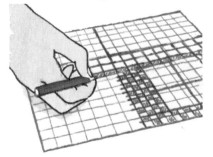

Draw the area to be tiled on a squared grid – if you are using square tiles, one square can equal one tile. Plot your pattern on the grid and colour it in so that you can see the effect and work out accurate quantities.

Centre your design on any feature it relates to, and use your plan to foresee awkward cuts. Whether you choose the middle of a tile or the meeting of two tiles as the mid-point of the area to be tiled (design permitting) can make a big difference to the amount of tile cutting you have to do.

Even if you have planned the design on paper, lay out the tiles in a 'dry run' on the floor. Number them, if necessary, so that you fix them in the right sequence.

How to tile walls

Tiling is perfectly feasible for the average handy person, but start with a small project, such as a basin splashback. Avoid anywhere that requires any tricky tile cutting, and choose tiles that don't need any complicated pattern matching.

Preparation

Tiles must be fixed to a dry, flat surface. Replaster any badly damaged areas and fill small holes and cracks with proprietary filler (see page 20). If the surface is uneven, it may be better to fix the tiles on chipboard or hardboard instead. Then either fix the board to the wall with battens and tile it, or tile the board while horizontal and then fix it with impact adhesive.

You can also tile over existing tiling, as long as the original tiles are firmly stuck down and thoroughly cleaned before covering. In addition, you must stagger the joints so that no new joint is on top of an old one. You will end up with a thick ledge where the tiling ends. Use a wooden trim or edging tiles with quite a deep profile to hide any visible joins between the two.

As long as you establish a perfectly horizontal base from which to work up the wall, your tiles should quite literally slot into place to provide a stunning and hardwearing surface.

Hanging tiles

Hanging tiles is not too difficult, but you need to work out exactly how to position the tiles before you start, otherwise you could end up with a tile cut in an awkward position – or a mismatched pattern. Use a profile gauge to help you, if you like (see page 56). Also make sure that you have to hand all the necessary tools to cut and fix your tiles (see pages 56–57).

1 Tiles are hung from the bottom of the wall working upwards, so you first need to establish a perfectly horizontal base. Measure one tile depth up from the floor, skirting (baseboard) or worktop, allowing a gap for grouting. Attach a temporary batten to the wall with long nails, without hammering them fully home. Use a spirit level to check that it is horizontal and aligned to your measured point. Fix a vertical batten in a similar way, so that its inner edge marks the edge of the last whole tile. If your design has a mid-point, mark the centre of the wall and measure out from this.

2 Apply tile adhesive to the wall using a serrated spreader. Cover only about a square metre or yard at a time, and keep the container covered.

3 Press the first horizontal row of tiles firmly into place against the edge of the batten, inserting spacers

Handy hints

- Try to avoid having to cut pieces of tile less than a quarter of the tile width.
- If planning your tile layout on a squared grid, use a tracing-paper overlay to see where tiled edges will fall best – this is more convenient than having to keep marking and erasing alternatives.
- When using plain or marbled tiles, open the boxes and 'shuffle the pack' to mix the tiles before use to minimize the effect of any colour variation between boxes.
- Remember that tiling will raise the level of the floor and you may have to make adjustments to cupboard doors and the door opening into the room.

A simple basin splashback with an easy-to-follow design that requires no tricky tile cutting or pattern matching is the perfect project for a first attempt at tiling.

between each tile. Work up the wall, row by row, and when you reach an obstacle such as a window, carry on fixing the whole tiles, leaving tile cutting until later. Keep checking that the tiles line up in both directions and that any pattern matches.

4 Once all the whole tiles are fixed, remove the battens. You can now cut tiles as necessary (see right) and hang them around the edges. You may have to put adhesive on the backs of these part-tiles, rather than on the wall.

5 Once the tiling is done, leave the adhesive to dry for at least 12 hours and remove the spacers if necessary (see page 57). Prepare the grout according to the manufacturer's instructions, then press it into the joints using a sponge or small spreader. Remove the excess with a clean, damp sponge, then use a grouting tool or equivalent (see page 57) to compress and finish the joints. Allow the grout to dry, then polish the tiles with a clean, dry cloth.

Cutting wall tiles

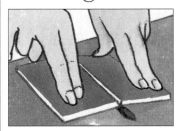

1 To cut straight edges, use a chinagraph pencil to mark on the glazed side of the tile where it is to be cut. Score the glazed surface firmly using a pen-like tile cutter. Place a matchstick beneath the scored line. Press the tile down firmly either side and it should snap cleanly. Alternatively, use a mechanical tile cutter.

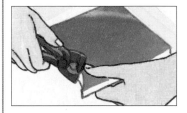

2 To cut a curved or angular shape, score the shape in the same way, then nibble away the excess with tile nibblers or use a tile saw. Smooth rough edges with a tile file, coarse sandpaper or carborundum stone.

Tiling tricky areas

With good tools and patience, tiling around tricky shapes is reasonably straightforward. If you are planning a room from scratch, consider the size of tiles to be used to avoid too much tile cutting and shaping later on.

Tiling around corners

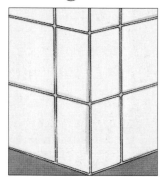

1 Finish external corners with tiles cut to the same size for both walls. Ideally, use tiles with glazed edges so that there is no raw edge facing out, or use a trim strip. With patterned tiles, it is better to use whole glazed-edged tiles and work away to where a cut tile and a pattern break will be less obvious.

2 Internal corners should also, wherever possible, be completed with either whole tiles or part-tiles of the same size meeting in the angle between the two walls. With patterned tiles, use the offcuts from tiling one wall to begin the opposite wall so as to maintain the continuity of the pattern.

Tiling around fixtures

1 Try to tile around washbasins and other fixtures in a symmetrical fashion. Ideally, plan to use a row of whole tiles above fixtures, since cut tiles can look untidy. Where this is impossible, be sure to fill any gaps with a good-quality waterproof sealant.

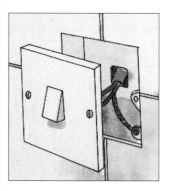

2 In bathrooms, light fittings must be fitted with a cord pull. Elsewhere, to produce a neat finish around a light switch, turn off the power at the mains, undo the screws in the faceplate and ease the front of the switch clear of the wall. Tile up to the edge of the mounting box so that the faceplate will cover the cut edges of the tiles. Carefully screw the faceplate back into place, hiding the tile edges.

Tiling around doors and windows

The door into a room or the window in a wall are both very much focal points, so try to maintain a visual balance by tiling around them evenly, using whole tiles whenever possible. The door or window frame may not be vertical, so don't use it as a guide for your tiling rows unless it is completely true.

Tiling a recess

For a professional appearance, ensure that the tiles are 'balanced' across a window recess – in other words, that the cut tiles either side of the recess are in symmetry. Tiles that protrude on external corners must have glazed edges or be covered by a trim strip.

1 To produce the neatest effect, tiles lining a window recess should project to overlap those on the wall. The tiles in the recess should therefore have glazed edges.

2 Make sure that the tile spacings in the recess are in line with those on the wall. Use cut tiles with glazed edges at either end of the ledge to finish it off neatly.

3 Start the sides of the recess with a cut, glazed-edge tile so that the spacings remain exactly consistent with those between the horizontal rows on the wall. Tile the underside of the recess last and tape the tiles in position until the adhesive has fully dried.

Drilling tiles

1 Apply a criss-cross of masking tape over the spot to be drilled to stop the drill bit slipping and cracking the tile during drilling.

2 Drill the hole using a pointed tile bit or a sharp masonry bit, but don't drill right through the tile into stone or brick. If you have a variable-speed power drill, run it at its slowest setting to maximize your control. Never use a power drill set to hammer action – the vibration will shatter the tile.

3 Peel away the masking tape afterwards and the small pieces of drilled tile should come away with it.

To drill holes in a tiled wall for accessories like a towel rail or shelf, use a tile drill bit and drill very slowly and carefully to avoid cracking the tile.

How to tile floors

The technique for laying floor tiles is much the same whether you are using flexible or hard tiles. They require a stable, level surface, so unless the subfloor is perfect, it is best to put down plywood or flooring-grade hardboard or chipboard before laying the tiles.

Handling floor tiles

Flexible floor tiles like vinyl, linoleum, cork and rubber are easier to handle and cut than hard tiles. Some are self-adhesive (see page 83) and simply have a peel-off backing.

Before laying flexible tiles, remove the packaging and leave them in the room where you are going to lay them for 24–48 hours so that they can acclimatize. Make sure that you have the right kind of adhesive for the tile. Most of the harder types of tile are best left to the experts to lay – they are heavy to handle and hard to cut, and some have to be set in a bed of mortar.

Flooring patterns

Floor tiles can be used creatively to make whatever pattern you like and an imaginative design can help to improve the proportions of the room. For example, stripes laid widthways will make a floor area look less long and narrow, while a chequerboard effect will create an impression of greater space.

Like wall tiles, floor tiles need to be centred on the middle of the floor or a dominant feature like a fireplace. If you are creating a pattern, draw up a scale plan of the room on graph paper, so that you can see how best to position the pattern and how many tiles of each colour you will need.

However sure you are of your calculations, always buy a few extra tiles as a contingency measure to allow for accidents and wastage.

Laying floor tiles

It is usual to lay tiles from the centre point and work outwards, towards the edges of the room, but in bathrooms and kitchens where much of the floor space is occupied by cupboards, you may have trouble finding the centre of the area to be tiled. It's important to lay tiles by working from a right angle you have marked on the floor because very few rooms have walls and corners that are completely true. If you started by working from one wall, you could soon find your tiles out of alignment.

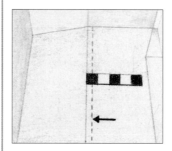

1 Mark the centre line in chalk and check that the edge tiles will be at least half a tile wide. If not, move the line a half-tile width to one side.

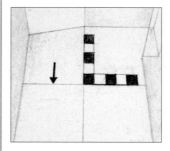

2 Mark a centre line in the other direction, at right angles to the first. Check as before and again move the line if necessary to avoid having small strips of tile at the sides of the room.

3 Spread tile adhesive along the floor, on either side of the centre line chalk marks.

4 Lay the central tiles first, either side of the chalk line, using floor tile spacers if necessary.

5 Continue working outwards from the centre tiles towards the edges until the floor area is completely covered. Lay a spirit level across the tiles periodically to check that they are level. Adjust the amount of adhesive beneath a tile if necessary to bring it in line with the others. Lay the tiles around the edges of the room last, cutting the tiles to fit as necessary. If using hard tiles, finish off by grouting in the same way as wall tiles (see page 63).

Hard tiles like quarry tiles are a popular choice for kitchen floors. They are more difficult to lay than flexible floor tiles, so probably best laid by experts for the best result.

Cutting flexible tiles to fit

1 To cut flexible tiles for edges, place a tile that needs cutting squarely on top of the last tile, then place another tile on top of it, butted up against the wall. Mark a pencil line along the edge of the top tile on the tile below. Cut the marked tile to this line – it should fit in the gap. Apply adhesive and firm into position, aligning the cut edge against the wall.

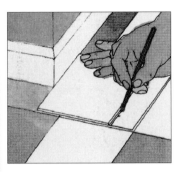

2 Use the same technique to cut around an external corner where a single tile is involved. Place two full tiles over the last laid tile and slide the top one over to butt up against the skirting (baseboard) or wall. Mark a pencil line on the tile beneath.

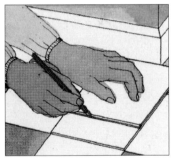

3 Move both tiles around the corner, without turning them, and position over the last laid tile on that side. Mark a line as before, cut out the part that is not required and glue the part-tile into place.

4 To tile around a pipe, first cut an edge tile to fit the space, then push the tile against the pipe and mark the centre of the pipe. Next, move the tile against the wall and mark the pipe centre on the edge of the tile. Draw light pencil lines from both points and cut a hole where the lines bisect. Make a slit in the tile to allow you to feed it around the pipe.

Flooring basics

Which type of flooring?

There are plenty of different types of flooring, but before you commit yourself to stripping old floorboards or laying a marvellous expanse of marble, consider objectively the look you want, what demands you will put on the floor and how much you can afford.

The options

There are three basic types of flooring:
Hard, which includes brick, stone, wood and tiles (see pages 74–75).
Resilient, such as vinyl, linoleum and rubber (see pages 80–81).
Soft, which includes carpets and rugs, as well as natural floorcoverings like sisal and jute (see pages 86–89).

Practical considerations

Floors take a lot of punishment – feet tramp across, bringing in water, mud and grit from outside; fidgety feet scuff the same patch in front of chairs or sofas; tracks are worn where there is only one possible route across a room. It's heartbreaking to spend time and effort laying a floor, only to find it looks scruffy after a few months or requires tremendous upkeep. To make sure you get it right, think about the following practicalities before making a decision:

- How much wear and tear will the floor receive? Floors that get a lot of heavy use or through-traffic, for example those in halls, on staircases and landings, need to be hardwearing.
- Does it need to be washable? The flooring in halls, bathrooms, kitchens and children's rooms all need to be easy to clean and resistant to dirt and stains.
- Do you want a permanent flooring – a wise choice for kitchens, conservatories and halls – or something that will not be too difficult to change?
- What is the condition of the existing floor? Is it easier to remove it or cover it up? This could affect the options available to you.
- Do you need a type of flooring that you can lay yourself, or is there enough in the budget to pay for it to be fitted? Large pieces of sheet material or rolls of carpet can be quite difficult to handle if you are inexperienced in these matters.

Design considerations

The flooring needs to relate to the style of the room or area. It is a significant proportion of the room's surface decoration, so will not go unnoticed. You therefore need to think carefully about the look that you want in addition to the practical considerations of the flooring:

- Strong colours and bold patterns may appear to hit you in the eye, so avoid these where you want to create a relaxing mood.

Soft flooring is warm underfoot and a comfortable choice for living rooms, although a pale colour is impractical in homes with small children and pets.

In areas like hallways that inevitably take a lot of through-traffic, wood is a good choice for a hardwearing flooring that is easy to keep clean.

Handy hints

- Always buy the best-quality flooring you can afford. Price is usually a good guide – if a good carpet is beyond your budget, look at cheaper alternatives to carpet rather than very cheap carpets.
- To create an impression of space, use a similar flooring, or the same colour flooring throughout – particularly effective in small flats or apartments, and bungalows.
- Where two different floorings meet at a doorway, you may have to install a threshold. Metal ones look very utilitarian; wooden ones are easy to stain or paint to match the floor or décor.
- Try to use manufacturers' own adhesives or sealants where possible.
- Check aftercare instructions, to avoid spoiling a floor with the wrong treatment.
- Some floor tiles need sealing after laying to protect them from stains – check with the manufacturer.

- Where much of the floor is covered up with furniture or equipment, it may be a waste of effort to design a patterned floor that will not be appreciated. Interesting patterns often look best in halls, kitchens, corridors and large, sparse rooms where they can best be shown off in all their glory.
- Consider the size of any pattern in relation to the scale of the room. Pale colours and small patterns can look disappointing on a large floor.
- When planning a patterned floor, whether in tiles or as a painted design on wood, think about how the pattern relates to the room's configuration – you don't want an awkward 'break' in the design at a doorway or in front of a fireplace, for example. Working out the design on a scale plan first is always helpful.
- Collect samples of your proposed flooring and look at them in the room where you plan to use them, to gauge their effect in situ.

Use stencils and floor paints on stripped wooden floorboards to reproduce the effect of a patterned carpet without the expense of the real thing.

Preparing your floor

Whatever type of flooring you choose, it will need to be laid on a subfloor that is smooth, clean, level and dry. Problems such as damp or rot cannot be ignored and will need to be cured before you start work.

Remedying an uneven subfloor

There are several ways to level an uneven subfloor:
- Cover it with panels of plywood or flooring-grade chipboard or hardboard. Lay hardboard smooth side uppermost if covering with adhesive, or rough side uppermost to give a better grip for underlay and soft floorings. Stagger the joins of the new panels so that they do not align with existing boards.

- Use a self-levelling screeding compound. This comes as a powder that is mixed and poured over the floor, where it fills any imperfections and finds its own level to set in a perfectly flat finish.
- Dig up and rescreed. These are both jobs that should be carried out by a professional.

Removing old floor tiles

To remove old cork and vinyl floor tiles, start with a loose tile or one in the middle of the floor and lever it up using a flexible scraper. Remove all the tiles one by one, then use a hot-air gun to soften the adhesive left on the floor so that you can scrape it off. Linoleum can be removed similarly – you may find a floor scraper, which has a long handle, useful for the job.

To remove old quarry and ceramic tiles, you will need to lever them up using a hammer and bolster chisel. Wear protective clothing, such as gloves and goggles, as shards of tile can be very sharp.

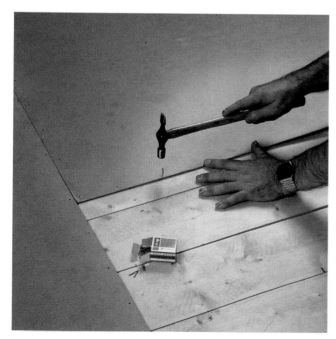

Covering an uneven subfloor with sheets of flooring-grade hardboard or chipboard will ensure the best result when it comes to laying the final floorcovering.

It's a shame to remove or cover up original flooring tiles unless you really have to. Consider restoring them if possible using proprietary restoring products.

If you find that draughts and dust are coming up through the gaps between your old floorboards, try filling them with fillets of wood or with papier mâché.

Dealing with old floorboards

Old floorboards are likely to be uneven and may have been cut to install or repair pipes and cables. If they are to be covered with a soft or resilient flooring, the ridges will show through as wear in the new flooring, even with an underlay. So, whether you intend to make a feature of the floorboards or to conceal them, you need to make sure that they are in good condition.

Fill wide gaps with fillets of wood glued on either side and tapped into position, flush with the neighbouring boards; and fill narrow gaps with papier mâché made from paper and water-based PVA (white) glue. Repair or replace any damaged boards. Punch down nail heads – at least 3 mm (^{1}/$_8$ in) below the surface if you are going to sand the boards – and screw down any loose boards (first checking beneath for cables and pipes that you might damage).

Old floorboards can also be turned over – the underside should be smooth and the wood aged and seasoned. Call on professionals to do this, as it is a major job and will involve removing and replacing skirting boards (baseboards).

If leaving floorboards uncovered, you could either sand them back for a light finish or coat them with a dark stain or paint to suit your room scheme.

Hard flooring

Laying a hard floor is usually a job for the professionals, since the materials are heavy to handle and require precise fitting. However, there are some types that you can fit yourself.

Brick

Similar to terracotta tiles but thicker, floor bricks are effective in halls, living areas of period homes and conservatories, but impractical for kitchens. Brick is crumbly and porous if not sealed.

Tiles

Floor tiles come in a huge range of colours and styles:
Ceramic These work well in halls, kitchens, dining rooms and conservatories. They are thicker and stronger than ceramic wall tiles, and without the high glaze that could make them slippery. Never lay wall tiles on floors.
Quarry Made of tough clay in various earthy colours, quarry tiles look good wherever a rustic look is required. They are usually porous, so may need sealing. If left untreated, quarry tiles have a matt appearance. For a more shiny finish, treat them with linseed oil and turpentine.
Terracotta These tiles can be brittle. Buy recycled ones from a salvage company or new ones with a slightly 'distressed' surface.
Encaustic Also called Victorian tiles because they are often found in entrance halls of this period. Encaustic tiles have a pattern, usually geometric or heraldic, constructed as an inlay that never wears.
Mosaic Like mosaic tiles intended for walls (see page 59), these tiny clay tiles are supplied in sheet form and require sticking down and grouting.

Concrete

Concrete is common as a subfloor in modern homes, but makes a hardwearing top floor if smoothed (but not polished) and sealed with special non-slip resin. Concrete can be laid as slabs or poured on site, coloured with pigment and even embedded with decorative materials.

Stone

Original stone flagstones are still found in period homes, which proves their long-lasting quality. Nowadays, many types of stone make practical and attractive flooring. A cheaper option is reconstituted stone slabs, as sold by garden centres, which can be used indoors if they are adequately sealed.

Slate

Slate floor tiles are usually slightly roughened (riven), making them non-slip and safe underfoot, but not so easy to clean. Ensure that the slate is flooring quality, not for roofs or wall cladding.

Wood

The hardwearing resilience and warmth of wooden flooring makes it ideal throughout the house.

In many older houses, suspended floors are made from floorboards fixed to the joists. Old boards can be restored or turned and relaid (see pages 76–77). Wooden flooring can also come as wood-block, laid in herringbone patterns, or intricately patterned parquet. If you have either type, you may have to do some restoration work, but do not paint an old parquet floor.

Wood strips, panels and veneers are alternatives to solid wood. They are constructed of a thin layer (veneer) of wood on a cheaper backing. Some are suitable for laying by non-professionals. Do not confuse manmade 'wood laminate' with real wood.

Laminates

These include woodgrain effects (often called wood laminate), marble and a host of individual designs. They need no sealing or polishing and can be cleaned with a damp cloth. They are suitable for most rooms except bathrooms, where they could swell and start to lift if saturated – check with the manufacturer.

Plywood

Plywood can be used as a floor in its own right, as well as for levelling a subfloor. A birch- or maple-faced plywood can look very stylish in a modern living room.

Some of the options

Tiles offer lots of decorative options and are probably the most commonly used type of hard flooring, but it's always worth considering some of the alternatives.

Terracotta is a popular traditional flooring. In the summer it is cool underfoot and, with the help of underfloor heating, in winter the tiles absorb the heat and keep warm for hours.

These rich red antique terracotta tiles have been rescued from chateaux, manor houses and even barns in Burgundy, France, and are understandably expensive.

Reminiscent of Roman mosaics, these stone mosaic tiles are ideal on both floors and walls – the greater the surface area covered, the better they look.

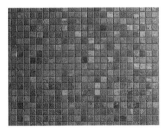

Stone flooring is ideal for a traditional rustic-feel room, but can easily be made more formal or contemporary with the addition of oversized modern stone pieces such as raw chunky side tables.

Pros and cons of hard flooring

For
- Hardwearing
- Looks good and even improves with time
- Easy to maintain and stain-resistant if sealed
- Good option for allergy sufferers, as less dust build-up than carpet

Against
- Unyielding underfoot
- Noisy, so rarely practical in a flat or apartment – many leases stipulate soft floorcovering for all floors in flats or apartments
- Heavy. The subfloor or joists may need strengthening, especially upstairs. May become slippery when wet

Real slate is formed over thousands of years, and no two pieces will be the same. Use it in the kitchen for a long-lasting flooring. Varnish regularly to keep it looking good.

Oak, a straight-grained wood characterized by distinctive flecking, is a particularly good material for flooring. Wood floors are a healthy alternative to carpet, as there is no build-up of dirt.

Facelift for a wooden floor

If you find a wood floor concealed beneath old floorcoverings, why not refurbish it? Sanding is an option for old wooden floors in good condition, while modern paints can bring new life to wooden and other types of floor.

Sanding

If you have an existing wooden floor, you may be able to restore it to its natural beauty by sanding. This involves a lot of physical work, but the rewards can be tremendous. A sanding machine makes the job much easier, although you may need to finish off some areas by hand.

Ensure that the floor is solid wood, since your sander could pulverize wood strip, wood mosaic panels, parquet or other veneers. Sanding is a very noisy business, so work only during sociable hours and warn your neighbours – they would probably prefer to go out for the day!

You can hire sanders for the day or weekend. Insist that the staff show you how to use the machines and provide a full set of instructions. You will need:

- A large drum sanding machine
- A small belt or orbital sander for the edges, or a sanding attachment on a drill
- Coarse, medium and fine abrasive strips for each machine
- Ear defenders, goggle and face mask

Before you start, deal with any loose and damaged boards (see page 73). Sweep up all dust and debris, seal the door with masking tape to keep dust from the rest of the house and open the windows so that you are working in a well-ventilated space.

1 Wearing the safety equipment, begin with the big sander and a coarse abrasive sheet. Work diagonally across the floor, taking care not to knock the skirting boards (baseboards).

2 Change to a medium-grade abrasive and sand parallel with the direction of the boards.

3 Finally, switch to a fine-grade abrasive, again working in the direction of the boards. Empty the dust bag frequently, and vacuum up wood dust from the floor.

4 Use the small sander or the sanding attachment on your drill to deal with edges and corners. Again, work through the different grades of abrasive paper.

5 Finish by sanding by hand any areas that the machines have missed.

You'll need to hire a sanding machine to strip original floorboards. It's noisy, dirty and very hard work, but the end result will be worth all the effort.

6 Vacuum thoroughly, including window sills and crevices between the boards. Wipe over the floor with a soft cloth soaked in white spirit, to remove any remaining dust.

Painting

Floorboards that are in poor condition or heavily filled can be coated with floor paint, which presents endless possibilities for colour use and design. You can create a geometric pattern or an eye-catching border, simulate a rug or give your floor a faux marble or tile effect – the possibilities are limitless. You can also transform a concrete or cement floor with industrial floor paints. A chequerboard effect can look stunning.

1 Start in the corner furthest from the door so that you have a clear exit and begin by applying a thinned coat of varnish to the whole floor. Allow to dry.

2 When marking out your design, use chalk and a rule to keep lines crisp, and low-tack masking tape to delineate the edge of each colour as you work.

3 Now apply the paint, making sure that your brushstrokes go in the direction of the wood's grain (length of floorboards). For any special effect, such as marbling, do a trial run on some spare wood first.

4 Remove the masking tape carefully and allow each coat to dry thoroughly before protecting all paintwork with several topcoats of varnish.

Reviving vinyl and linoleum

Special makeover paint can bring new life to tired vinyl or linoleum. Why not paint a simple border to outline architectural features, or define kitchen units or bathroom fittings? Alternatively, you could try using stencils and stamps (see page 32) – for example, print a Victorian tiled look with sponges cut into geometric shapes or stencil a unique patterned flooring. There is no need to varnish the floor.

Staining and sealing

You can alter the colour of timber floors using a woodstain followed by a varnish to seal, or a varnish stain, which combines the two jobs. Both oils and varnishes darken or yellow wood slightly, and woodstains can come up much stronger than on a shade card, so test first on a spare piece of sanded wood or in an unobtrusive corner. An undercoat of varnish thinned with white spirit will dilute the effect of a stain. Each finish will need several coats. Lightly sand between coats to key the surface for the next coat, and wipe away dust with a cloth and white spirit. For safety reasons, always use a matt or semi-matt varnish on floors to avoid creating a slippery surface.

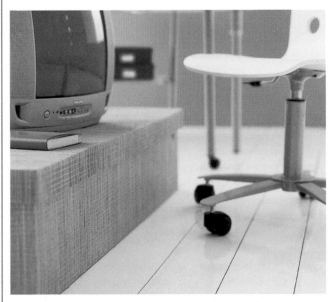

Painting old floorboards is an easier option than sanding and is recommended if the boards are in poor condition. Choose a light, neutral colour for a modern look.

Laying a laminate floor

Real wood is an expensive option, so if you don't have an existing floor that you can restore or you don't want the bother of stripping floorboards, a popular and relatively inexpensive alternative is a wood laminate, or floating, floor.

What is laminate flooring?

Laminates are formed by fusing an image of a natural material like wood on to a backing board of MDF (medium-density fibreboard) or condensed chipboard and protecting it with layers of transparent laminate so that it looks like a hard floor finish. Laminate boards resemble floorboards, but are only about 1 cm ($\frac{1}{2}$ in) thick. They are tongued and grooved to fit together easily and are glued or clipped to each other, but not to the surface beneath – hence the term 'floating floor'. Wood laminates come in a wide variety of colours, from blonde ash to dark oak or rich cherry, and in different grades to suit different areas of the house.

Laminate flooring boards are supplied in packs that cover a calculated area, usually in the region of 20 sq m (215 sq ft). Divide the area of your floor space by the area covered by the pack to give you the number of packs you need, remembering to allow for wastage.

The pros and cons

A laminate floor is easy to clean and non-allergenic, but as noisy underfoot as most hard floors. If you are laying over a hard floor, and especially in an upstairs apartment, the insulating membrane usually supplied with the laminate may not be enough. You will need to consider some form of extra insulation beneath the laminate or even insulating between the joists. Laminate can be laid over almost any other type of flooring, and if you predict problems with noise, you might consider laying it over a ready-fitted carpet, although this is not ideal.

Although laminate flooring is tough and hardwearing, it should not be saturated with water. Mopping the floor with excess water could cause the boards to swell and push up. Use a damp mop or cloth only, and always wipe up any spills immediately.

Tools and materials
- Strong pair of scissors
- Foam insulating membrane
- Steel measuring tape
- Laminate boards and 2 cm (¾ in) quadrant moulding (quarter-circle wood trim)
- Adhesive, in a glue gun
- Wooden or plastic spacers, 6 mm (¼ in) thick
- Set square or straight edge
- Pencil
- Craft (utility) knife
- Hand tenon or crosscut saw
- Tamping block and hammer
- Damp cloth

A modern alternative to a real wood floor, wood laminate flooring is relatively easy to lay, comprising tongue-and-groove boards that are glued or clipped to each other.

How to lay laminate

The following instructions are for boards that need gluing, which is the traditional method, but the principles are the same for the newer glueless installation systems.

Before you start, ensure that the subfloor is sound and flat (see page 72). You can remove the skirting boards (baseboards) if you like, although this is not essential and they can be fiddly to replace.

1 Cut the foam insulating membrane into lengths and cover the existing floor with it, abutting lengths as necessary and fitting it neatly into the corners.

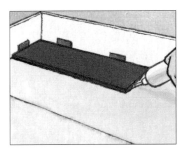

2 Start in the corner furthest away from the door. Lay the first laminate board against the wall, at right angles to any existing boards. Spread glue sparingly on the end tongue and slot on a second board. Work down the length of the room, inserting 6 mm (¹⁄₄ in) thick spacers between the boards and the wall to create an expansion gap. This allows the wooden floor to expand and contract with temperature changes, without buckling.

3 To complete the first row with a part-board, carefully measure and mark the board with a set square or straight edge and pencil. Score the mark with the craft (utility) knife and cut cleanly with the saw.

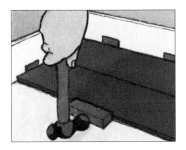

4 To lay the second row, apply the adhesive along the length as well as the end tongue. Tap the boards home with the tamping block and hammer. Wipe off any excess adhesive with a damp cloth immediately.

5 Continue laying the floor, length by length, across the room, staggering the joins to help strengthen the flooring and to avoid a noticeable line of board ends. You can achieve this by always using an offcut from the previous row of boards to start the next row. You may have to cut the last row of boards lengthways to fit – remember to allow for the 6 mm (¹⁄₄ in) space between the floor and the wall. Use a profile gauge (see page 56) if necessary to mark and cut boards to fit around pipes.

6 Leave the floor to dry, and make sure that no one walks on it before the glue is completely set.

7 Remove all the spacers. Cut lengths of 2 cm (³⁄₄ in) quadrant moulding (quarter-circle wood trim) to fit around the perimeter of the room, to cover the expansion gap between the laminate and the wall. Glue or nail the skirting (baseboard) face of the quadrant only, so that the floor can still move, or 'float'.

Handy hints

- Atmospheric moisture affects wood, so after buying your laminate flooring, lie the boards flat in the room in which they are to be laid for at least a couple of days.
- A damp-proof membrane should be laid to provide a barrier to residual moisture in a solid subfloor, but must not be used over existing wooden flooring.
- Many manufacturers supply quadrant moulding (quarter-circle wood trim) to match their boards.
- Laminate boards that are locked together can be easily removed.

Laying vinyl

Rigid vinyl is available as tiles and complementary borders, but can be quite unyielding underfoot. Flexible vinyl, available in sheets and tiles (including self-adhesive), can be 'lay flat', which moulds itself to the subfloor and needs sticking down only around the perimeter, or cushioned, which gives extra bounce.

Vinyl tiles

Vinyl tiles are easy to lay yourself – they simply butt up against each other and don't need grouting. Some require fixing with adhesive, while others are the 'peel-and-stick' type. Being smaller, vinyl tiles are easier to handle than vinyl sheeting and can be easily cut to fit (see page 67). They allow you to create borders and panels for design interest, but the joins between tiles mean that dirt and spillages can penetrate. Vinyl tiles range in quality from vinyl coated to solid vinyl.

Vinyl sheeting is more practical than vinyl tiles in a bathroom, as it provides fewer joins through which dirt and spillages can penetrate and ultimately lift the flooring.

Tools and materials

- Steel measuring tape
- Chalk line (see page 66)
- Self-adhesive vinyl tiles
- Pencil
- Straight edge
- Cutting board
- Heavy-duty craft (utility) knife
- Profile gauge and sharp scissors

Floor tile planning

Make a scale drawing of the room to be tiled on squared graph paper, where each square represents one floor tile. Mark on the plan all alcoves and other irregularities. In a kitchen, much of the floor may be taken up with fixed units, so include these to scale as well. Shade in the design for the pattern and to determine the quantities of tiles of each colour required.

How to lay self-adhesive vinyl tiles

The following instructions are for self-adhesive vinyl tiles, which are the easiest to lay, but the principles are much the same for all tiles (see page 66).

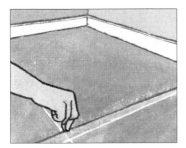

1 Find the true centre of the room by marking the centre of two opposite walls. Snap a chalk line between the two points. Do this between the other two walls and the centre of the room will be marked as a cross.

2 Lay out the tiles in a 'dry run' so that any problems come to light before you start sticking them down.

3 Peel the backing off the first tile and, aligning it with two arms of the central cross, press it down lightly and firmly. Work from the centre out towards the edges of the room, carefully butting the tiles up to each other and checking pattern matches.

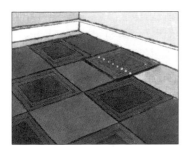

4 You will probably have to cut tiles to fit around the edge of the room. Lay a tile (do not remove the backing) over the last full tile, and mark with the pencil and straight edge where the cut is to come.

5 Cut the tile on a cutting board, using a heavy-duty craft (utility) knife against the straight edge.

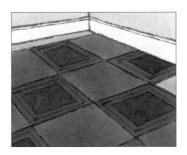

6 Peel off the backing and lay the tile in position with the cut edge against the wall, not abutting the last full tile.

7 To cut a non-rectangular part-tile (walls are seldom perfectly straight), you will need to turn the tile over to mark the fit, or carefully measure the space. For awkward shapes, especially curves, use a profile gauge (see page 56) to template the outline. A fiddly shape may be easier to cut with scissors than with a knife.

Available in styles that provide the look of hard tiles or wood for a fraction of the price of the real thing, most vinyl flooring can be laid by a competent do-it-yourselfer.

Soft flooring

The appeal of soft floorcoverings is their warmth and quiet. As well as conventional carpets and rugs, this category includes 'natural' floorcoverings made from a variety of grasses and vegetable fibres.

Natural-material floorcoverings

Grasses and vegetable fibres have been used for centuries as floorcoverings. Most are strong and hardwearing, and provide a smart alternative to carpets, used wall to wall or in rug form on wooden or concrete floors. They are, in the main, rather harsh or roughly textured, and not really suitable for floors where children might be crawling, or where it is usual to walk around barefoot.

Environmentally friendly and durable, natural floorcoverings are a popular choice in modern homes for their textural interest, neutral colours and ability to work well with other natural materials.

Natural-material floorcoverings stain permanently if water is spilt on them, so they are not suitable for use in the kitchen or bathroom. In addition, some can be slippery and are not suitable for use on stairs. These types of flooring (apart from rush matting) are laid in a similar way to carpet, although some cannot be stretched. They may need to be laid over underlay, although some of the latex-backed varieties can be stuck to the subfloor. When used as a close-covered carpet, all need to be professionally laid.

Coir

Formerly used only for doormats, coir, or coconut fibre, is now woven into coarse but interestingly textured flooring, which is sometimes dyed bright colours. It is usually supplied backed, but some narrow widths – for use in halls or corridors – are unbacked.

Seagrass

A tough tropical grass spun and woven into coarse matting, seagrass is usually left undyed, in a random mix of yellow, beige, green and a hint of russet. It is hardwearing, but not suitable for wet areas. Extra loose-laid mats are recommended as protection from furniture castors and wear in front of sofas. If laid on stairs, the grain should run parallel to the tread.

Sisal

Sisal is a whitish, stringy fibre that is woven into attractive herringbone, bouclé and ribbed patterns in a range of colours; it can also be stencilled. Sisal is not recommended for kitchens or bathrooms. A sisal and coir blend gives a finer weave and texture that is less scratchy underfoot. Sisal can stretch (even with a backing) and may need occasional refitting (a light spray of water can help shrink it back into shape). Like rush matting, sisal is inclined to shed fibres.

Other natural materials

Hemp is soft underfoot, but not as strong as some natural-material floorings. It is woven into attractive herringbone, ribbed and chevron textures. Jute looks similar to sisal and has a textured, ribbed weave. It is best used in areas of fairly light wear. Paper twine is twisted from strong, unbleached paper and made into rugs. Rush matting is hand- or machine-plaited into strips, sewn together to form standard-sized mats.

Some of the options

Coir, seagrass and sisal are widely used, but there are others. The natural fibres are often woven into a variety of patterns and can be bleached or dyed.

The 100 per cent coir from which this herringbone flooring is made is softened by soaking it in lagoons for up to ten months before it is woven.

A seagrass floorcovering is a fantastic alternative to carpet. For best results, do not use in the kitchen or bathroom, but hallways and stairs are fine, as it can take constant traffic.

This sisal flooring looks fantastic in any room, but for the living room or bedroom, combine with sheepskins or rugs to protect bare feet from the rough texture of the fibres.

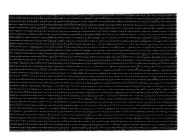

A bitter chocolate version of sisal flooring, this dark woven matting makes a striking change in any room and would look stunning as a rug on a wooden floor.

Pros and cons of soft flooring

For
- Warm and cosy underfoot, although some natural-material floorcoverings are much less so
- Good insulation from cold and draughts
- Helps deaden sound
- Natural-material floorcoverings are environmentally friendly

Against
- Vulnerable to staining, even if treated
- Impractical in areas that get wet
- Some natural-material floorcoverings can be slippery, so are highly unsuitable for use on stairs

The abaca fibre from which this natural-material flooring is made comes from the banana plant family, and gives a much smoother and more comfortable surface than many other woven natural-material floorcoverings.

This softly golden natural-material floorcovering made from jute gives a warm glow to any room, although it will not withstand very heavy wear and tear.

Working with colour

Since adjacent hues on the colour wheel are pleasing on the eye, blues and violets are always a successful combination. Touches of pale gold prevent the scheme appearing too bland.

Choosing colours

Since the visible spectrum is huge, colour possibilities are almost limitless. A red scheme, for example, can vary from crimson to scarlet, or terracotta to shocking pink, and it can be adapted for very different living areas. In a formal dining room, lit for evening entertaining, rich crimson tones can be very romantic; in a sunny family kitchen, the warm tones of terracotta tiles and accessories can appear friendly and inviting. In a bedroom, combining deep and pale tones of rose in the wallpaper, bed linen and curtains can produce a country-style look. For something unusual in the bathroom, consider teaming shocking pink flooring with wallpaper decorated with a sugar pink design.

For a colour scheme that is strikingly different, you could decorate areas of your home in bold blocks of primary colours. Think how a spacious and sparsely furnished sunny hallway or living room could look warm and welcoming with a daffodil yellow ceiling, bleached wood floors, orange curtains and turquoise cushions and accessories set against deep blue-grey walls. But this is only one possible colour scheme from the myriad of options available – use the colour wheel to help you select suitable hues.

Colour inspirations

You might like to build an entire colour scheme around a single item such as a rug, a ceramic piece or an item of furniture. For example, your imagination could be inspired by a wooden dresser painted a rough apple green colour. This colour contrasts wonderfully with brown, and so is ideally suited to a room with a natural wood floor, unpainted brick walls and bare pine furniture. The effect could be further heightened with additional touches of the same green – in cushion covers, painted window frames and a few accessories.

Stencilling lively motifs on walls and furniture (see page 32) is another way to add bold splashes of colour. The subject of the motif could vary from room to room – to reflect changing uses and activities – or stay consistent throughout, and so act as a unifying decorative element. In this way, you can begin to build bridges between the decorative approaches in different areas of your home.

The importance of your choice

As you will see in more detail on the following pages, your choice of colours can make a room seem smaller or larger, cramped or spacious, warmer or cooler, brighter or more subdued. Knowing which colours to

Warm red suggests intimacy and can be effective in the bedroom, providing you take care not to swamp the room and make it too dark. Neutrals always provide a good counterbalance.

intimacy, use plenty of deeper tones and bold colours in the warm red range, and include lots of colourful patterns and textures.

Decorating in rich, bright colours will help to enliven a dull, predominantly shady room, while subtle colours and darker tones will subdue a sunny room. Warm, bold colours and strong contrasts create a lively and inviting atmosphere, while cool tones and subtle contrasts create more of a calm and relaxed environment. The effects created, however, do depend on the intensity of the colours you choose, and whether you use them in their pure forms or mixed in the form of tints, tones and shades.

Despite all these 'rules', using colour is a very personal matter. Like music, colour varies enormously in the mood and atmosphere it creates. So allow your own style to develop as you plan and have confidence in whatever colours appeal to you. Remember, if you have a favourite colour you are wary of using all over, there are plenty of less obtrusive ways to incorporate it in your scheme – in the accessories (see page 103), on a feature wall or by way of an item of painted furniture.

select to achieve the precise effect you want is not always an easy task. The secret of success lies in the skill with which the fabrics, furniture, wallpaper and decorative objects are selected and mixed. So it is vital to look, learn and plan as much as you can before actually buying anything.

Developing a colour sense

Everyone can develop an eye for successful colour combinations, and become aware of how colours and patterns work together. To explore your own taste and style, observe the colour combinations in everyday life. Take note of your surroundings wherever you are and study window displays, magazines and advertising brochures. Make notes of what you like and don't like, cut out interesting pictures and make yourself a mood board (see page 106). You will soon develop your own personal preferences and discover the combinations you find most pleasing.

Learning the 'rules'

Besides consulting the colour wheel (see page 92), there are some broad guidelines to help you choose colours for your home. For example, to emphasize spaciousness, choose a limited range of white, pale pastels or shades of blue. To suggest cosiness and

Whites and neutrals suggest spaciousness and airiness, but can be too dull used alone. A touch of pale blue in the accessories can lift and complete the colour scheme.

Greens, blues and violets

Greens, blues and violets are on the cooler side of the colour wheel, but often appear far from cold. Many of them mix easily together and there are fewer strident colours, making them a 'safe' choice on which to base a scheme.

Green

Lively and invigorating, true green is the 'balance' colour, halfway between the warm and cool colours of the spectrum. In all its many guises, which allow it to be a restful, vibrant or zingy colour (compare grey-

The use of grey-green in a decorative scheme results in a subtle and sophisticated look that can help soften the potentially stark appearance of some modern décor.

Use greens, blues and violets to...

• Alter the perceived dimensions of an area. Cool colours, especially the paler values, make walls and ceilings appear to go away from you, thereby making the space seem larger than it actually is. For example, you can make a narrow area such as a landing look wider.
• Balance the feel of a room that receives a lot of light (see page 104).
• Create a feeling of airy spaciousness.
• Add contrasting cool accents to a warm colour scheme.
• Make an unattractive feature fade into the background.
• Create a calm and relaxed environment.

green with apple green or citric lime), it's an easy hue to live with and a popular choice for decorating schemes. The colour of nature and guaranteed to suggest 'landscape', green is a particularly good choice for a city apartment or dull townhouse to bring a sense of the outdoors indoors.

Some greens can be very cold, but mix well with contrasting warm colours to ward off the chill. Pale tints of green create a very spacious look for small bathrooms and bedrooms, and when green goes towards blue, it becomes minty, which is very refreshing in a kitchen, and on to aqua, an ever-popular colour for marine-themed bathrooms.

Dark shades of green are rich and sophisticated – deep malachite or forest green work well with both traditional and modern styles of décor. Used as the main colour, they work best in rooms with bright natural light and need balancing with white or cream and reflective surfaces. Alternatively, use dark green as an accent colour with pastel pink or yellow.

Blue is a versatile colour that combines well with most other colours. Here, pale blues are teamed with white and pale violet for a particularly pleasing scheme.

The greyed tones and shades of blue can be subtle. They look very effective teamed with crisp neutrals – white or cream – or warmed up with contrasting accents of orange, yellow or bright pink.

Violet

In its strongest value, a regal purple, this is a vibrant and demanding colour that needs neutral contrasts. It works particularly well with its adjacent colours on the wheel – blue, blue-green and pink-violet.

The pale tints of lilac and lavender are delicate and feminine, giving a romantic feel to a bedroom or bathroom, or a sophisticated look to a living room. Lilacs teamed with lime greens are very contemporary.

Greyed lighter tints give subtle heathers that are changeable in different lights, while the darker shades yield rich plums and aubergines. These work well in period settings, teamed with golds, cream and pale yellow in drawing rooms or dining rooms.

Greyed values of green can be very subtle. Sage, rosemary and olive greens are elegant, and look good in country-style drawing rooms, classic hallways or modern dining rooms, but, like yellow-greens, need to be well lit so that they don't look gloomy at night.

Blue

Blue is the colour of harmony, peace and devotion. It is associated with the sky and wide vistas, and creates an impression of space. But because it is basically a cool colour, use it with care. Blue is fairly low in reflective value, and will diffuse and soften strong sunlight, calming down over-bright rooms. However, clear bright blue can also be very cheerful, which makes it ideal for children's rooms or a basement kitchen.

The pure values of blue, especially as it starts to go towards green – peacock or turquoise – can be very demanding, so use these in small amounts if the room is small. Lightened to aqua, its freshness works well in bathrooms and kitchens.

Another decorative scheme proving that pale lilac and lavender combine well with blue, working together here to give a romantic yet sophisticated feel to a seaside room.

Using harmonious and contrasting colours

The colour wheel is an invaluable tool. It illustrates the natural neighbours and contrasts of each colour, and is a ready reminder of how colours relate to each other. Use it as a guide to which colours work well together as well as to inspire unusual combinations.

Harmonious colours

Harmonious colour schemes are based on neighbours on the colour wheel and always create a relaxing atmosphere. Pick colours that directly adjoin each other on the wheel or are very close to one another, for example choose three or four colours of a similar value that originate from a single primary colour, or team a primary colour with one or more of its neighbours for a more striking effect.

A monochromatic scheme

A tighter variation of a harmonious colour scheme is to work within a single colour, from its lightest tint to its darkest shade, but without straying sideways into the next segment of the colour wheel. This results in a monochromatic, or tonal, scheme.

Paint colour cards comprising strips of different tints and shades of the same colour are readily available from paint stockists and are ideal for helping build a monochromatic scheme. There will probably be several cards to cover all the possibilities within a single colour, giving you a wide range to play with. You might use the palest value of the colour for the woodwork, a slightly stronger tint for the ceiling, a mid-tone for the walls and a deeper shade for the floorcovering. Look for patterned wallpapers or fabrics that use different tones of the same colour to help prevent the potential boredom of using just one colour.

Most monochromatic schemes need brightening up with accent colours – look back at the colour wheel to see which hues contrast with your chosen colour.

A harmonious colour scheme like this one relies on using colours that are neighbours on the colour wheel – in this case soft, pale violets and blues.

Contrasting colours

Contrasting colours (also known as complementaries) produce colour schemes based on direct opposites on the colour wheel – red and green, yellow and violet, blue and orange. The special relationship between contrasting colours means that each brings out the best in the other. By choosing contrasting colours, you automatically team a warm colour with a cool one. If you focus on a strong hue of a warm colour (red, for example) for a minute or so, then look away, you will see an 'after colour' of its opposite (green) – the eye plays strange tricks!

When used together, contrasting colours can be stimulating and exciting, but if the colour values of both are too bright, the effect can be disturbing.

The pink of the sofa and scatter cushions are a welcome addition to this living room, bringing some life and warmth to the understated, neutral colours.

However, contrasts include the colours in all their tints, tones and shades – so a scheme based on sugar pink and olive green or turquoise and terracotta, for example, will still be contrasting, but not overpowering. Such 'diluted' contrasts work well in rooms where you don't want to relax too much, but don't want to be continually stimulated either.

A contrasting colour scheme can also be split-complementary, by using a primary colour combined with the two tertiary colours that flank its opposite on the colour wheel – such as blue with red-orange and yellow-orange.

You can create an equally successful contrasting colour scheme without actually using direct opposites, as long as you combine warm and cool colours, for example blue with yellow, orange with jade-green or green with violet.

Don't use your chosen contrasting colours in equal amounts in your decorating scheme, as the colours will all fight for attention and cancel each other out. Instead, feature on one predominant colour. You could introduce a third colour if you like, but certainly use no more than three.

A few accessories in hot orange introduce interest and touches of warmth to this monochromatic room scheme, which showcases the tints and shades of violet.

Working with neutrals

The only three true neutrals are white, black and the in-between of grey, but there are also the accepted neutrals – or naturals – the broken whites, creams, beiges, taupes and soft browns of stone, wood, undyed linen or wool.

Just neutrals

Neutral colours on their own create a harmonious scheme, and are often effective for a very contemporary look, but the pale neutrals can be difficult to keep clean. A basic rule of interior design is to consider the practical alongside the aesthetic. It would be impractical, for example, to decorate a hallway in pale cream with an off-white carpet and expect it to cope with a regular flow of visitors, muddy feet and even bike or baby-buggy wheels. On the other

Subtle contrasting accent colours in warm orange and brown, teamed with rough-textured fabrics and smooth, polished wood surfaces, add interest to this neutral living-room scheme.

hand, such a pure and simple colour scheme would provide a pleasing atmosphere of calm in a main bedroom or a living room.

Black and white, especially used together to create a bold pattern, is the most stimulating and disturbing of neutral combinations, and needs to be used with care.

Importance of texture

If using neutral colours on their own in a room scheme, a good mixture of contrasting textures – smooth, rough, hard, soft, shiny and matt – is important for providing visual interest and stimulation (see page 124), to avoid the scheme looking too bland.

Untreated wood, bamboo, rattan, wicker, bare brick walls, rough plaster, natural floorcoverings, unglazed terracotta, paper lampshades, undyed muslin and faux fur are just some of the rough textural possibilities in neutral colours. Smooth and shiny textures available in neutral colours include stone and ceramic surfaces, glossy leather furniture, laminate flooring, polished wood and gloss paint.

A pale neutral room scheme, devoid of any other colour, creates a very contemporary, elegant and restful grown-up look, not really practical for homes with small children and pets.

Neutrals in a mixed scheme

On their own, neutrals can be rather boring, so it is a good idea to enliven them with contrasting accents (see box, right). Follow the basic rule of relating these accents to the overall style of the room, and choose warm or cool colours according to the atmosphere you want. With a predominantly neutral scheme, you could use several different accent colours, mixing cool and warm, or different tonal values of one or two colours.

Neutrals can also make a major contribution to other colour schemes. They form a good background for printed fabrics and wallcoverings, and provide an element of quiet contrast, often unnoticed.

When is white not white?

Many so-called neutrals, especially near-whites, are actually very pale tints of a 'proper' colour. Creams can have hints of yellow or pink, while 'greiges' (grey/beige) and browns can have undertones of warm orange or cool blue. These origins may not be apparent until placed near another colour or neutral. Pure white woodwork can make an off-white carpet look dirty, for example, and a subtly shaded tweed fabric can just appear grey in the wrong company.

A neutral room scheme of whites and off-whites provides a good backdrop for dark wood furniture and a subtly coloured but boldly patterned wallcovering.

Adding accent colours

Use accessories to add accent colours, together with a patterned fabric, wallpaper or flooring. The dominant colour should relate to the main scheme, with secondary colours echoed in other plain accessories. To vary the tonal balance, add warm accents to a cool scheme, cool ones to a warm scheme and strong, rich colours to neutrals. Look at the following ways to introduce accent colours:

In a living room or dining room:
- Cushions or padded seat covers
- Rugs
- Plants and flowers
- Pictures, prints and photographs (including interesting frames)
- Glass or china
- Lamp bases and shades
- Table covers
- Candles and holders

In the kitchen:
- Pots and pans
- China, pottery and glass
- Tea towels, aprons and oven gloves (pot gloves)
- Pictures
- Herbs in decorative pots

In the bathroom:
- Towels and bath mats
- Boxes and baskets for toiletries
- Pictures
- Plants

In the bedroom:
- Bedside lamps and shades
- Pictures
- Rugs
- Cushions and throws

Note: Bed linen should be an integral part of the scheme, not an accessory.

Creating a colour scheme

The colour scheme you choose for decorating any area of your home may be determined by a number of factors – the size and shape of the room, existing furnishings and furniture, the lighting, the mood you wish to create and the function of the room.

Working with existing features

In an ideal world, you would be able to create a colour scheme with absolutely no constraints at all. But in reality, most of us have to work a new colour scheme around an existing item such as a bathroom suite, kitchen units, curtains, a fitted carpet or a sofa. As a basic rule, choose one colour of an existing item and use it as the cornerstone of your colour scheme.

Room size is an additional factor when it comes to colour schemes. For example, small rooms can feel claustrophobic if strong colours are used (see pages 96 and 110). The room's architecture and natural light may also affect your colour decisions.

Architectural factors

The architecture can suggest an overall style for the room, and any interesting features should be enhanced. An attractive fireplace is worth emphasizing – colour the chimney breast a rich dark tone to 'bring forward' a light-coloured fire surround, or paint it a pale colour to contrast with a dark surround. If the room has beautiful windows, give them a very simple treatment to enhance their natural shape.

Natural light

The orientation of the room will affect the daylight that it receives. For rooms without much morning light, choose colours from the warmer side of the colour spectrum. For those bathed in morning sunlight, which are cooler and darker in the evening, try to work with the paler 'sunshine' tones. Rooms that are bright and warm from midday onwards work best with cooler, receding colours.

Function and mood

When choosing a colour scheme, you must also consider the purpose of each room and how much it is used. For example, do you have a busy living room with a lot of activity in a small space? If so, lots of strong colours may add to the confusion. Instead, plan a scheme that provides a fairly neutral background.

Don't make the mistake of picking a colour simply because it is currently in fashion – you'll soon regret it! Whatever your colour scheme, think about how

Attractive architectural features can be emphasized by clever use of colour. For example, this pale painted chimney breast contrasts with the darker walls around it to make the fireplace stand out.

If you have to work around the colour of existing furniture, make this colour the basis of your scheme and match it with various closely related tones for a harmonious look.

Handy hints

- Take fabric swatches with you when shopping. It is surprisingly difficult to recall exactly how light or dark a colour is.
- Colours look different according to the light, so it helps to examine all colours under the same lighting conditions.
- Large expanses of a single colour can look dull. Instead, match a variety of closely related tones with an occasional splash of contrasting colour to create a lively harmony.
- For successful colour schemes, incorporate a balanced mixture of primary colours, darker tones and lighter tones.
- Suit your colour scheme to the function of the room being decorated. A home office or study, for example, decorated in clean, simple tones offers the least excuse for eyes to wander, while a hall decorated in warm colours immediately appears welcoming.

frequently you may need to redecorate and how much you may want to spend. If you want to be able to create an entirely different atmosphere at minimal cost, keep the room's basic colour scheme fairly light and simple. You can then introduce strong colours with the curtains, blinds, painted woodwork, prints, plants, accessories and lighting – all of which can be changed with minimal disruption.

Balancing colours and patterns

It is important to relate the strength of colours and size of pattern to the scale on which they will be used. Bright colours and bold, jazzy patterns will look at least twice as strong on a large expanse of floor or wall, so always try to see as large a sample as possible. If you plan to paint a wall, use a tester pot on a length of lining paper and hang the painted paper in the room for a few days, moving it on to different walls to see how it looks in the varying light at different times of the day. If in doubt, choose a shade lighter than you think you want, since paint colours are notoriously difficult to assess from a small sample.

Combining patterns is one of the trickiest schemes to pull off successfully (see page 122). Use a mood board (see page 106) to experiment with different effects. It is often too broad a colour range that makes

pattern mixes look messy, so carefully study the colours that make up each design and aim for patterns that share a common palette.

Don't overlook the important part that neutrals play. Even a rich, colourful scheme will rely on white or cream for contrast, to emphasize the brilliance of the other colours and to avoid sensory overload.

It can be tricky mixing different types of pattern, so it helps to restrict pattern to small areas like cushions and to choose patterns with a common colour palette.

Making a mood board

Good colour schemes don't just happen –
they have to be planned. Since nobody
can 'carry colour' successfully in their
eye, this means taking samples of
decorating materials home, placing them
together and looking at them under both
daylight and artificial light.

What is a mood board?

A mood, or sample, board is a collection of visual ideas,
colour swatches and sample materials for a specific
project. The display of samples of all the proposed
furnishings gives the effect, almost, of a miniature
room, and professional interior designers always work
up mood boards for their clients.

It is both useful and fun to create one yourself when
planning a decorative scheme, as it's an excellent way
to visualize your colour scheme and design concept,
and avoid costly mistakes. The process of collecting
ideas allows you to crystallize your thoughts and
decide which ideas to develop further and which ones
it would be best to drop.

How to make a mood board

Start by browsing through magazines and catalogues
(they don't have to be on interior design) for
inspirational images and design ideas you might like to
consider for the room you intend to decorate. Cut out
any pictures you like and once you have some ideas in
mind and a starting point – this could be a main base
colour, a favourite texture or a patterned fabric that
you like – start collecting images and samples. Include
colour cards from paint charts; swatches of fabric for
curtains, upholstery and scatter cushions; pictures of
lighting and key furniture; pieces of wallcoverings and
samples of carpets or other floorcovering.

Handy hints
• Time spent planning the room to be
decorated and making a mood board allows
you to budget properly and helps avoid
making decorating mistakes.
• Bear in mind that a room scheme you've
seen somewhere else – perhaps in a plush
hotel or restaurant – that you want to copy
might not be totally practical for your lifestyle
or size of room. It is important to be
prepared to modify and come up with your
own ideas rather than slavishly copying
someone else's.
• Don't rush into completing your mood
board. It's not something to be compiled in
one go, but needs to evolve over a period of
time as your ideas take shape, and you'll
invariably change your mind several times
along the way!
• Don't think about one aspect of the room
in isolation. Everything should work together
– colours, pattern and texture; furniture; wall,
window and floor treatments; lighting.

Get yourself a large piece of cardboard or mounting
board and lay your samples next to each other on the
board. Arrange them in the general order of the room
– with floorcovering at the bottom of the board and
ceiling colour near the top – and make sure that the
materials are more or less in the correct proportion.
For example, wall and floor samples should be larger
than the upholstery or curtain fabric.

Paste your pictures and samples on to the board
and review how all the elements look together. If you
decide that something doesn't work, remove it and
keep looking for alternatives. It is better to spend a lot

Elements of a mood board

1	Picture	7	Blinds: overall look	13	Tub chair 2
2	Main wallcovering	8	Sofa 1 fabric	14	Wood flooring sample
3	Covering for chimney breast	9	Sofa 2 fabric	15	Rug sample
4	Ceiling paint	10	Single chair fabric	16	Contemporary fireplace
5	Curtain fabric	11	Wood sample for shelving	17	Cushion fabric
6	Blinds: wood type	12	Tub chair 1		

of time and effort at this stage, researching and compiling a mood board, rather than completing your decorating and then deciding that you don't like the overall look of your new room.

Putting it into practice

The starting point for the mood board pictured above is the rug (15), which is striped in golds, greens, yellows and terracottas. It will be placed on a wooden floor in a living room and is the centrepiece around which the rest of the mood board is put together, providing an inspiring starting place for colour and texture.

Wallcoverings The main wallcovering is soft yellow-green, with a bolder texture and deeper colour on a grasscloth for the chimney breast to enhance the modern fireplace. A paler yellow-green paint has been selected for the ceiling.

Window treatment A silky, textured fabric for curtains in a darker tone than the walls, but of a similar colour, has been chosen to enrich without distracting. Rattan blinds are to go under the curtains, providing privacy and filtering the light.

Upholstery Upholstery fabrics in contrasting textures in green, gold and terracotta complete the theme. The tweedy texture for the sofa cover contains some turquoise, which could be used, with terracotta, for extra accessories.

Decorative elements Paintings or prints can be introduced at the final stage to pull together all the elements of the design, including any accent colours.

Visual tricks with colour

As well as creating an atmosphere with colour, and visually heating or cooling a room, you can use colour together with pattern and light to play decorating tricks that appear to alter a room's proportions (see also pages 110 and 112) – far less expensive options than any structural alteration!

Lowering a ceiling

There are several options here. The most obvious is to paint the ceiling with a warm, advancing colour and the walls with a cool contrast. Alternatively, consider colouring the floor to match the ceiling, or at least use the same tonal value for both. If you prefer darker walls, then you can team them with a brilliant white ceiling, as it is a misconception that white ceilings always make a room look higher.

When choosing wallpaper, you could put a horizontal pattern on the walls, or divide them up using a picture rail or frieze. Other options include tenting the ceiling with fabric, investing in tall items of furniture or simply painting the cornicing/coving to contrast with the ceiling.

Adding height

There are a variety of useful techniques for dealing with a low ceiling, which can be tailored to suit any colour scheme. The first is to paint the ceiling a pale receding colour – sky blue is particularly effective. Otherwise, you could colour the ceiling a slightly lighter tone than the walls or match it to the background of any wallcovering. Using a light-reflecting finish on the ceiling may also be effective, but only if the plasterwork is in good condition. In addition, vertical stripes for wallpaper, curtains and blinds will help to direct attention away from a low ceiling, as will low-slung furniture.

Making a small room look larger

Pale receding colours – such as blues, greens and lilacs – and small patterns will create an illusion of space, as will light-reflecting shiny surfaces and textures, and mirrors. It is wise to avoid too much contrast in colour, choosing instead a harmonious, neutral or monochromatic scheme. Keep accessories simple, and avoid clutter.

Making a large room cosier

Warm, advancing colours and contrasting textures will bring instant warmth to a large room. Emphasize different features or areas with lighting and use bold patterns. Using complementary colours, for example curtains that contrast with walls and furniture that contrasts with the floor, will also draw a room in.

A large room can appear a little unwelcoming, but the use of warm, advancing colours will help to bring the walls in and make the room seem cosier.

This faux fireplace, in a contrasting colour that makes it stand out from its surroundings, helps distract attention away from the boxy shape of the room.

Camouflaging or enhancing features

Paint or colour unsightly features the same colour as the background – this is a particularly effective trick if you have unattractive radiators or built-in cupboards in your home. Otherwise, use receding colours, from the cooler half of the colour wheel, on both the feature and the background.

Conversely, paint or colour attractive features like an imposing chimney breast to contrast with the background, for example use neutral colours against a white background or an advancing colour (warm) against a receding one (cool). Otherwise, simply light your favourite feature dramatically.

Tricks with lighting

• A free-standing lamp placed in a corner can make a room appear larger, as can a central ceiling light because of reflection from the ceiling.
• Using table lamps makes a room feel more intimate and therefore smaller.
• Varying the height of the light sources in the room creates separate areas of light and shade, and makes rooms more interesting.

Adding width

This will really depend on the size, shape and purpose of the narrow area. Try painting or papering end walls in a bold, advancing colour or with a strong pattern, or if it's on a long corridor, paint or paper both long walls in a cool, receding colour. If there is a window at the end, try dressing it with floor-to-ceiling curtains in a companion fabric to the wall treatment. Other options include putting a well-lit, eye-catching piece of art or group of pictures on an end wall; hanging mirrors on one of the long walls; and making the floor seem wider with widthways stripes or a chequerboard effect. You could also try painting the skirting board (baseboard) to match the floor.

Making a box room less square

This is a very common problem, for which there are several very easy solutions that will instantly transform the space. Consider colouring one wall in a bold contrast to the other walls. Or, create a focal point in the centre of one wall such as a fireplace, opulent window treatment or bold drapes above a bed to contrast with the surrounding wall. You could even create a trompe l'oeil effect. If storage is an issue, use one wall for built-in furniture, decorated with contrasting panels or beading.

Painting the walls to match the sofa allows the natural light that floods through the imposing windows during the day to be the room's focal point rather than the angle of the walls.

Using colour in small rooms

Whatever main colour you choose for your decorating scheme, it will have greater dramatic impact on a small space than on a large one. Since you can decorate a small room relatively quickly and inexpensively, why not take the chance to experiment? You can try more unusual colour combinations and explore different textures and effects.

Illusion of spaciousness

Very pale colours make surfaces recede and seem less noticeable, thus creating the impression of spaciousness. The simplest way to ensure an uncluttered and restful effect in a small room is in fact to limit the basic décor to white. However, although white is a good choice for small rooms, it can be stark and boring if used in its pure form. By using just a hint of colour, you can begin to make more of a statement. For a start, walls can be painted in one of the enormous range of almost-whites – from warm peach to soft, cool blue. Texture, too, can add interest and warmth to a predominantly white scheme.

Incorporate curtains, cushions, small prints or rugs as a way of introducing colours in occasional splashes. When the time comes for a new look, all you need to do is change the accessories. However, if you have a lot of furniture and accessories, it may be better to select plain fabrics and muted tones to avoid a cluttered feeling in the room.

Experiment with colour

Despite the oft-quoted guidelines, don't feel limited to white and the off-whites just because you are decorating a small room. Why not start with a strong colour, such as yellow or purple? There is no basic problem here, as long as you limit your colour scheme to different tints and shades of a single colour. You can

Using just a hint of colour in a small room can offer a good alternative to plain white. The use of muted colour can be continued in the soft furnishings to add additional interest.

even incorporate dark colours in a small room as long as you avoid creating too many colour contrasts or combining them with bright, richly coloured or patterned furnishings. If you include too many competing colours, a small room can end up looking cramped and cluttered.

If you are painting in bold colours, be aware that when applied to the wall, paint always ends up darker than the colour suggested by the swatch card.

Limiting the colour of walls, floor, furniture and accessories to a restricted range of pale neutral tints helps make a small room appear larger than it is.

Handy hints

- If you use dark colours in a small room, make sure that the lighting is adequate so that they don't look dull and gloomy.
- Since mirrors reflect both light and space, they can dramatically transform a small room. They are particularly effective in small spaces and corners, where they can be angled in specific directions to highlight interesting features.
- Experiment with indirect lighting, such as uplighters. It influences subtle colours and gives an extra dimension to small spaces, particularly in alcoves and corners.
- Introducing texture in accessories and fabric can help give a small room interest, dimension and depth.

Illusion of height

In a small room with a low ceiling, you can use colour to give an illusion of height and space. If the room has no decorative features, simply paint the ceiling white and the walls a single almost-white tone, from skirting boards (baseboards) to ceiling. If the room has traditional mouldings such as dado or picture rails and cornicing, you can 'shade up' by painting each horizontal 'band' of the wall a subtle tonal variation of white.

For example, start by painting the ceiling white, the cornice an almost-white tone, then the area above the dado rail with a touch more colour. Carry on like this down to the skirting board (baseboard). The floor will be the deepest colour, but avoid very dark floor tones, which will make the room look even smaller than it is.

Rooms within rooms

Many houses have deep alcoves, recesses or even a box room, which can be modified to make a room within a room. If the area is reasonably large, then an en-suite bathroom may be a possibility. If the space is smaller, it could be ideal as a walk-in closet. Natural light will probably be restricted, if not absent altogether, so you need to pay particular attention to the colours used and the way they are affected by different types of artificial lighting.

You don't have to stick to pale colours in small rooms. Here, a feature has been made of a deep alcove by wallpapering it in a bold, contrasting colour.

Using colour in large rooms

Making a small room feel spacious is a familiar decorative challenge. Less frequent, perhaps, but equally challenging is the task of making a large, high-ceilinged room feel warm, intimate and friendly. In both of these situations, colour is a vital tool for producing the particular effect you want to achieve.

Which colours?

Warm-coloured surfaces in reds, oranges and yellows and other deep, rich colours give the illusion of advancing towards the viewer, so they are an ideal choice when you want to make distant walls seem nearer and the room cosier. In contrast, cool blue and green tones give a feeling of movement away from the viewer, so they can make an already large space seem larger and airier. There are a few decorating solutions that will guide you towards a suitable scheme.

A large room can withstand – and even benefit from – strong, pure colours. In fact, decorating it in a variety of colours, tones and textures is another way of making it appear intimate. All the vivid contrasts break up large expanses of space, creating interest and a feeling of warmth. It is a good idea to link these strong colours with their very pale equivalents – or neutral colours such as tones of grey – to avoid garish results.

Using pattern

Not only can large rooms take strong, pure colours with ease, they can also be made to appear more welcoming or intimate by the use of bold, flamboyant pattern on the walls (see page 120). This might be desirable in a large living room or bedroom.

Lowering a high ceiling

As outlined on page 108, there are a number of simple decorating tricks for reducing the apparent height of a high-ceilinged room. One method is to paint the ceiling in a dark, rich hue such as a brick red, possibly matching the floorcovering, then paint the walls in a complementary pale colour. This will add intimacy and, at the same time, avoid a drab or dull atmosphere.

If the walls of the room have dado rails or other interesting features, paint these in a contrasting lighter tone to create another focal point and further break up the expanse of wall surface. Alternatively, paint both the ceiling and the top of the walls down to the dado rail in one dark colour, the dado rail in the contrasting pale colour and the lower walls in a medium hue that complements the ceiling and upper walls. This will give the illusion of a much more compact room with a lower ceiling height.

The dark-painted pillar in the middle of this large, light room helps both anchor the high ceiling and break up the space so as to unite the distinct living and dining zones.

Creating zones

In a very large, open-plan room that needs some sense of unity, one way to achieve a feeling of intimacy is to create several clearly marked zones within the space. Each zone can be defined by its colour scheme, as well as by its furniture and lighting (see page 196). For example, to break up a large living room, choose perhaps four complementary hues – one for a dining corner, another for studying, another by the windows for daytime work and another for evening relaxation and socializing. (In the same way, you can use colour to define the separate functions of dual-purpose rooms like a home office/spare bedroom or a kitchen/diner.)

Choose colours that are sufficiently different to provide the desired contrast, but not so far apart in the spectrum that they clash and compete for attention. Link the colours together by interspersing some neutral tones, such as beige or grey, and use these in shared features such as the flooring or colour of the skirting board (baseboard).

Orange-coloured walls evoke a sense of warmth and intimacy – ideal for a cosy dining area – while the contrasting picture rail helps break up the walls and creates a focal point.

Creating areas of Interest

• If you decorate a room in predominantly warm, dark colours, introduce furnishings in pale tones of the same warm colours for harmony and brightness.

• Create two well-defined but linked areas in a large room by decorating one area in subtle warm tones and the other in similarly subtle cool tones.

• Try painting just one wall of a large room in a dark hue to create an area of intimacy.

• Experiment with lighting to create different areas of interest in a large space, such as recessed lights for soft background effects, spotlights for work surfaces and display lights for focal points.

Working
with
pattern

Choosing patterns

Like colour, pattern can serve many functions. It can be a focal point or a backdrop, and it can alter the apparent dimensions of a room. Patterned wallcoverings can soften angles and are more effective than plain surfaces for disguising an odd-shaped room or covering uneven walls.

Where to start?

When choosing a pattern, consider how effectively a large area of pattern will complement or contrast with your existing or proposed colour and design scheme.

Mood and function Patterns influence the mood of a room, so start by considering the room's function. Large, bold patterns – like deep, rich colours – are intense and active, and work best as focal points in large living areas. Small, subtle patterns – like pale, neutral colours – offer a less-challenging backdrop.

The predominant colours in a pattern influence mood. Reds have a warm, welcoming quality; blues are restful and calming; browns and oranges are warm; yellows are bright and reflective; greens have a natural, cool and spacious feel. Patterns in pale, neutral tones of these colours act like texture to add depth and interest.

Colour scheme Think about the basic colour scheme with which the pattern will contrast or harmonize. If the pattern is fairly small, it will take on the appearance of a single colour when seen from a distance. You can then use the principles of colour matching (see pages 92–95) to decide whether it works with the rest of the décor.

Style of pattern A common mistake is to choose a pattern that is either too dark or too fussy. Don't forget to consider both the lighting and the size of the area you are decorating. If you are not certain what pattern to choose, opt for one with an off-white background and colours that subtly contrast with or complement the other colours in the room.

Striped patterns in just two colours can appear quite formal, whereas multi-coloured stripes like those on the walls, bedding, floor and furniture in this bedroom produce a fun, modern look.

Balance Balance is all important. Too little variety of colour, tone, pattern or texture and the room may seem lifeless. Too much and it can look chaotic. If you already have a neutral colour scheme, you have the scope to balance it with a variety of rich patterns. But if you have heavily patterned, richly coloured curtains, consider a plain wallcovering in a muted tone of the predominant colour in the pattern.

Types of pattern

Just as there is an infinite range of colours, there is a huge variety of patterns from which to choose. Patterns are either static or dynamic. A static pattern is one where the eye remains within the design – geometric checks, or a small repeated motif, for example. With a dynamic pattern, the eye is constantly travelling from one part of the design to the other – these include damasks and twining floral and leaf designs.

If you are mixing patterns, think carefully about the combinations (see page 122), and consider the atmosphere and style you are after.

Stripes

Depending on the combination of colours used and the width of the stripes, a stripy design can be subtle (like candy stripes) or bold (like deckchair stripes). Stripes work well on walls, floors, upholstery and soft furnishings, but take care to use the correct width of stripe for the size of the area. Regimented stripes in two contrasting colours can provide a formal look, which is particularly effective on blinds, and on walls

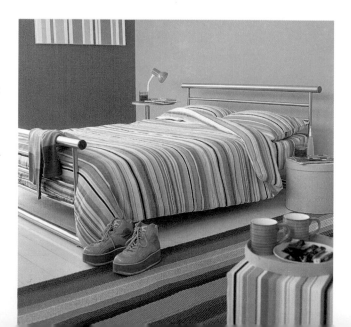

A mix of gingham and different-sized floral patterns is typical of a country-style look and these fabrics work well together as long as they share similar colours.

up to dado rail height in period rooms. Stripes with blurred lines in muted colours create a very different, more relaxed effect.

Use stripes to alter the perceived dimensions of an area. For example, bold, vertical stripes increase the apparent height of a room, while horizontal stripes make walls appear longer and lower, and can make a narrow room appear wider. A striped hall carpet looks smart and welcoming as it draws the viewer in.

Checks

Checks can vary in size from large-scale bold squares and trellis-style diagonal checks to a minuscule two-colour check design that gives the appearance of a solid colour from a distance. Checks work well with both solid colours and stripes.

Gingham (white checked with another colour), once regarded quintessential country kitchen and often mixed with florals, looks equally good in modern rooms with surfaces not usually associated with it, like a stainless steel kitchen or in a bathroom with a steel basin and towel rails.

Spots

Bold spots work particularly well on rugs and furnishings, while spots in subtle colours can look good on walls in children's rooms and family rooms. Spots and stripes in the same colours work well together.

Geometrics

Regular geometric patterns (any regular design other than spots, stripes and checks) tend to be static and formal. They are particularly good on rugs and in

masculine rooms. They make a great style statement, but need careful use if you want to mix and match patterns on walls or fabrics.

Pictorials

Pictorial designs range from French-inspired toile and traditional chinoiserie patterns to oriental-style prints and fun themed designs for bathrooms and children's rooms. Pictorial fabrics make fantastic roller blinds, as the lack of pleats and folds allows the design to be seen fully when the blind is in use.

Florals

Floral patterns range in size from busy sprigs to huge dramatic flower heads, and encompass both traditional and modern styles. Traditional interior floral fabrics have recently enjoyed a resurgence of favour. Contrast them with ginghams for country cottage appeal, or team them with modern accessories and metal surfaces for a contemporary take on a classic look.

Floral wallpaper needs more care. It can be overpowering, but is ideal on feature walls (see page 126) or used on the lower half of walls, up to dado rail or even picture rail height. Repetitive floral patterns give a sense of movement and flow.

Abstracts

This is a huge category, which includes ethnic patterns and faux animal skin prints, plus anything else not covered above. An abstract pattern often works best if it's the main attraction, so it could be a good starting point for a room scheme.

The large, bold pattern of the wallhanging set against a white wall provides a focal point for this living room yet works well with the fabrics in smaller patterns in coordinating colours on the sofa.

Size and scale

The scale of patterns in your decorating scheme can vary from enormous repeats that dominate a room to minuscule mini-prints that are hardly visible from a distance. Don't be afraid to use pattern – sometimes the bolder the pattern, the better the effect.

Large patterns

Large-scale patterns require careful planning. If the design contrasts highly with its background, it will create a basic colour scheme around which the rest of the décor must revolve.

Usually, large bold patterns emphasize form and movement. They look most attractive when used on open, uncluttered areas such as on the floor of a large

A balance of large- and small-scale patterns is achieved here by the use of boldly patterned floor-length curtains teamed with a few cushions in relatively small-scale patterns.

A large-scale pattern on the wall makes an effective background for plain white furnishings in this apparently simple yet striking monochromatic colour scheme.

room, on bed covers or on the walls of a stairwell, where the design can be seen in its entirety without the interruption of furnishings.

Like dark colours, large-scale patterns can also be effective on a single wall of a large room, as a focal point (see page 126) or a backdrop to plain furnishings. Large patterns advance and tend to make a small room look smaller; they are less suitable for rooms with lots of windows, doors or alcoves, since the motifs will be constantly interrupted. However, you can emphasize the intensity of a small space by using a large pattern in a small room.

Avoid placing a large-scale pattern on furniture where it will not be seen at its best and where it may need to be kept a specific way up. It is more practical to choose a small self-pattern for a sofa, which allows the cushions to be used any way up, thereby distributing wear evenly. If you are unsure which size of pattern will best suit your room, err towards the larger side, as this will at least look distinctive.

Mixing small, regular patterns, such as this pretty floral sprig, with a larger, bolder pattern can be successful as long as the patterns share a common colour.

A small-scale patterned wallpaper is ideal in rooms with multi-angled surfaces. Here, it helps to visually enlarge a small attic bathroom with a sloping ceiling and inset alcove.

Small patterns

Small patterns suit smaller spaces. They are particularly effective in rooms with lots of surfaces at different angles, such as an attic room. Here, miniature designs, such as small floral patterns, can give the illusion of a larger, cohesive space.

Small patterns produce a quite different effect when seen on a large scale, so don't judge them from a small swatch. Some mini-prints lose all definition and give the impression of a texture rather than a distinct pattern when applied to the whole wall. Others take on a single colour and tone when viewed from a distance, making them the easiest to match to a colour scheme. Whether you allow a small-scale pattern to function as a single tone throughout a room or coordinate it with areas of plain colour is a personal decision. In general, however, a balance of plain colour and patterned areas is easy on the eyes. Avoid using small designs in large rooms, since they can look spotty. Steer clear of a fully coordinated look that uses only small-scale patterns. Instead, mix small patterns with larger-scale prints or weaves to add depth and variety.

Texture

Like pattern, texture is often used for decorative effect – adding interesting focal points to a room. Using textured surfaces, whether on walls, furnishings or flooring, is also a useful device for concealing underlying imperfections.

Texture and colour

Texture affects how we see colour – for example, a rough, bobbly fabric will appear different from a light-reflecting gloss paint in the same hue. Texture is therefore an important element to consider when planning a decorating scheme. Rough surfaces scatter light and make colours appear darker and duller. They can, in some situations, create interesting shadows. By contrast, smooth surfaces reflect light and make colours look lighter and livelier. Even black or very dark walls reflect some light if decorated in gloss paint.

Tricks with textures

• Warm, rich colours and hard, shiny surfaces appear to come towards you. Cool colours and soft, matt textures give the appearance of receding more into the background. Bear this in mind when you want to alter the apparent size of a room.
• Break up large spaces into intimate areas by using different-textured hangings or rugs.
• Create contrasts by regularly interspersing textures and patterns with areas of plain, smooth colour.
• Combine 'rough' textures like untreated wood, stone or sisal flooring, wicker and hessian (burlap) with 'smooth' textures like glass, metal, polished wood, finished leather, shiny chintz and fine silk.

Creating a mood

Smooth surfaces create a cold, clean, sparse mood – especially when combined with black or white. Such a scheme may be ideal for the bathroom, and could also work well for a business-like study. However, for a bedroom or candle-lit dining area, textured surfaces such as brick, wood and coarse fabrics create a more preferable, softer mood. When used with a combination of muted colours and tones, they impart warmth and intimacy to the overall scheme.

Combining colour, pattern and texture

Just as the predominance of a single colour can be overwhelming or dull, so, too, can the overuse of a single texture or pattern. However, it is best to limit the variety of textures and patterns if you have a complicated colour scheme or the effect can be confusing. For example, if you have a range of textured surfaces in natural tones of brown and beige, it is a good idea to limit the patterns and stick to smooth surfaces for the other colours in the room.

Achieving a balance of colour, pattern and texture can be tricky, but you can rarely go wrong with natural colours and textures as the basis for your scheme.

Examples of texture

Texture is not only found in woven fabrics, but on virtually any surface from walls to floors, from rough, natural fibres to smooth tiles, as can be seen from the small selection of different products below.

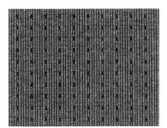

This geometrically textured flooring is a great modern alternative to traditional woven carpeting. Precise and structured patterning gives a contemporary feel to any interior style.

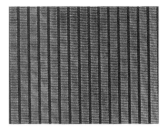

Here, raised geometric stripes have been created in a synthetic suede by using clever textile techniques to give a simple weave that would look impressive as cushion covers or curtains.

This textured natural flooring is almost like tweed, with many different colours woven together. Because it is undyed, the soft greens and toffee colourings are completely organic.

The interesting surface quality of this wallcovering adds a new dimension to simple décor. It offers a gentle background, which would be particularly suited to a living room or bedroom.

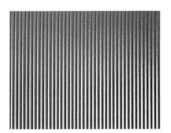

This ribbed-lead-effect resin flooring gives an industrial look. Durable and ultra-modern, it works well in open-plan flats teamed with metal staircases and polished bare concrete.

The intricate surfaces of fabrics that are pleated, naturally creased, ruched, smocked or embossed are either woven in or permanently set, keeping fabrics textured even after washing.

Ceramic tiles introduce a smooth, glossy texture into a scheme. This reproduction Delft tiling can be used in the kitchen or bathroom, or, for a more traditional approach, around a fireplace.

This matt, delicately textured wallpaper is highlighted with soft pin-stripes of pale gold. When using metallics, choose tarnished or antiqued finishes for a subtle, more elegant feel.

Focal patterns

If the room you are decorating has a special feature, you might like to use pattern to emphasize it and treat it as a focal point. Alternatively, use pattern to *create* a focal point – single walls, alcoves, sofas or even curtains can all become focal points if patterns are used carefully.

Focal features

The most obvious focal points are fireplaces, large bay windows and unusual recesses. Any of these could be accentuated by applying pattern to the surface. For example, to highlight a fireplace, use a wallcovering in a warm-toned complementary style on the surrounding wall, or turn an alcove into a focal point with a splash of colourful pattern; offset these by keeping the adjacent walls plain. Similarly, you could enhance a bay window with richly patterned, floor-to-ceiling curtains.

In this bedroom decorated in mostly plain colours, the eye is immediately drawn to the room's focal point – stunning, full floor-length curtains with a rich gold-on-blue pattern.

The opposite can also work successfully. For example, you could decorate most of the room in subtle patterns, leaving an expanse of plain colour in a featured alcove as a backdrop to shelves of attractive ornaments or a display of framed pictures.

Feature walls

Unlike older properties with striking fireplaces, interesting nooks and crannies, exposed beams, panelled walls or doors, ceiling roses or bay windows, most modern homes lack any architectural detail on walls and ceilings. In any size room, deliberately creating a feature wall by using an interesting patterned wallpaper, for example, is a good way of overcoming this plainness. Similarly, a feature wall in a room can help draw the eye away from an unattractive view outside the window.

Using patterned wallpaper on one wall in this large room makes a feature of the bed, while also helping to zone the room by visually separating the sleeping and washing areas.

These striking bold curtains are the focal point in a plainly decorated room with its large, relatively pattern-free areas of walls and furniture.

Handy hints

• For maximum effect, use the most flamboyant patterns in the most simple settings – bold floral prints, for example, give a boost in plain, uncluttered rooms. Enhance a sense of coherence by picking out a single colour from the pattern design for the plain colours in the room.

• Patterns look more exciting than plain fabrics, but bear in mind that you can quickly tire of a powerful design, and it can soon dominate a room or start to look dated.

• Upholstery and curtains need to last a long time – it would be a shame if your curtains still have many years of use left in them, but you are simply tired of an over-dramatic or outdated design. Use simpler, neutral patterns on the more expensive items such as curtains and large pieces of upholstery, and save the exciting patterns for items that can be changed more frequently such as cushions and tablecloths.

In large rooms, consider using patterned wallpaper on one or two walls or on a chimney breast. Large-scale geometric patterning, for example, could become a real feature within a room.

Similarly, multi-functional rooms such as kitchens, living-dining areas or bedrooms that double as studies can be divided into sections through clever use of patterned and non-patterned surfaces. Keep the colours simple and coordinated, and use pattern to direct the eye away from work areas.

Working with existing patterns

If you have no choice but to work with a fitted patterned carpet, patterned curtains or a patterned suite of furniture, stick to plain contrasting-coloured walls as the backdrop and plain-weave fabrics for the other soft furnishings.

Too many large-scale patterns in a room can appear to fight, but this boldly patterned piece of furniture looks great teamed with plain walls and accessories.

Soft
furnishings

Furnishing your home

Before you decide what to put in a room, think about what you will use it for, and create a scheme that is specific to you and your lifestyle – not other people's expectations and conventions. Don't choose furniture and furnishings until you have decided on the final effect.

Design versus fashion

Interior design is a careful decision-making process. It is not just a question of filling a room with bits and pieces, however chic and modish. Distinguish between design, which is creating interiors that work for you, and fashion, which is the latest craze in styles and colours (and which may not last very long).

Bear in mind that different fabrics have different textures, qualities and price tags, which make some fabrics more suited to a particular project than others.

You may not want to go to the trouble of creating a mood board (see page 106), but it is well worth finding an undisturbed spot like the corner of a noticeboard where you can arrange your choices of fabrics together and live with them for a while. Something that appealed to you at first sight may not be so attractive after a couple of weeks, and you will be able to see easily whether your colour and pattern choices are working, or whether they need some adjustment.

Choosing fabrics

Not all fabrics are the same, so the first thing you need to know is whether or not the fabric you have fallen in love with is suitable for what you are trying to achieve.

It is well worth buying just a metre or yard of a fabric you like before committing to a soft furnishing project. You can then find out how it drapes and handles, how the pattern suits your home and whether the colours work for you and in their intended space.

Draping

Some bulky fabrics, such as velvets, tapestries and chenilles, make excellent insulating curtains, while very heavy fabrics are best used flat for seating and cushions, because they can be too stiff for curtains,

List your activities

Perfection is probably impossible, but you will get a lot closer to designing the ideal room if you first list all the activities that will be taking place there. The list may be longer than you think! It won't just involve the obvious things like eating, sleeping, watching television and cooking, but also many other varied activities such as storing things, exercising, listening to music, reading and indoor gardening.

The bulky fabric used for these curtains drapes well and has good insulating qualities, retaining the room's warmth when drawn for the night.

which then won't fold or drape easily. Heavier fabrics give presence to roman blinds, however, where a lightweight fabric can appear flimsy.

Letting light through

You can vary the amount of light that comes into a room by the thickness or translucency of your curtain fabric – achieving either a soft light to work by, or a comforting twilight to help you go to sleep.

Reaction to light

Some fabrics, such as silks, are particularly susceptible to strong sunlight, and will fade quickly. For bright rooms, consider fitting additional roller blinds or inner sheers to protect a precious fabric from the light.

Price

A budget fabric tends not to last as long as a more expensive one. But you can create great effects with a large quantity of a cheaper fabric, then replace it later.

Fabric widths

Most furnishing fabrics are approximately 140 cm (54 in) wide. Some cottons and voiles may be wider than this, and some fabrics – eastern silks, for example

– are appreciably narrower. A bargain price per metre or yard of fabric may not always prove to be an economy if the width is very narrow, as you may have to buy a greater total length of fabric to get the effect you want.

Colour variation

Just as with rolls of wallpaper, dye colours may vary slightly from roll to roll of fabric, so make sure that you buy enough fabric to complete your project, as it could prove difficult to match later.

Sheer fabric at the window allows light in, while still providing daytime privacy from the outside world. The textured fabric in the room adds interest to the all-white colour scheme.

Soft furnishing fabrics

The range of soft furnishing fabrics is extensive – just make sure you work with your chosen fabric appropriately to avoid shrinkage and damage. You *can* use dress fabrics for soft furnishings, although they are not usually as strong as proper furnishing fabrics, and certainly not as resistant to fading in strong sunlight.

Natural fabrics

The range of natural fibres used for furnishing fabrics is limited, but the textures and patterns they produce are almost infinite. Weavers have developed many ways to create texture, making the same fibre soft, stiff, supple, smooth or rough.

Cotton

Cotton is relatively economical to produce and extremely versatile. It can be dyed; it produces a smooth surface for printed patterns; and it can be woven into heavy upholstery cottons or light transparent muslin. It is not luxurious, but very tough and practical.

Linen

More expensive than cotton because it is more time-consuming to produce, linen has a distinctive crisp, dry texture that is very appealing. Despite being strong and smooth, it creases easily, so you have to appreciate the rumpled look that is its hallmark. It is often combined with cotton to make 'linen union', which gives you the best of both fabrics. It is commonly used for tablecloths and napkins.

Wool

Sheep's wool has a natural crimp to it, which makes a soft, warm fabric that drapes beautifully and can be dyed in deep, rich colours. It has a tendency to shrink, so is often blended with other fibres for furnishings.

Silk

Silk is the finest and smoothest natural fibre and is derived from the cocoon spun by the silkworm larva. The fibre has a unique sheen that gives the woven textile, also called silk, an instant look of luxury. The cheaper, lightweight Indian or Chinese silks make beautiful curtains, although they usually come in narrower widths than European silks.

Hemp and jute

Jute is quite a rough fibre, while hemp has very similar qualities to linen in that it is lustrous and extremely hardwearing – it also tends to crease. It is not in common supply, although manufacturers are currently experimenting with mixing it with cotton or silk for both curtain and upholstery fabrics.

Natural fabrics like cotton and linen (woven from the fibres of the cotton and flax plants, respectively) are good, hardwearing options for soft furnishings.

Artificial and unusual fabrics

Synthetic fabrics, such as nylon and polyester, are often easier to maintain than natural ones. Polyester sheers, for example, will wash without shrinking, and require little or no ironing compared with sheers made from cotton muslin or lace.

Manmade fibres were originally devised as a cheap substitute for expensive natural fibres, such as silks. However, a new generation of textile designers has exploited the strength and versatility of these fibres in producing fabrics that are magnificent and innovative in their own right. Furthermore, many natural fibres benefit from having a small percentage of strong nylon or polyester added to the weave, which produces a fabric that will withstand modern wear and tear.

Markets and antique outlets are good sources of unusual fabrics. You can quite easily transform an old tapestry or beautiful silk brocade panel into a blind or cushion. Just remember that older fabrics need to be treated with care, and they may not be strong enough to withstand a lot of stretching and stitching.

Leather and suede

Leather and suede make excellent furnishing fabrics. Covering a sofa or armchair in real leather is a job for the experts, but the beginner can successfully cut hides into simple shapes to make a smart cushion cover. Artificial leather and suede are now almost indistinguishable from the real thing, and you can buy them by the metre or yard, which makes them much easier to use.

Real, or artificial, leather and suede can be tough to sew, so keep the construction simple. Use a heavy cotton thread and a specialist three-sided hand or machine needle for leather, designed to make a clean hole. Hold the pieces together at the seams with masking tape or paper clips and press with a warm (not hot) iron.

Plastic-coated fabrics

Plastic-coated fabrics make cheerful and robust table covers for multi-purpose rooms. They need the minimum of stitching – just cut and use in simple shapes. All puncture marks made by pins and needles remain permanently, so it is impossible to pin or tack these fabrics together, and any alterations show. Always use a long machine stitch, as these fabrics can tear where perforated by a stitching line. Pressing with a hot iron will melt the fabric, so use your fingers or a heavy book to flatten any seams.

Handy hints

Natural fabrics bring few problems in sewing, either by hand or machine, and are excellent materials for beginners:

• Use a matching cotton thread and a medium needle for cottons and linens. Polyester thread is better for wool, as it has a slight stretchiness.

• Many silks are easy to handle, but fine ones can prove difficult – use a new, sharp needle and a fine polyester thread rather than cotton.

• Press pure linen carefully on the wrong side, as the iron can leave shiny marks.

• Press silks with a dry iron, as the fabrics can watermark badly if pressed damp, leaving a permanent stain.

Furs

Fur fabrics make fun cushions, or cosy and luxurious throws for beds and armchairs. Like all unusual fabrics, pattern pieces need to be kept to simple shapes. Pin all the seams first, as the fur pile will make the layers slip as you stitch. You may have to use a pin to pick free the fur pile along the seam lines on the right side.

Faux fur cushions are currently a popular choice in decorating schemes, introducing thick, luxurious texture as well as a sense of fun into contemporary living rooms and bedrooms.

Since cushions require only small amounts of fabric, they are ideal showcases for luxurious or unusual fabrics that would be both visually overwhelming and too costly if used in larger quantities.

Unconventional and techno fabrics

Contemporary textiles may incorporate extraordinary weaving and knitting techniques, giving openwork or three-dimensional effects, sometimes interlocking several layers, or interweaving small, delicate items like beads or sequins.

Use a synthetic sewing thread when stitching these fabrics, as they can fray badly when cut and, because of their structure, can be difficult to sew inconspicuously. Allow for wide turnings in seams and for deep hems, and avoid any design that needs a lot of detailed and tailored structuring.

Voiles and sheers

Sheer fabrics can be light and airy, billowing in the breeze, or darkly dramatic in rich jewel colours. They have a magical quality of filtering harsh light and can provide screening from the world outside during the day, but when lights are on in a room at night, most sheers offer little privacy and you may need additional curtains in a heavier fabric, or a roller blind. Sheers should be used simply, without complicated stitching and construction, in order to emphasize the intrinsic beauty of the fabric.

Working with sheers

• The transparency of sheers means that all seams and turnings show through on the right side, so simplicity in making up is best. Neat selvedges at the side of the fabric mean you can usually avoid side hemming.

• A deep, double bottom hem of at least 15 cm (6 in) looks better than a narrow, single one. Allow for this in your fabric calculations.

• Some sheers are available wider than ordinary furnishing fabrics so that vertical seams can be avoided. If you need to join panels to fit a large window, hang them side by side rather than stitching them together.

• You can buy transparent heading tapes, which avoid an opaque or coloured band showing at the top.

• Always use sharp scissors to avoid ragged edges and snags on these lightweight and slippery fabrics.

• Use fine dressmaking sewing thread (polyester or polyester-cotton) to match the lightness of the fabric and avoid puckering.

• Always fit a new machine needle before stitching these fabrics, and loosen the top tension slightly. Ease the fabric gently through the machine with one hand in front and one behind the needle to stop it bunching and becoming damaged. Experiment on a scrap of the fabric, doubled over, until you get your stitching right before you attempt any long seams or hems.

Sheers vary tremendously in the effect they have, and it is advisable to check them against the light before you buy, as this affects both their colour and their apparent texture. Some have shine and polish, and are available in bright colours and daring techno weaves. Others have the soft, gauzy look of traditional cheesecloth and cotton. Linen-based sheers always have that appealing, slightly rumpled look. Many contain artificial fibres, such as polyester, which makes them easier to maintain.

Lace

Often based on historical patterns, laces are most effective when hung almost flat, as panels, so that the design can be seen clearly. Heavier laces can be stiffened and made into roller blinds, while other sheers can be treated this way if they have a high cotton or linen content.

Hanging sheers

These lightweight fabrics are simpler to handle than formal curtains, as they don't need lining or robust tracks and fixings. Avoid old-fashioned expanding wires for hanging, however, as these tend to sag. Tensioned wire cables are perfect or you can sew a simple casing along the top of a sheer fabric and ruche it on to a pole or a café-style rod. Sheers on any type of pole attached with rings, ties or clips always look good, and are an economical way of dressing up a window.

Sheer care

Some, but not all, sheers are washable. Loosen gathering tapes, remove any clips or hooks and use a delicates machine wash with a limited spin to avoid creasing. Artificial fibres absorb little water, which means you can hang the curtains back up straight from the washing machine. Laces and sheers of cotton and linen may shrink, need ironing and can become limp with repeated washing. You can wash the fabric before you make it up, to preshrink it, but always test-wash a sample to be sure. If in doubt, have it dry cleaned.

Available in a huge range of colours, sheer fabrics filter light and provide some privacy. For night-time screening and warmth, use an additional window dressing like a roller blind.

Muslin

Muslin is the name given to soft, matt, sheer fabrics, often embroidered or printed. Inexpensive butter muslin and cheesecloths are excellent when used in large quantities and can create a dramatic effect, although they are not very durable. They shrink and become limp on washing, and are best used for short-term effects and thrown away when shabby. Slightly more expensive cotton and cotton mix sheers will have a longer life.

Organza

Stiffened sheers are known as organzas. They have an appealing party-dress quality and often include metallic or opalescent threads, which shimmer as they move in the breeze of an open window.

Sheer fabrics are not only useful at windows. This length of embroidered muslin makes an effective room divider in an open-plan area without detracting from the available daylight.

Bed and table linen

Most bed and table linen is very easy to make, because it is largely based on rectangles and squares with straight seams and simple hems. If you feel like it, you can add embroidery – by hand or machine – or decorative edges.

Suitable fabrics

Although the terms bed and table 'linen' are frequently used, the more suitable fabrics for these items are, in fact, cottons or polyester-cottons – anything that will withstand constant washing and ironing.

Some materials shrink slightly when washed for the first time, so it is a good idea to prewash any fabric before you cut and hem it. Dress fabrics, shirting or polyester-cotton bed linens can also be made up into tablecloths and napkins.

Sheeting It is possible to buy sheeting fabric by the metre or yard in cotton or polyester-cotton, and in widths larger than usual. This means that you can make up sheets and duvet (comforter) covers without seams for both single and double (full-size) beds.

Cotton fabrics Although these are very good and hardwearing, like linen, they need ironing.

Cotton/synthetic blends These have very similar qualities to cotton, but are easier to care for, needing little or no ironing.

All-synthetic fabrics These wash and dry very quickly, but they can feel unpleasantly slippery, and tend to build up static electricity. They are non-absorbent, which makes them hot and uncomfortable to sleep in when the weather is warm.

Patterned or plain?

Bed and table linen have traditionally been white for a good reason – white fabrics can take constant washing and ironing best, particularly the high temperatures to which tablecloths may need to be subjected. And pure white linen, well cared for, has a look of effortless class.

Because linen fibres were traditionally used to make the fabric used for sheets and pillowcases, 'linen' has become the generic term for such bedding today.

It is worth taking the trouble of cleaning and starching small pieces of beautiful white linen or embroidery to accessorize tables and beds.

You can make bed linen out of dark-coloured or patterned fabrics, but these will inevitably fade with frequent washing, and may look drab before they have reached the end of their useful lives. Brightly coloured borders on pale fabrics may be a better option.

Making bed covers

A throw-over bedspread is easy to make and can cover either the whole bed, touching the floor at the foot and sides, or just the duvet (comforter). Bedspreads can be made of less durable fabrics than other bed linen, because they don't need to be cleaned so often.

You can use standard-width fabrics for bedspreads, but this means that you will almost certainly have to make a join or two. If this is the case, use fabrics of similar weights, and position the seams so that there is a whole width in the centre of the bed, with smaller part-widths on either side. If you do have to join fabric, either try to make the seams unobtrusive or use them to contribute to the design in some way – a band of lace can be sewn between the edges, for example. Just make sure that these decorations will wash as well as the fabric itself.

Heavier-weight fabrics make the best bedspreads, but if you want to use fine fabrics, you can make the cover more substantial by backing it with a lining fabric and finishing the edges with a binding.

You can leave the corners of a bed cover square, or you can round them off to stop them trailing on the floor. Leave them plain or finish with a decorative trimming as a border or edging.

Making a three-panelled cover

The generous cover pictured here (right) is made from three contrasting fabrics, stitched together lengthways along the selvages, with a double-hemmed edge.

You will need
- Basic sewing kit
- Three contrasting fabrics for the bands
- Matching thread for each of the fabrics

1 Take measurements over a made-up bed, using a fabric measuring tape. Measure from the top of the mattress at the head, over the pillows to the desired length at the foot. Add 8 cm (3¼ in) to your calculations to allow for the two seams. Measure the width from the desired length on one side over to the same length on the other side. Add 5 cm (2 in) all round for the hem.

A bedspread allows you to disguise bed linen – perhaps a duvet (comforter) cover and pillowcases – that doesn't match the décor of the room, but is too costly to replace.

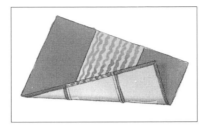

2 Cut the fabrics to the desired lengths. With right sides together, machine stitch the panels together, allowing for a 2 cm (¾ in) seam. Make a few reverse machine stitches at the beginning and end of each seam to give a strong finish. (If you cut off the fabric selvages, you will need to neaten or finish the raw edges.) Press the seams open.

3 Press under a double hem with 2.5 cm (1 in) in each turning, and stitch. Note: If the fabrics contrast strongly, it is worth taking the time to stitch each one with its own matching thread for a professional finish.

Table linen

Since eating and entertaining have become more informal activities, it is no longer necessary to load the table with the traditional intimidating collection of matching cutlery and formal accessories. Instead, dressing up a table for entertaining now offers a great opportunity to be creative, even if you're on a budget. Your table can be cool, crisp and white, or bright, fun and funky. One easy decorative option is to place a second cloth over a plain one. This can be a piece of dramatic fabric that is not big enough to be a tablecloth or a long strip of silk, embroidery or brocade used as a runner down the centre of the table.

Measuring up

Tablecloths There are no standard sizes for tablecloths. If you already have a cloth that fits your table, simply measure its width and length. Otherwise, measure the dimensions of your tabletop and, for a short cloth, sit on a chair and measure from the top of the table to your lap. If you want a floor-length tablecloth (handy for disguising very ugly table legs), measure from the top of the table to the floor. Remember to add the measurement to each side of the tabletop measurement, plus a further 5 cm (2 in) all round for a narrow hem, or 10 cm (4 in) for a wide hem.

Napkins Napkins are usually square, and can be anything from 30 cm (12 in) square for tea size to 60 cm (24 in) square for dinner size. A good all-purpose size is 40 cm (16 in) square, which fits economically into standard-width fabrics, allowing for turnings and avoiding wastage.

To estimate fabric for a napkin, take the finished size and add 1 cm (⅜ in) on each side for the hems. A metre or yard of 90 cm (36 in) wide fabric will make four 40 cm (16 in) square napkins. A metre or yard of 140 cm (54 in) wide fabric will make six napkins of the same size.

The contemporary dining experience is likely to be funky and informal, mixing bright colours and patterns in the glassware, crockery and table linen.

Mix and match

A stunning piece of printed fabric over a table can have the effect of decorating an entire room. It's not necessary for all the table linen to match perfectly, and you can buy remnants of relatively expensive designer fabrics for a good price, because they are too small for regular curtains and upholstery. A mixture of colours and patterns can be charming, and will act as a perfect foil for plain plates and dishes. There is an art, however, to putting different fabrics together successfully, no matter how casual the finished effect appears to your guests. For a scheme to work, there has to be a common denominator behind your selection:

- Base all the fabrics on a similar design motif, such as stripes, checks or florals.
- Choose fabrics that are closely related in colour, even though the designs may be different.
- Purchase fabrics that have a similar texture – all rough matt linens, all light and gauzy or all with a metallic trim.
- Opt for fabrics of a similar tone – rich, vivid, jewel-like colours or soft, pretty and pastel.

Making tablecloths and napkins

Making tablecloths and napkins with your own chosen fabrics and colours is a good way of using small pieces of attractive fabric not large enough for other projects.

You will need

- Basic sewing kit
- Fabric
- Matching thread

1 Cut out your chosen fabric to the calculated size, including the hem allowance (see box, opposite), making any joins that may be necessary in order to make the fabric wide enough. Finish the edges with a double hem.

2 To make neat hems on tablecloths and napkins, cut off each corner diagonally. This cut should be about 10 cm (4 in) long. Press a narrow turning to the wrong side along this diagonal, and then press the edges over all round so that the corners meet in a neat mitre.

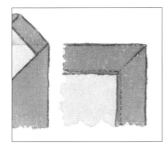

3 Open the pressed sides, and turn the raw edges under by about 1 cm (³⁄₈ in). Press. Turn the hem back in place all round and pin or tack if necessary. Machine stitch close to the inner edge of the hem.

Pastel-coloured striped fabric used for placemats and table runners is a perfect foil for plain coloured crockery with a pearlized finish, as seen here.

Protecting the table

Wood and other surfaces can be badly marked by hot plates and pots, or left with unsightly rings from drinks glasses. If you have a quality table, it needs protecting. Traditional, individual table mats can look fussy and are often quite slippery or awkward. Softer, woven mats are more attractive, but a really hot casserole straight from the oven will need significantly thicker insulation, perhaps even a metal trivet.

A layer of padding under the cloth gives a calmer, less-cluttered table. You can buy table-saver materials, often with a rubberized base, by the metre or yard.

Oilcloth and similar plasticized fabrics wipe down in seconds and need no sewing at all – just cut them to the required size.

Cushions

Getting the seating right goes a long way to making a home feel more comfortable. Cushions of various kinds are important, not only for decoration, but also to soften functional and uncomfortable furniture.

Making cushion covers

Making a cushion involves creating a cover to fit over a ready-made cushion pad – an inner cover containing the filling – so that the outer cover can be removed for cleaning. The best pads contain a feather filling, but budget cushions stuffed with acrylic wadding or polyester are almost as good. Decorative cushion covers can be made using any fabric, while seating and floor cushions need a robust and washable furnishing fabric. You will need a zip (zipper) or buttons, unless you're making the envelope-style cushion cover (right).

An easy sewing project, most cushions require only small quantities of fabric and some method of opening and closing such as a zip (zipper) or buttons.

Making an envelope-style cushion

All you need for this simplest of cushion covers, which involves no fastenings, is fabric for the front and back.

You will need
- Basic sewing kit
- Square cushion pad
- Covering fabric and matching thread

1 The width of the fabric should be the cushion pad width plus 4 cm (1$^1/_2$ in) for seam allowances, and the length of the fabric should be twice the pad width, plus 18 cm (7 in) for the overlap.

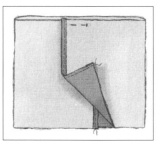

2 On the two short sides, press a turning of 2 cm ($^3/_4$ in) to the wrong side, and then fold it again to make a double hem. Stitch along the edge.

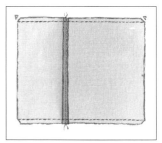

3 With right sides together, wrap the fabric around so that the short edges overlap by 10 cm (4 in) on the back. Pin together the raw edges of the cover at the top and bottom. Stitch along these edges with a 2 cm ($^3/_4$ in) seam allowance. Trim the corners diagonally. Turn the cushion cover the right way out and press. Insert the cushion pad through the slot at the back.

Making tie-on cushions

Tie-on seat cushions add comfort to wooden dining chairs and are tied to their backs to keep them in place.

1 To make each tie, cut a strip of fabric approximately 33 x 9 cm (13 x 3^1/$_2$ in). Press in half lengthways, right sides facing out. Turn the raw edges into the fold and press again. Stitch the open edges together on the right side. Trim the raw ends neatly with scissors.

2 Next, measure the size of your cushion pad from seam to seam. You will need to cut two pieces of fabric per cushion, each one the size of the pad, plus 2 cm (^{3}/$_4$ in) on one side (for the zip/zipper), and plus 9 cm (3^1/$_2$ in) on the three other sides (for the decorative Oxford edges).

3 Lay the two pieces of fabric together, right sides facing. On the zip (zipper) side, pin together two short seams of 12 cm (4^3/$_4$ in) from each corner, along the seam allowance.

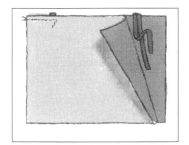

4 Slip two pairs of ties into the seams, each one 10 cm (4 in) in from the edge, then stitch the seams with two lines of stitching to secure the ties.

5 Press the short seams flat, and press the seam allowance to the wrong side along the opening between the sewn seams.

6 Lay the zip (zipper) face up on a flat surface and centre the cushion cover on top of it, with the wrong side of the fabric to the zip (zipper) and with the opening over the zip (zipper) teeth. Tack the zip (zipper) in place.

7 On the right side of the fabric, stitch along both sides of the zip (zipper) and across the top and bottom ends using the zipper foot of your machine, or as close to the teeth as you can get with a normal sewing foot.

8 Remove the tacking and open the zip (zipper). With right sides facing, fold one side of the cushion cover on top of the other, and pin and sew the three raw sides. Trim the corners diagonally for ease of turning.

9 Turn the cushion the right way out and press the seams flat, using a pin to pick out the corner points.

10 To make the Oxford edges, stitch around three sides of the cover (the ones without the zip/zipper) 7 cm (2^3/$_4$ in) from the edge. Insert the cushion pad, close the zip (zipper) and tie the cushion to the chair.

It's worth adding ties to seat cushions, as it stops them sliding around uncomfortably and always ending up on the floor. Either make matching fabric ties or choose a plain or contrasting colour.

Window dressings

Curtains and blinds can perform a variety of functions and the success of any window dressing relies on careful planning. Where is the window? What shape or size is it? What is your aim? What is your budget? Is there an outside view to enhance or obscure? These considerations give you plenty to think about before making a decision on style.

What are you trying to achieve?

Think about what it is you want your intended window dressing to do in each situation. A style or design you admire may not give you what you need, so decide first what function the dressing is being asked to perform and the options will become clearer:

Insulation Curtains keep heat in and control draughts, particularly in older homes without double glazing.

Excluding light Controlling the amount of light in a room is important, especially for a bedroom.

Protection from light Sunlight can damage your furnishings, so a light screen is useful in sunny rooms.

Camouflage An ugly view will be less conspicuous if you have a bold and imaginative window dressing.

Privacy To avoid being overlooked, some form of permanent filter between you and the outside world is often essential.

Decoration Curtains and blinds can tell a fashion and style story better than almost anything else in a room.

Budget considerations

Curtains can be an expensive item, but a room will look unfinished without them. Grand curtains are labour- and skill-intensive, so making your own can bring very useful savings. It is preferable to use more of a cheaper fabric than to skimp on an expensive one, and plain fabrics are more cost-effective than patterned ones. You can always buy ready-made curtains – they are more limited in range and style, but are less

A high window like this one, where privacy is unimportant, can be left permanently uncovered and just softened with a pretty yet unobtrusive frame of draped sheer fabric.

expensive than having them made to measure. If they are longer than your windows, you can easily hem them to the correct size.

Soft blinds like a roman blind add colour, folds and softness to a window and are economical with fabric at the same time. A money-saving roller blind can be used to give privacy, perhaps with an inexpensive sheer curtain draped to one side to add style.

Other considerations

Curtains give an additional frame to a window – as you think about how and where to hang them, look at the window in the context of the wall around it. Use the wall space above and around the window to improve the shape and proportion of your window dressing.

Wall space Always make sure that there is enough wall space for the curtains to stack back on – curtains don't look good if they are cramped and will block light. If you have wall space only on one side, a single curtain may be a good solution. If there is no wall space for stacking, then a blind may be a better idea.

Eyelet headings hold curtains in perfectly even folds and are particularly suited to more formal curtains, like those in this high-ceilinged room with period architectural details.

Shape Round or unusual-shaped windows are often difficult to dress. By hanging a conventional set of curtains, you ignore the shape completely, while shaping the top of the curtains to fit means that you won't be able to pull them back and forth. One solution is to hang a length of voile or sheer fabric over the window, so that the architectural shape and light still shine through.

Size Very small windows may need no dressing at all except, perhaps, for a painted border around the outside, or a piece of lace or embroidery as a tiny valance across the top. Tall windows are easier to dress than horizontal ones – some treatments don't work when stretched out across a whole wall, while modern 'picture' windows need modern styling. In both cases, emphasize the beauty or texture of the fabric you have chosen, and keep construction to a minimum.

Curtain length Short curtains – to the sill or slightly below – are appropriate if you are trying to create a cottage or rustic-style interior. Otherwise, try to fit floor-length curtains if you can, as they are more elegant. However, if you have a radiator beneath the window, you must decide whether you want to cover it with full-length curtains, or whether short curtains or a blind would be a better choice.

Framing Curtains can frame a beautiful view, or obscure an ugly one. To keep the eye in the room, use strong colours and a bold design. To make the most of a splendid outlook, frame the window with unobtrusive colours and a plain or simple pattern, and allow the view to take centre stage.

Toning colours Use fabric colours that tone with the wall colour if the windows are awkwardly placed, not balanced, or if you have used very different treatments on other windows in the same room. The lack of symmetry will then be less obvious.

Formal or informal? High ceilings and large rooms can take very formal curtains, while in smaller, irregular rooms, the focus should be on a beautiful fabric used simply for a cosier effect.

Avoiding problems Check for pipes, power points and switches when you are planning your window dressing. Their presence indicates that electricity cables run in the wall, which could be a safety hazard when fitting curtain track. In addition, think about whether you will be able to open the windows once the dressing is in place. Blinds often get in the way of casement or tip-and-tilt windows, in which case curtains will be the preferable option.

When drawn across the full width of the bay, these eyelet-headed curtains threaded along a tensioned wire completely disguise the shape of the bay window behind.

Curtains

Few things have greater effect on a room's appearance than what is put at the window, so your choice of curtain style, colour, accessories and method of hanging are important. If you are making your own curtains, an essential part of the process is measuring up accurately.

Curtain poles and tracks

There is a huge range of curtain tracks and poles to choose from, and you may find the task daunting at first. Hardware (houseware) and department stores will have a basic range of tracks and poles, but it's also worth locating a few of the many specialist companies making individual poles and finials in varied and often exquisite materials.

Curtain tracks

Most curtain tracks are made of metal or white plastic and tend to be inexpensive and functional, but not particularly attractive. Generally, however, they sit behind some kind of pelmet, swag or valance and are not on show. A plastic track is often the only budget option for a bay or bow window, as it can be bent into the correct shape at home.

Curtain poles

Not only do poles support the curtains, but they also make a style statement in their own right. You will find poles in all styles from traditional to modern – and at all prices. You can commission exquisite handmade poles for your windows, in beautiful decorative finishes including gilding. You can also create such effects yourself by decorating a budget pole from a do-it-yourself store.

When buying a wooden pole, always opt for the largest diameter you can afford – narrow poles look flimsy and can sag across a large span. Metal poles are stronger, and so can be thin and slender with small, delicate finials, in finishes from wrought iron to brass and stainless steel. The sleekest system of all involves tensioned wires, which carry the curtains on fine rings or clips.

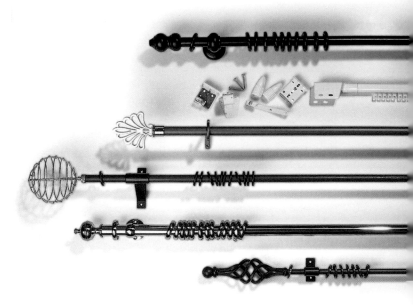

You'll encounter a tremendous range of styles and materials for curtain tracks, poles and finials, which means that making your final choice can be quite difficult.

Calculating length and position

It's important to get right both the required length of the curtain pole or track and its position in relation to the window. Too short a pole/track will leave the curtains covering the sides of the window and blocking daylight when drawn open, while one that is fixed too close to the top of the window will let light shine through along the top edge of the curtains.

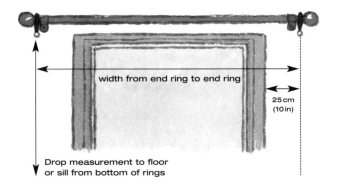

width from end ring to end ring

25 cm (10 in)

Drop measurement to floor or sill from bottom of rings

For an average window, make the track or pole approximately 25 cm (10 in) wider than the window on either side, and take the track or pole at least 10 cm (4 in) above the window.

Measuring for curtains

For each pair of curtains you need two measurements: the complete length of the curtain track or pole, and the finished length (or drop) of the curtains. Use a long steel measuring tape, as fabric tapes can stretch and are unlikely to be accurate. This is a job that is much easier with two people. Sketch your window first, and fill in the measurements around it.

Fix the pole or track in position, then measure its length. Next, take the measurement for the length of the finished curtains from the bottom of the rings, where the curtain hooks will go. (For tab-top curtains, take the measurement from the top of the pole – see page 147.) The curtains should end 1 cm ($^3/_8$ in) from the floor or window sill.

Fullness

Curtains always have more fabric in their width than the length of the track or pole, so that they fall in folds when drawn closed. Depending on the look you want, you can allow different quantities of fabric:

One-and-a-half-times fullness This is where the fabric is one-and-a-half times as wide as the finished width of the window dressing and is useful for contemporary styles where you want only a little folding. It is also the most economical option.

Curtains have a major impact on the look of a room. Without a window dressing of some sort, a room can look unfinished and the windows appear cold and stark.

Calculating fabric requirements

1 Once you have decided on the curtain fullness you want, multiply the width of the track or pole by the required fullness (two, or two-and-a-half, for example). This will give the total width of fabric you need to make the curtains.

2 Divide this figure by the width of the fabric (see page 131), and then round up to the nearest whole number. This will give you the number of widths you need to sew together to make the curtains.

3 Add 25 cm (10 in) to the finished length of the curtains to allow for turnings at the top and a hem at the bottom. (Don't forget to allow for pattern matching – see below.)

4 Multiply this figure by the number of widths obtained in Step 2 to give the total amount of fabric required.

Twice fullness This is a good average amount, especially for contemporary curtains. It gives fullness, while allowing the pattern of the fabric to be displayed.
Three-times fullness This gives an extremely full, designer look, but may be too bulky for smaller rooms.

Pattern matching

When you join two widths of a patterned fabric, you must match up the design at the seam, adding an allowance for this to the fabric calculation (see box, above). Common pattern repeats are 15 cm (6 in), 30 cm (12 in) or 64 cm (25 in) and they are given on the fabric labels. Add the amount of the pattern repeat for each width of fabric you are putting in the curtains. This will also allow you to place the design so that it looks good along the hemline of the curtain, without cutting through an important motif or design element

No-sew curtains

It is possible to make good-looking curtains simply by hanging a number of single, unstitched panels of fabric on a pole, using accessories like curtain clips or by attaching ties. You can use instant hemming web to strengthen the top edge of each panel and to make a hem at the bottom. Hemming web can also be used to shorten ready-made curtains without sewing.

Making tie-top curtains

Tie-top curtains (see top picture, page 135) are simple to make – you could even use lengths of ribbon if preferred instead of making your own ties. They suit an informal look and the style can be used for heavy-duty fabrics as well as for voiles and sheers. You could attach the ties directly on to a curtain pole or track, but they work much better tied on to curtain hooks or rings, which allows the curtain to be drawn freely.

You will need

- Basic sewing kit
- Fabric for the curtains (see page 145 for calculating quantities)
- Fabric for ties and a facing
- Matching thread

1 Cut the fabric into lengths for the curtains, remembering to allow extra fabric for hems and turnings. If the curtains require more than one width of fabric to achieve the desired fullness (see page 145), machine stitch the widths together and press the seams open. Press a double hem on the outside long edges of each curtain about 4 cm ($1^{1}/_{2}$ in) wide, and machine stitch.

2 Next, make enough ties to fit along the top of the curtain, bearing in mind that they need to be spaced approximately 15 cm (6 in) apart and that you need two tie lengths at each point. For each one, cut a strip of fabric 40 x 10 cm (16 x 4 in). Press each strip in half lengthways, wrong sides facing, fold the raw edges into the centre and press again. Machine stitch along the open edge to secure.

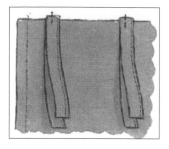

3 To attach the ties, pin, then tack pairs of ties along the top of the curtain, pointing downwards, making sure that there is a set at each end.

4 To make the facing, cut a strip of fabric 8 cm ($3^{1}/_{4}$ in) wide and the same length as the top of the curtain, plus an extra 4 cm ($1^{1}/_{2}$ in) (you may need to join strips). Press under 1 cm ($^{3}/_{8}$ in) along one long edge.

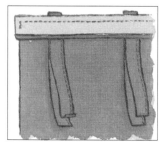

5 With right sides and raw edges together, pin the facing along the top of the curtain, enclosing the ties, and leaving a small turning at either end. Machine stitch through all the layers, 2 cm ($^{3}/_{4}$ in) from the top of the curtain.

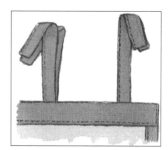

6 Remove any pins and tacking stitches, turn the facing to the wrong side of the curtain and press flat. Turn in the short edges and press again. Stitch along the short edges and the bottom turned edge of the facing to hold it in place.

7 Check the final length of the curtain. Turn up the spare fabric into a double hem and machine stitch.

Making tab-top curtains

If you make up the curtains entirely by machine, add the tabs following the instructions for the ties in the project opposite.

1 Cut the fabric into lengths for the curtains, remembering to allow for hems and turnings. Machine stitch any widths together and press the seams open.

2 To make the lining, cut the lining to the same length as each finished curtain, and the same width, less 10 cm (4 in). (Sew panels together if necessary.) Sew a double hem by pressing under 2.5 cm (1 in) along the bottom edge, followed by a further turning of 2.5 cm (1 in). Machine stitch.

3 To attach each lining, place it on the curtain fabric, right sides together, so that the top and left-hand edges are level. Machine stitch down the left-hand edges to within approximately 5 cm (2 in) of the bottom edge of the lining, making a 2 cm (³/4 in) seam. Rearrange the two fabrics so that the right-hand edges are level, and sew in the same way. Turn each lined curtain through to the right side and press so that an equal margin of the front fabric shows on either side.

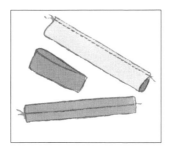

4 The tabs need to be spaced along each curtain top with 15 cm (6 in) gaps between them. To make each tab, use a piece of fabric 36 x 14 cm (14 x 5¹/2 in). Fold each piece in half lengthways and, with right sides

facing, machine stitch the long edges together to make a 'tube'. Press the seam open, then turn the tab through to the right side and press flat with the seam running down the middle of the tab.

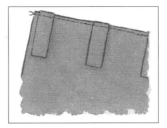

5 Fold the tabs into loops (with the seam to the inside of the loop) and place them along the top edge of each curtain, on the right side, hanging downwards. Make sure that there is a tab at each end of each curtain. Pin, then tack in place before machine stitching 2 cm (³/4 in) from the top edge of the curtain to hold everything in place.

6 To make the facing, cut a strip of fabric 15 cm (6 in) wide and the length of the top of the curtain, plus 4 cm (1¹/2 in). Press under 1 cm (³/8 in) along one long edge and stitch. With right sides together, pin the raw edge of the facing along the top of the curtain, sandwiching in the tabs, and machine stitch through all the layers, 2 cm (³/4 in) from the top.

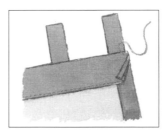

7 Turn the facing to the back of the curtain and press. Turn in the side turnings and stitch in place carefully.

8 Check the final length of the curtain. Turn the spare fabric to the wrong side, making a double hem. You may need to unpick a little of the stitching holding the lining in place to do this neatly. Fold in the ends of the hem to make a shallow angle and hem stitch.

Blinds

Blinds can be divided into two basic types: soft blinds, which include roman and similar blinds, and hard blinds, such as roller and venetian blinds. Soft blinds can be made at home out of many different fabrics, while hard blinds are usually ordered from specialist manufacturers or bought ready-made from decorating stores.

Ready-made blinds

Buying ready-made blinds is cheaper than having them made to order and they are available in many different sizes to fit most windows. There are a few different options, all of which are readily available.

Roller blinds These are most often used on their own, but they can look rather plain. For a more effective treatment, use one behind a set of curtains – the blind can be functional, while the curtains add style and grace. If your window is an unusual size and you cannot find a ready-made roller blind to fit, it may be possible to cut one down to size using a ruler and sharp scissors on the fabric, and a hacksaw to cut the roller. Light-resistant, or blackout, roller blinds are made of a plasticized fabric that blocks light. They are useful in bedrooms, particularly for young children, and for sloping roof windows in attic rooms, and come in a range of plain colours and patterns.

Lace or voile fabrics Blinds made from stiffened lace or voile can be very useful for blocking just a little light in a room and providing privacy.

Venetian blinds These are extremely flexible and can control the light coming into a room in a number of ways. There is a huge range of colours and widths available for the slats. It is advisable to avoid dark colours, which show the dust quickly and are difficult to clean.

Vertical blinds These are most commonly used in offices, although they can be a good solution in a very modern house. They are made of stiffened fabric and, like venetian blinds, offer various options for controlling the light at a window.

Custom-made and do-it-yourself blinds

Several large manufacturers produce pattern books, which you can find in department or curtain stores. They showcase a range of stiffened fabrics that can be made up into a blind of your choice.

Alternatively, you can take your own fabric to a manufacturer, who will make up a roller blind by stiffening the fabric in some way. In most cases, this involves gluing the fabric to a backing, which can be light-resistant if you want. You need to supply the fabric on a roll, not folded, as any creases will show up in the finished blind.

It is also possible to make your own roller blind from a kit, which comes with a spray-glue stiffener. This can be quite successful, but the spray has a powerful solvent and has to be applied outdoors, and it is not always easy to apply an even coating of stiffener to the fabric.

Customizing roller blinds

• Stick a length of lace, braid or ribbon along the lower edge using craft glue.

• Replace a nylon and plastic centre pull cord and acorn with either a length of coloured cord, string or ribbon, or a leather thong, attached to a tassel, a pebble or a shell. Remove the existing cord holder from the back of the lath. Thread the new pull cord through, tie a knot to hold it and replace the holder.

• Cut small designs and shapes from the blind fabric using a sharp craft (utility) knife on a cutting mat, or use decorating stencils to create quite complex designs. Even a simple pattern of small 2 cm (¾ in) cut-out squares spaced around 20 cm (8 in) apart is extremely effective.

Blinds are useful in bays where the room shape can cause problems for curtains. They also help keep window sills and floors uncluttered and are ideal in rooms with limited space.

Measuring for blinds

You can fix a blind either inside or outside the recess of a window. Whether you are making a blind from scratch, ordering a ready-made one or cutting one down to size, accurate measuring is paramount – measure twice, cut once.

Fitting a blind inside the recess

Using a steel tape, measure the width at both the top and bottom of the window (these measurements are often different, even in modern buildings), and use the smaller measurement. Similarly, measure the drop in several places, and use the smaller measurement. This

will give the recess size (width is customarily given before drop). Deduct 1 cm ($^3/_8$ in) from the measurements for clearance.

Most recessed blinds work well only if the recess is straight and true, and you can check this by measuring the diagonals of the recess. If there is more than 1 cm ($^3/_8$ in) difference between the two diagonals, then the recess is likely to be skewed and you will have to fit the blind on the outside of the recess.

Fitting a blind outside the recess

If the blind is to hang outside the window recess, add 5–10 cm (2–4 in) to the measurement of the width of the window each side and at the top. The bottom edge can touch the window sill, but if the sill is not level, overlap it by 5–10 cm (2–4 in). (The overlap can be larger or smaller. For example, if you want the fabric of the blind to show, but not obscure too much light, the blind can be fitted well above the window, and left partly pulled down.)

Hanging roller blinds

With a roller blind, the fabric is always narrower than the overall width (pin width) at the top, because you need space for the brackets and mechanism. When buying a roller blind, therefore, it is important to find out whether the measurement you need is the fabric width, or the pin width.

Also decide whether you want a spring-pull mechanism, the classic centre pull cord under tension, or a sidewinder mechanism. This operates the blind by a continuous chain or cord at one side, and tends to be harder wearing.

Find your style

Contemporary style

The combination of light and space, form and function and colour and texture are the basic elements that create contemporary style. A refreshing alternative to overcluttered nostalgia and outdated period style, contemporary is a look that can be adapted to streamline and create the illusion of space in almost any home.

The look

You don't have to be an architect or design purist with cutting-edge furniture and intentionally bare surroundings to live with contemporary style. Although it can be minimalist and stark, it can also be adapted to combine all the basic elements but with a softer approach. With its trademark open-plan layouts and seamless looks, it is an ideal choice for anyone wanting an organized home that is practical, hardwearing and clutter free, yet without sacrificing a sense of comfort.

Contemporary style does, however, demand a certain rigidity for pared-down living and banished mess, so that the overall effect is neat and precise. This requires clever space planning in every room to ensure that there is a place for everything. This can create a practical and easily maintainable living solution, even for a busy family. With effective storage facilities, belongings can be quickly and easily hidden away behind closed doors in built-in cupboards so that the carefully planned silhouette of the room and its contents are both comfortable to relax in and soothing to the eye. Accessories are kept to a minimum and the few that are used are large scale and given room to breathe by a zone of empty space around them.

The materials used in the fabric of the décor are of paramount importance. The focus is on surfaces – walls and worktops, floors and furniture. These should be carefully chosen to create areas of uninterrupted space with hardwearing contemporary surfaces throughout teamed with sleek-styled furniture with clean, smooth lines.

The streamlined shapes and bold statement pieces in this living room are typical of contemporary style, but the neutral colour palette with its accents of ice blue helps create a more relaxed approach.

Why it appeals

The clean lines and attention to detail of contemporary style appeal to those who love sleek, sculptural design and light, airy open spaces, and who want to streamline their possessions with structure and organization. Not suited to those who love intimate surroundings, a hotchpotch of possessions on display or a cluttered mix and match of furniture, contemporary style is a careful and considered approach to design and decorating where every item and surface in the home is included only after much thought and planning.

Although often considered unsuitable for a busy family home with its hard surfaces, sharp edges and minimal style, contemporary interiors can be surprisingly practical, durable and luxurious. Hard floors can be non-slip and underfloor-heated, well-planned storage allows for plenty of space for toys and other kids' paraphernalia and innovative contemporary surfaces are often surprisingly durable and easy to keep clean. There's no need to compromise on comfort or colour either – sculptural lines and hard

Key characteristics

- **Seamless open-plan living** Keep colours and surface styles consistent to link the spaces together and make the rooms look larger and lighter. With a neutral scheme, complementary colours will blend and harmonize throughout the space; if using a bold colour palette, introduce bright splashes of colour from zone to zone.

- **Hard surfaces** Surfaces should blend seamlessly, so think of them as planes of colour and texture and choose hardwearing materials such as glossy granite, reflective glass, shiny marble and practical hardwood, Corian or stainless steel.

- **Good storage** Storage must be well organized to hide away clutter yet make it easily accessible. Choose floor-to-ceiling built-in cupboards with flush doors or neat free-standing units with sleek lines.

- **Streamlined and simple fireplace** Opt for a contemporary hole-in-the-wall-style fireplace complete with remote control. Or, room permitting, invest in a huge statement piece such as a state-of-the-art log burner, which can be placed in the centre of the room.

- **Spotlights** These are a versatile and practical solution for a sleek, minimal scheme, as they are unobtrusive yet provide a variety of lighting effects – mix low-voltage spot lighting with statement pieces like geometric-shaped side or table lamps.

- **Less is more** With plenty of fitted storage, furniture can be streamlined and elegant and accessories minimal but right – a few dramatic-shaped vases grouped together or a statement piece of art to add that all-important personal stamp and final flourish.

- **Neutral-coloured walls** Walls should be in soothing neutrals, either plain painted or with sophisticated finishes from polished plaster to wood veneer panelling.

Floor-to-ceiling glass windows and a neutral wood floor contrast with contemporary statement furniture in a vibrant colour to create a slick, fresh dining area.

surfaces can be softened with luxurious cushions, throws and soft tactile rugs, and contemporary furniture and furnishings come in a riot of vivid colours guaranteed to add a dramatic look for those who want to make a bold statement in their home.

Key colours

With modern open-plan living, the immediate reaction is to use pure brilliant white as a background colour, although this can look cold and rather clinical, especially if combined with huge floor-to-ceiling glass windows. It is more effective instead to use softer shades of white or a tonal colour palette on the walls to create a flow of colour throughout the space rather than to complicate the look by adding a variety of colours. Architects often use a monochrome palette of greys to define different areas in a room and create a sophisticated finish. This trick also works with colour palettes of white, taupe, eau de nil and sand.

A neutral background colour allows a wide choice of furniture and accessories. Choose furnishings in tones that blend and harmonize with the background colours, and add interest around the room with texture and interesting surface details. Alternatively, a neutral background is the ideal contrast for statement pieces of contemporary furniture in bright, vibrant colour or as a backdrop for colourful pieces of modern art.

Walls

In a contemporary decorating scheme, walls are generally in soothing neutral tones or natural materials such as limestone, marble, slate, concrete or polished plaster, all of which create a gentle, easy-on-the-eye background for sleek, streamlined furniture and furnishings or bold, colourful works of art. Although pure white is a popular choice for contemporary style, it can look cold and jarring on the eye and a softer off-white or creamier tone may be more effective, especially in a living room or bedroom. Wallpaper is an unusual yet worthwhile consideration for contemporary style, but it must be in big, bold and eye-catching designs that make a statement or luxurious textured surface papers like metallics, suedes, leather or wood effects (see page 40).

Kitchen and bathroom walls should mix practicality with style. Identical horizontal and vertical surfaces create the most effective and cohesive-looking finish. Choose sleek-looking surfaces such as limestone, marble, slate, concrete, glass or even mirrors and use liberally for walls, floors and bath surround. Although expensive and heavy, these choices create a sophisticated and seamless blend with adjacent surfaces for cutting-edge contemporary style.

A vibrant blue glass lampbase adds a splash of colour to a sleek white decorating scheme. Accents of colour in your accessories will lift a neutral scheme.

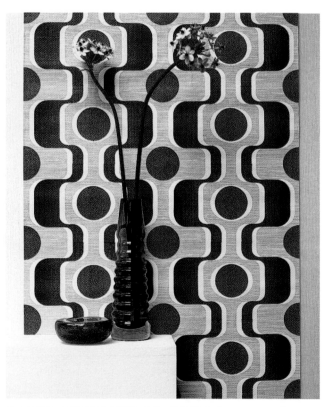

Geometric-patterned wallpaper in vibrant contemporary colours makes a bold statement in any room. Use large patterns carefully, interspersed with plain neutrals to subdue the effect.

Window dressings

Contemporary style is ideally suited to houses with large, floor-to-ceiling feature windows or architect-designed houses where glass plays a major part in the design of the house. Since this is not always possible, make the most of the available light by keeping window styles as simple and streamlined as possible.

Blinds (see page 148) are the most obvious choice, as they are unfussy and come in a variety of styles, designs and fabrics to suit all shapes of window, from practical rollers to roman, venetian and vertical blinds. Where privacy is paramount, frosting the windows is an effective solution – you can do this yourself with a transparent film or spray, or you could replace the glass with special frosted or sandblasted glass panels.

If curtains are a must, opt for very simple structured styles – floor length are best to make windows look longer and more streamlined, and choose tactile fabrics such as felt, suede or wool with simple tab or eyelet headings or sheer curtain panels in luxurious voiles and silks.

Lighting

Plenty of natural light is key to creating a contemporary look, and in open-plan homes with their clean-coloured walls and reflective surfaces, light will bounce around, making the space look larger and feel lighter and brighter. Artificial lighting should be introduced in clever ways in a contemporary home to provide either unobtrusive or feature lighting, both of which need to create a cohesive look around the space without compromising on style.

You should always try to include the lighting requirements in the basic room plans from the very start so that the lighting can be incorporated into ceilings, floors, walls, fitted furniture and decorative features in the room. With free-standing lighting, go for floorstanding and table lamps in wood, metal or leather, and team with simple-shaped plain shades in suede or linen. Sleek chrome and glass lighting or bold coloured perspex are also good choices for modern lighting materials.

Floors

Flooring materials can be very varied, but the required effect is of a seamless space where even if two contrasting floors meet they blend harmoniously and are a major part of the decorative scheme. Hard flooring could be soft, creamy limestone, black, shiny granite or glass, colourful resins, utilitarian cement or ceramic or porcelain tiles. Soft flooring could be luxurious shag pile carpets, rugs, leather or rubber. All types of wood are a possibility and can be sleek, shiny and lacquered or soft-toned and mellow.

Check that the surface you choose is practical for your needs. Limestone and granite, for example, look and feel wonderful, but are extremely heavy and floors

For a more informal touch, hard contemporary lines can be softened by adding accessories in tactile textures such as silks and satins – a luxurious throw can add a welcoming touch to a sofa.

may need strengthening. It is also worth considering underfloor heating with cold, hard surface floors – this not only makes the flooring warm and much more appealing underfoot, it also eliminates the need for other more visually prominent forms of heating such as radiators.

Soft furnishings and fabrics

With such a strong emphasis on simple outline and structured shape, soft furnishings are often disregarded as a major element of contemporary style. However, a few clever touches can make all the difference to a scheme by injecting a splash of bold colour into a neutral room with cushions or throws, drawing attention to a window with a decorative fabric panel or softening the lines of a sculptural bed with beautiful toning linens in tactile textures. The addition of soft furnishings can add those all-important touches that bring your character and personality to the room to prevent it from looking too cold and clinical. Fabrics should be soft suedes and leathers, felt and wool or silks and linens. Alternatively, in contrast, they can be bold, flamboyant prints or textures with oversized pattern and vivid colour.

Wooden boards with a shiny polished surface look streamlined and sophisticated, yet provide a mellow modern backdrop for furniture and furnishings.

Country style

A perennially popular look, country style draws its inspiration from the simple elements of the countryside and its eclectic mix of natural materials, functional furniture and soft earthy colours. All of these elements combine to create a cosy and informal look with timeless appeal.

The look

It's a mistake to think that country style should be confined only to homes in the country. Generally, because the style mixes antiques, comfortable furniture and fabrics with lots of natural materials, it is easily adapted to suit almost any home. Country style is the perfect solution if you want a homely and welcoming look and, with its mix and match of old and new and collection of cosy clutter, it captures the nostalgia of a bygone era. It is a look that should constantly grow and evolve over time, absorbing a kaleidoscope of pattern, colour and texture as you pick up an antique at a sale, collect a pretty fabric or set of linen or paint and transform an old chair.

Depending on the colour schemes you choose and how you interpret the theme, country style can be pure classic country with its traditional lived-in rustic feel, rich, earthy colours, cosy clutter and mix of floral prints and patterns, or given the more simple approach of a fresh modern country look. The latter has the essence of country style, but without the frills and flounces, incorporating a lighter, softer colour palette and a more streamlined approach to furnishing.

Why it appeals

Comfort, simplicity and relaxation are the key words that explain why country style is so appealing. Country style is a way of life with a mix and match of possessions collected over the years, which you love and feel comfortable living with, and the style is constantly evolving. Unlike many of the other popular styles, the country look is timeless. It can work equally well in a small cottage or a large house, as the aim is

With its warm, golden tones, pine furniture is the perfect traditional choice for a country-style bedroom. A fresh colour palette of pastel blues and creams prevents the room from looking dark and dingy.

the same – to create a haven where you can retreat and relax. It is also particularly well suited to large families and lots of pets, since its cluttered style and no need for perfect finishes or surfaces needs little order to look good. Clutter and mix-and-match possessions simply add to the character of the look and are an intrinsic part of its charm and popularity.

Key colours

The ingredients that make up the colour palette for country style are instinctively taken from nature. Soft earthy tones such as terracotta, ochre and burnt umber are a popular choice and create a warm, cosy decorating scheme, while fresher shades inspired by leaves and flowers such as pretty pale greens, pastel blues and primrose yellows and creams add a lighter and softer background colour – particularly good for bedrooms and kitchens or to prevent small rooms looking too dark.

For a modern country approach, opt for the same elements that make up traditional style – wood furniture and simple furnishings, but with more streamlined lines and a softer colour palette.

Since traditional cottage-style proportioned rooms have small windows and low ceilings, it is tempting to paint them white to try to make them look larger and lighter. Always avoid bright whites, which can be stark and cold, and instead opt for off-whites, creams and neutrals for a softer effect – the key to success is to create a calming, relaxing environment by building up layers of soft colour, pattern and texture.

Fresh flowers are essential for adding a pretty touch to a country-style room. Display them in simple ways to make them a feature – a row of flower heads in night light holders has immediate impact.

Key characteristics

- **Wooden furniture** Country furniture has always been made from materials local to the area, so wood is the most obvious and practical choice. Choose chunky pieces of stripped pine or darker oak furniture, or pieces with a matt painted finish in deeper rustic shades for a more traditional look or fresh pastel colours for a modern country look. A dresser (hutch) is a key piece for adding country style to your home – use it to display a selection of china.

- **Natural-looking floors** Keep floors in simple natural materials in soft earthy colours, for example pine and oak planks, rustic terracotta or stone tiles or matting. Rugs are ideal for adding a soft and colourful touch – choose traditional rag rugs and runners for an authentic look.

- **Fabrics** Patchwork, embroidery, natural weaves and pretty cottons all add to the homespun character of the style.

- **Open fire** The open fire and kitchen stove with their cosy glow were traditionally the hub of the home. Incorporate an open fire or woodburning stove into rooms for an authentic country touch – as well being a practical feature, they also provide a focal point and character. A range cooker is also an essential ingredient of country style and adds to the romantic image of the nostalgic country kitchen.

- **Simple lighting** Since country cottages were often small with tiny windows and low ceilings, lighting was limited to wall lights and candles. Overhead lighting is generally too harsh for a soft and shadowy country look, so keep styles simple with the soft glow from wall lights throwing light from wall to ceiling. Position table lamps in dark corners and add a mix of decorative candlesticks and lanterns in a variety of materials.

Walls

For country-style walls, forget pristine finishes and perfect corners – any imperfections and uneven textures only add to the character and charm of the style. While plain painted walls in soft country colours are the most obvious choice, it is worth considering a variety of alternative wall finishes that add traditional character and charm to a room.

Textured plaster finishes painted with a soft wash of colour mimic traditional country-style walls – plastering over a layer of lining paper will make it easy to strip off when you fancy a change. Stencilling and stamping (see page 32) are also a simple and effective way of bringing classic country feel to a room in an instant, while floral sprig wallpaper is a perfect choice for a pretty country-style bedroom or living room.

Panelling is another option – as well as adding character, it adds impact and proportion to a room, while being tough and durable. Panelling is especially versatile for adding country style to a bathroom.

Window dressings

In large houses, chintz has long been the quintessential fabric of choice for creating country style, with its bold printed floral patterns made up into sweeping and elegant curtain styles. In smaller country cottages,

Wooden flooring is a versatile choice for a country-style interior. Avoid laminates and wood-effect flooring with their uniform appearance and opt for wood boards, which mellow with age.

A creamy white painted wall provides the perfect backdrop for country collectables. It creates a fresh, bright look which is ideal for a small room.

however, simple-styled curtains predominantly made from patterned cotton or natural yarn fabrics have always been more practical.

To replicate such country-style window dressings, choose lightweight cotton fabrics with either small sprigged patterns or checks, or opt for fashionable yarns such as natural textured linens or soft wools. Keep curtain styles unfussy with short curtains for tiny windows, and long flowing curtains for larger windows with simple gathered or slot headings. Fabric blinds are also a suitable country-style solution to prevent windows from looking too fussy or cluttered and are a practical solution in kitchens and bathrooms where privacy is paramount but curtains would be impractical. Keep blinds streamlined and simple in a lightweight cotton fabric either with a roman, roller or pretty roll-up blind with fabric ties.

Lighting

Country-style lighting is primarily about creating a warm and welcoming atmosphere, with soft mood lighting provided by the romantic glow of a real fire and the flicker of candlelight and lanterns. Although practicality must be taken into account, not only for safety but also to provide lighting for specific tasks around the home, you don't need to install bold spotlights or powerful bulbs in every light source. With modern technology, it's easy to install dimmer switches for controllable light and to create an even and effective balance of light around each room by

A junk shop lamp has been revamped with a lick of cream paint and a matching pleated fabric shade to add a pretty touch to a country bedroom scheme.

combining a mix of wall uplighters or downlighters, pendant fittings, table and floor lamps. Choose fittings in traditional country-style materials such as wrought or cast iron, brass, tin, carved wood or chunky stoneware and team with pretty fabric or card shades. Junk shop ceramics and collectables are often converted into decorative table lamps that have an original and unique look.

Candles and lanterns add the perfect finishing touch to a country-style interior and exterior. For a special occasion, a row of lanterns lighting up a pathway or on an outside wall adds an effective display.

Floors

Country-style floors need to cope with a constant stream of children and pets to muddy boots, wet shoes and a wide variety of household paraphernalia without showing the dirt, so hardwearing and practical materials are the only solution.

Natural flooring such as quarry or stone tiles are an effective choice for a kitchen and hallway, as they are both highly durable and easy to keep clean, although

You can mix and match patterns and textures for country-style furnishings, but stick to a complementary colour scheme using two or three contrasting shades to make it work.

they can be cold and hard underfoot. Wooden floors are also a practical and popular choice. If you are lucky enough to have good-quality floorboards in a good condition, you could restore them to their former glory (see page 76) by sanding and sealing. Reclaimed timber is also a worthwhile although relatively costly investment, but can add instant character and personality to a home.

Natural floorcoverings such as coir and sisal (see page 86) are particularly hardwearing and durable, and do add the perfect rustic touch, but they are not suitable for some areas in the home such as stairs (as they can be slippery) or bathrooms and kitchens (as they don't like getting wet). Being a natural colour, matting also provides a neutral base for colourful rugs.

Soft furnishings and fabrics

Country-style fabrics are all about simplicity. Patterns and textures are cleverly mixed together to create an eclectic yet inspirational result. Plain slubby cottons and tactile linens are mixed with polka dots, ginghams and checks, with a printed floral cushion or two thrown in for good measure – and it works! Providing you stick to a simple palette of two or three harmonious or contrasting colours, a mix of pattern and texture in these colours will work together successfully.

Homespun furnishings are essential for adding character and those all-important personal touches to the scheme – beds and sofas covered with patchwork quilts and cushions, crisp embroidered linens, warm woollen throws and handmade lace will add to the quintessential country feel.

Classic style

Drawing inspiration from periods in history, the key to successful classic-style decorating is to create rooms that are formal, symmetrical and elegant, and to achieve the feeling that they have been lived in for generations.

The look

The quality of the interior elements is paramount – from architectural details to furnishings and finishing touches. Each item has a specific place and so there is little flexibility once the scheme is designed.

Symmetry is vital in each room scheme – a bed flanked by two matching side tables and lamps with a mirror or print centred over the bed; a living room with two identical chairs opposite a contrasting sofa yet with matching proportions or a display of prints perfectly balanced on a wall. The effect is designed to be sophisticated and elegant, yet easy on the eye. Although a classic-styled interior can include contemporary elements, these are still incorporated using classic guidelines with the same end result.

Furniture can be a mix of antique, reproduction or more modern pieces. It must, however, be based on classic styles – imposing beds, huge-scale mirrors, ornate statues and elegantly shaped furniture with decorative detailing. Teamed with authentic period colour, plain and elegant furnishings and carefully chosen accessories, the elements combine to create a look that will add timeless appeal to your home.

Why it appeals

If you want to live surrounded by luxurious things displayed simply and elegantly, then classic style is the look for you. This very grown-up look demands not

Rooms inspired by a classic period should be sparing in their use of colour – take a look at the modern ranges of authentic heritage paint colours to recreate a classic look with ease.

You can't go wrong with classic white accessories. From simple white china to crisp linens and fluffy white towels, they can be an essential and timeless addition to the classic interior.

Painting your home in 'period' colours has never been easier thanks to the widely available special 'historic' paints – ranges of heritage colours based on archive material.

only structured living, formality and attention to detail, but also the very best quality in furniture and furnishings and strict discipline in arranging them so that they create a streamlined formal interior.

The classic look works better in homes with traditional features and roomy proportions. A larger room gives the space to appreciate how the scheme works, adding to the elegance of the look. In the classic interior, the home is divided into separate areas, each with a particular role to play. A classic living room may not be suited to the modern paraphernalia of busy family lifestyles, so consider establishing a separate informal family room, or incorporate clever storage solutions without compromising on either looks or practicality.

You don't need to confine the décor to a single period to create a classic style that works. Rather than create a pastiche and overworked interior, it's best to choose a mix of pieces that work well together, but also suit the requirements of your lifestyle. A children's bedroom, for example, could include a wonderful antique bateau-style bed and traditional soft furnishings, but be teamed with modern built-in cupboards painted in classic colours to provide practical and concealed storage for clothes and toys.

Key colours

If you are a purist, you might want to track down and use exact replicas of authentic paint colours to capture and recreate a particular period. However, most paint

Key characteristics

- **Symmetry** Balance each room by using matching chairs or side tables and add impact with pairs of pictures in matching frames or a duo of table lamps.

- **Architectural features** Use architectural features to add proportion and definition as well as instant classic style – from a main focal point such as a fireplace to mouldings such as a dado, picture rail and cornice.

- **Good-quality materials** Essential for classic looks that will stand the test of time – solid hardwood or marquetry floors, marble fireplaces, polished wood furniture and luxurious fabrics from damasks to silks.

- **Grand windows** Opt for generous curtains in traditional styles such as swags and tails or with pelmets to emphasize and accentuate window proportions, making them look larger and longer.

- **Elegant furniture** Upholstered furniture is generally in straight-backed styles with wooden frames. Choose pieces with elegant proportions and team with furniture in dark polished woods, leather or glass.

manufacturers now produce special 'historic' ranges with specific period colours based on archive material, which makes the job of choosing colour much simpler.

Originally, the classic colour palette was derived from natural pigments and so produced a diverse range of tones from soft creamy whites and golds through to shades of taupe, brown, red, green and more exotic shades of blue. While the colour shades vary dramatically, rooms should be decorated extremely simply with a sparing use of colour on walls teamed with either white or a darker shade used to highlight skirtings (baseboards), mouldings and panelling. Contrast with beautiful ornate wood flooring to provide a sophisticated backdrop.

Paint effects (see page 28) can also be used – scumbling, dragging, sponging and marbling were traditionally used to make inferior surfaces look as good as the real thing and are still popular today.

Walls

From historic colours in rich matt paint shades to panelling and heavily patterned wallpaper, the colour and pattern you choose for your wall surfaces (which are often divided by mouldings and features) can make or break the decorating scheme. They not only pull a scheme together and set the tone for the furniture and furnishings, but are also key to creating the mood in the room by playing with the combination of colour and proportion.

It is easy to select a paint colour to suit your requirements from today's specialist heritage ranges. Classic-style painted walls are generally decorated in a plain, flat matt single colour, and the aspect of the room should be a major determining factor as to which colour to use – avoid using pale blues and greens in north-facing rooms, as they can look cold and rather harsh. Soft creams and richer golds make an effective background colour that works with most furniture styles and can be teamed with rich gilt or dramatic black accessories.

Authentic panelled walls and decorative features add instant character and classic looks to a room, but when adding features yourself to a room, pay attention to the divisions of the wall, adapting the proportions to suit the scale of your room. Authentic-looking panelled walls can easily be created using softwood or MDF (medium-density fibreboard) and painted in a colour to match your walls.

Wallpaper is an effective choice for classic style – opt for wide vertical stripes, large-scale damasks or classic-style prints in tone-on-tone colour.

Window dressings

Large windows are more suitable for classic-style window dressings, since they should be elaborate designs using generous amounts of fabric, which would swamp and overpower the proportions of a small window. Choose floor-length curtains topped with a pelmet, valance or elaborate swags and tails. Curtains can be used on their own or, for a really luxurious look, teamed with roman or roll-up fabric blinds in a contrasting fabric.

The classic-style window relies greatly on elegant shaping and correct formal proportions and in the most elaborate interiors were frequently left undressed, relying on shutters for keeping in the warmth. Traditional-style shutters are still a practical and effective choice today, left in natural wood or painted, and work especially well in rooms where you want to avoid fussy curtaining.

Although you can make windows look larger by disguising their proportions – for example, use a wider curtain pole and place it higher than the window – in small rooms, pare down the style and opt for simple floor-length curtains in a pale colour with a plain or subtle patterned fabric. Choose a complementary wall tone to avoid drawing attention to the windows.

Lighting

Good lighting is key to a successful classic-style interior – a badly lit room will distort the proportions and balance in the room, whereas good lighting can reapportion space beautifully and to great effect. With large rooms and big windows, natural light and soft colour schemes will only add to the feeling of space and elegance in a room, and it's easy to accentuate this with large mirrors above mantels and on walls.

Supplement the natural light source with flattering ceiling lights with dimmer switches to adjust the ambience in the room, then add task and feature lighting to complete the look. Symmetrical features are vital for adding grandeur and formality to the look – for example, a pair of carved wooden wall sconces, matching sleek glass table lamps with silk shades or ornate bronze candle holders.

Classic style demands quality materials for lighting – a decorative glass lampbase adds a smart, sophisticated touch to this classic scheme, echoing the unusual tabletop.

Wooden floorboards add character and there is often a glorious floor lurking under years of paint – renovate flooring to its former glory by sanding and sealing it, then adding a coat of varnish.

Floors

Most suitable for hallways, bathrooms, kitchens and dining rooms, the most classic-looking floors are black and white squares in tile or marble or terrazzo (marble chippings set in cement), all of which look wonderful and are extremely durable and hardwearing, but are heavy and expensive. If installing a traditional marble or tiled floor, consider installing underfloor heating at the same time to prevent the surfaces from being cold underfoot.

For living rooms and bedrooms, polished wood flooring, in dark rather than light tones, is also a wonderful-looking classic choice and looks just as good on its own or topped with rugs – hardwoods are the most effective solution and improve and mellow with age. Traditional parquetry and marquetry designs are as popular as ever and original designs can still be found in reclamation yards. Alternatively, existing softwood boards can be made to look fairly authentic (see page 76) by sanding, sealing, finishing with a stain and polishing to a glimmering sheen.

Wall-to-wall carpet is another option for a classic-styled room, but it must be of a really luxurious quality and the carpet colour needs to enhance the overall decorating scheme.

Soft furnishings and fabrics

There is no point in skimping on either quality or quantity with classic-style furnishings. Linens, cottons, silks, satin damasks, velvets, wools and tapestry are all classics and can be introduced in a variety of soft furnishings, from upholstery and curtains to cushions and bedcovers. In complete contrast to contemporary style, trimmings are key, too – finish upholstery with decorative piping or gimp, and curtains with fringing or braiding.

In general, plain colours and subtle patterns work best to emphasize the simple, elegant look of the interior and to highlight the quality of the fabrics – sweeps of plain fabric take the eye from plain to plain, while bold pattern will distract the eye. If you do want bolder patterns and colours, opt for richly coloured brocades, watered silks, embroideries and tapestries, all of which can add impact to a classic scheme, but they should work with it instead of overpower it.

In the bedroom, invariably the look should be luxurious white bed linen and in the bathroom piles of fluffy white or crisp linen towels and a luxurious deep-pile bath mat.

Crisp white linens add a luxurious touch to any room in the home. Look for the most exquisite linens, embroidered cottons and handmade lace.

Global style

It is easy to see why global-style decorating has become so popular. It is all about creating your own style and incorporating a multi-cultural blend of pieces. Whether you are adding a few simple touches or transforming entire rooms with different themes, it's fun to create rooms with personality and individuality that are unique.

The look

With the huge selection of global products in many high-street ranges and the ease of internet shopping, it has never been so easy to access a variety of furniture and furnishings from around the world. With its exotic looks and myriad of regional styles teaming unusual shapes, textures and colours, global-style decorating is a fascinating new world to us and breaks all the rules of creating a formulaic interior.

In complete contrast to the mass-produced items of the west, global style offers handmade and homespun designs using traditions that have been handed down through the ages, giving each item a unique personality of its own with individual characteristics.

Global style can work in any size and style of home and with any budget. A few simple 'ethnic' accessories teamed with a bold colour palette can add unexpected touches to a room without costing a fortune, while a complete transformation with furniture and furnishings creates an exotic mood and a very different way of living.

Why it appeals

Global style has widespread appeal that can be used to transform any style and size of interior, and is a look that can be introduced in a variety of ways. As well as the more confident and exotic approach of combining traditional design, vibrant pattern, bold colour and unique creations, the look can also be restrained and orderly with a more classic approach. For example, a beautiful carved Indonesian cupboard topped with a

An antique teak daybed makes a dramatic statement in this room and is topped with textiles in a medley of bright, tropical colour that picks out the primary tones of the dish.

collection of wooden boxes or a colonial-style four-poster bed are feature pieces in a room that can be juxtaposed with a more traditional decorating scheme.

The key to creating a successful interior is to choose items that you love and ensure that your home is practical for the way you live – think of inventive ways to use more unusual pieces of furniture. A Thai rain drum, for example, can make a dramatic coffee table, while a handcrafted pottery jug can be transformed into a lamp. With its fusion of styles from around the globe, you make of the theme what you want and, depending on your budget, you can add small inexpensive accessories or splash out on antiques.

As well as using dramatic colour, the style is also a great way of introducing a variety of textures to a scheme. Handcrafted items are often made from unusual materials and have a wonderful tactile quality to them – think of handmade woven baskets, rough carved wooden furniture and smooth marble. Differing textures keep a room visually rich and their natural

Look to incorporate accessories in interesting and unusual fibres. These woven grass baskets would make a decorative centrepiece as a coffee table.

looks are a wonderful foil for colourful paint tones and fabrics. If the idea of bringing the world into your home appeals, then global style is the look for you.

Key colours

Colours from around the world vary from country to country and from region to region. For example, deep reds are symbolic in China, you'll find vibrant pinks in India and vivid blues and whites are commonplace in the Mediterranean. So, pick a few colours that you like and try to stick to these throughout the scheme to allow colour and pattern to flow. Most global furnishings, however, are colour-led by the pigments and natural dyes available from their native surroundings, which is why natural tones of rich reds, browns and earthy shades are fairly widespread.

The best starting point is to choose an object that is to be a main part of the scheme – a bold patterned rug, colourful wall hanging or work of art, for example – and pick out a colour from the piece to use as a basis for the decorating scheme. For a more classic approach, which combines traditional western pieces with exotic touches, opt for a more neutral colour such as off-white, cream or beige, which will bring a sense of calm to a riot of colour and also provide a soft background to highlight the global features.

Decorative details add the unique and interesting features that make all the difference to a scheme, however small they are. Have fun searching out an eclectic mix of styles and patterns.

Key characteristics

- **Exotic style and atmosphere**
Whatever mix of furniture, furnishings and accessories you choose, the combination of exciting objects and global artefacts will add unique features to your home.

- **Textiles** Textiles can be used to cover walls, furniture, windows, floors and beds. They can add warmth to a scheme and, with their wonderful array of pattern, colours and textures, will have instant impact. Use them simply for a more streamlined approach or fill a room with a medley of layers.

- **Wooden objects** Wood lends itself to a variety of furniture and accessories, from the roughest-hewn timbers carved into naive bowls and sculptures to the mellow aged patina of an Indian cupboard.

- **Texture** Texture plays a vital part in a global interior and can include rattan, cane, wicker, bamboo or palm leaves made up into furniture, screens, window dressings or floorcoverings and accessories.

A wall painted in a soft colour provides a simple contrast to the vibrant patterns and colours of global style. Dilute the paint 50:50 with water and apply it with a large brush for a weather-worn look.

Walls

Paint is the most obvious material for decorating walls in a global-style scheme. Instead of a perfect painted finish, go for a much more informal approach with a soft wash of flat matt colour in earthy shades. This provides an ideal simple contrast to the riot of pattern and eclectic design, yet also makes a defining statement and sets the mood in a room. Textured stucco plaster is also a consideration and again painted in a wash of soft pastel colour works wonderfully well with rustic-styled furniture to bring a sun-faded mellow look to a room.

You could also consider wallpaper. There is a plethora of unusual and effective materials to choose from (see page 40). Look at grasses, bamboo and handmade papers for a natural textured approach, or vibrant large-scale designs with rich colour and gold or silver leaf – ideal for a feature wall teamed with elaborate dark wood Indian furniture or Moroccan accessories.

Window dressings

Avoid fussy traditional curtain styles or frills and flounces, and instead opt for a simple approach using the diverse range of textiles and materials available.

Although you could make plain floor-length curtains from a colourful patterned fabric or textured weave, be inventive and use traditional textures in clever ways. A silk sari, for example, makes a wonderful and dramatic curtain by stitching a casing through which to slot a curtain pole. Lightweight throws, rugs, scarves and linens can also be adapted into curtains or blinds. You could also use natural blinds to keep window styles simple and streamlined – wood, bamboo or paper are a good choice. Alternatively, look at wooden shutters and decorative carved panels. Be inventive with antiques, too, as there are many unusual and decorative carved panels that can be adapted and used to frame or add a decorative feature to a window.

Lighting

Although practical lighting is essential in kitchens, bathrooms and work areas, global style is all about capturing an exotic mood and creating a cosy and inviting atmosphere. For a more classic take on global style, introduce dramatic and unique pieces that enhance the furniture and furnishings you already have. For example, a pair of chinoiserie lamps in a living room or a decorative silver Moroccan lantern in a hallway can add impact and interest to a scheme. For a more dramatic approach, choose chandeliers and table lamps, candles, lanterns and sconces in eye-

Pretty coloured lanterns with a hint of Morocco are ideal for adding a colourful touch to an interior or exterior scheme and can be moved between the two depending on the season.

Natural textured flooring provides the perfect link for the basic ingredients of global style and is a neutral base on which to add layers of colour.

Soft furnishings and fabrics

Fabrics from exotic countries have long been a source of inspiration and a welcome treasure in interiors over the last few centuries – just think of the silks from China or paisley shawls from Kashmir, which are as popular today as ever. The charm of soft furnishings lies in their diversity and in the age-old techniques used to produce the batiks, appliqués, block prints or weaves that look surprisingly contemporary, yet are often traditional handed-down designs.

Soft furnishings from around the globe can be used to transform every room in the home, from wall hangings, cushions and upholstery to curtains and window dressings. Styles can be diverse, too, from a kelim-covered sofa in the living room and Japanese kimono in the bathroom to a silk sari in the bedroom and batik-printed linens in the dining room.

Ensure that the fabrics you choose are practical for your needs – many highly decorative, delicate or embroidered fabrics require careful laundering and are fragile, so in a busy family household, opt for more durable woven cottons and washable materials.

catching shapes, and let light flicker and dance around a space, accentuating the bold colours with shadows and highlights.

Floors

Rugs are an essential part of any global-style look and will instantly transform a room, reviving a plain decorating scheme by introducing fresh colour and pattern. Many of the designs such as kelims, dhurries and gabbehs have been around for thousands of years, and often the rugs are still produced using traditional methods and colours that vary from region to region. Rugs are one of the most versatile furnishings for your home and, since they come in every colour, pattern, shape, size and price imaginable, there will always be one to suit any size and style of home. As well as for the floor, consider using a rug on the wall – they make wonderful works of art.

For a neutral base beneath a rug and to provide a wall-to-wall flooring solution as an effective contrast for a bold colour scheme, opt for natural floorcoverings (see page 86) such as coir, sisal, seagrass or bamboo, or, for a softer approach, a natural textured carpet made from wool.

A mix and match of exotic textures and materials brings luxury and opulence to a global-style interior. Remember that flowers can be used to add an instant splash of colour, too.

Romantic style

A comfortingly familiar and accessible look to achieve in any home, romantic style is all about following your instincts and creating an interior in which to retreat and relax, surrounding yourself with things you love – be it faded florals, mismatched glass and crockery, handed-down treasures or an eclectic mix of antique and junk shop finds.

The look

Romantic style is never about the 'worth' of the objects. Instead, it is a style that is all about surrounding yourself with favourite things to create an environment that evokes memories of nostalgia and an embodiment of comfort and charm.

Romantic style doesn't have to look overcluttered and traditional. A pared-down approach with elegant white painted furniture or utilitarian enamelware, for example, has a simplicity of style that looks modern and would work in a more streamlined romantic scheme, yet has enduring appeal. Indeed, the renewed trend for vintage-style and retro homes has resulted in many manufacturers reintroducing contemporary fabrics and furnishings printed with vintage-style designs, reproduction furniture aged with vintage characteristics or updated traditional-style enamelware and kitchenware.

Classic romantic style has a used, faded charm and lived-in look. Fabrics are soft and natural textures – linens, cottons, muslins and chintz in faded and tea-stained patterns or scraps of antique linens, silks, velvets or tapestry. Vintage finds are key to the look – a wrought-iron bed in creamy white distressed paint, an overstuffed sofa with a faded linen cover or an antique glass chandelier hidden under decades of grime – left as they are or restored to their former glory.

Why it appeals

Comfortingly familiar to all of us yet uniquely personal, the appeal of romantic style is its embodiment of our nostalgia for days gone by. As well as being cosy and relaxing with its eclectic displays of mismatched objects, it can also provide a form of escapism by creating an environment that is far removed from everyday life yet remains practical and durable and suits almost any home.

You don't need to spend a fortune on creating a vintage-style interior. On the contrary, the style is achieved by instinct rather than fashion. Second-hand shop finds, visible patching and even chipped and slightly damaged items have their place in the interior, providing they have a use. This informal look is ideal for a family, as items need to be practical and not too pristine or precious. Although romantic style can be cluttered, it does not need to be untidy or inefficient – in a well-organized kitchen, every item should have its place and at least one purpose.

An old painted wrought-iron bedstead teamed with an eclectic mix of patterned linens gives a homely lived-in feel to a small bedroom. The soft colours and sheer drapes give it a feminine feel.

Sweet-scented fresh-cut flowers are an essential addition to a romantic-style decorating scheme, as they add impact with their bright splashes of colour as well as make a room smell delightful.

Decorating your home in romantic style and discovering new and exciting pieces from antique market or junk shop expeditions is an ongoing gradual process that develops over time and is as rewarding and enjoyable as living in the finished rooms. The key to success is to not take the style too seriously so that your home becomes a museum – instead, follow your individuality and instinct to create a home full of atmosphere, cosiness and warmth.

Key colours

Romantic style is a blend of past treasures, so even the colour palette has the faded, lived-in charm of a bygone era. No furniture or furnishing colours are bold, bright or jarring in the scheme; instead, everything should have the patina and aged looks of vintage finds. Accents of bright colour can be introduced to the scheme via a vase of pretty garden flowers, a tapestry cushion or a handpainted china bowl to add a little impact to the romantic scheme without being overpowering.

Painted furniture is key to the look, but tends to be creamy white or pastel shades in soft blues, greys and pinks painted with a distressed, worn appearance – paint finishes should not be perfect, as any imperfections only add to the character of the piece. Since fabric and pattern are vital to the look, walls tend to be soft blends of matt paint or rough plaster with a soft wash of colour that echoes the furniture tones.

Key characteristics

- **Treasures** A rich and varied mix of treasures, even with a little quirkiness and eccentricity, is key to the look – a kitchen dresser (hutch) full of mismatched china collected over the years, a comfortable battered sofa covered with a pretty floral throw or layers of mix-and-match rugs over a mellowed wood floor.

- **Countryside inspiration** The inspiration for a romantic-style scheme is drawn from the gardens and countryside with its old-fashioned roses, cottage garden floral prints and organic textures and surfaces provided by soft woven linens and antique cottons and muslins.

- **Flowers** These are a must. Forget styled arrangements or over-the-top displays – pick flowers fresh from the garden and arrange in glass jam jars, simple pieces of enamelware or pretty china.

- **Cosy and relaxing** The style requires comfortable upholstery, rustic wood or painted furniture in mismatched styles. A traditional fireplace can make all the difference, too – choose a cosy log fire or wood burner for living rooms and a large stove in the kitchen.

- **Aromatic fragrances** These are key to creating the ambience of a romantic-style interior. Whether it's the wood smoke of a roaring log fire, fresh flowers from the garden or scented night lights and candles, always inject finishing touches into a room scheme that tantalize the senses.

Pattern is an eclectic mix – never coordinated, but an inspirational collection of faded florals, vintage ticking, country-style ginghams and traditional plaids. Antique linens are also central to the scheme – for example, antique cotton or monogrammed linen sheets, sheer voile and lace panels at windows and crisp white table linens.

Walls

Textured chalky paint in soft natural colours seems to work best in a romantic-style interior, as these colours provide a subtle background for the mix of pattern and texture in the scheme. White or off-white walls or soft pastel creams, yellows and pinks are a versatile choice and bring a cohesive look to the eclectic mix of treasures – from painted to mirrored and gilt surfaces, china and coloured glassware. Bright colours are often avoided, as they would seem out of place and harsh with the vintage look. However, the resurgence in fashion for vintage-style interiors has led to a slightly brighter colour palette being introduced – clean minty greens, soft baby blues and pinks – although these bring a much more contemporary look to the end result.

Panelling, painted in a soft colour, can also be a key feature in a romantic-style interior and is decorative as well as being a practical surface. Wallpaper is a worthwhile consideration, too, as there are plenty of pretty floral sprig and tea-stained designs as well as faded stripes available, which are ideal for adding a romantic touch to a room. Teamed with vintage-style pickings, wallpaper can add instant character and charm. If the patterned paper could be too overwhelming, use it on just one wall as a feature (see page 126) and paint the other walls to coordinate.

Printed voile fabric in soft pretty colours allows sunlight to flood into the room, while providing a degree of privacy. This type of window treatment is more suited to a living area than a bedroom.

Window dressings

Window furnishings should be in unfussy styles, but not streamlined or clinical. Opt for flowing floor-length styles in faded florals or checked linens, cottons or muslins. Although simply styled, the curtains can still have attention to detail – pretty tied or tab headings, or button or ribbon trims – but incorporated in a subtle and clever way instead of being dressy and showy so that the curtains blend into the scheme in the room. Roman, roller or roll-up blinds are also a good choice and can be made from cottons and linens or from recycled materials, such as a simple blind made from a vintage tablecloth or tea towel.

Lighting

While the addition of soft candlelight, lanterns and mood lighting is the most obvious choice for creating a romantic atmosphere in your home, incorporating as much natural daylight into a scheme is a wonderful way of adding a soft haze to rooms. It also adds to the faded look of furnishings and décor, softening any harsh colours and fabrics with its natural bleaching qualities. Instead of using pendant fittings or harsh spotlights, introduce a variety of different size and style

Neutral walls in a soft white or off-white paint provide a plain canvas for treasured possessions in a mix of styles, and bring calm and cohesion to the scheme.

A chandelier is a must for romantic style. Scour junk shops and flea markets for wonderful vintage finds or opt for a modern reproduction and distress it with a coat of mottled paint.

of table lamps around room schemes to create scattered pools of soft light. In keeping with romantic style, choose complementary lampbases made from painted wood, brass, glass or ceramic, topped with pretty fabric shades.

Floors

Highlight the rustic charm of a romantic-style interior by using a mix of mellowed surfaces for flooring. Wood is the obvious first choice, but should be aged and distressed planks, or painted boards in soft pastel shades. These don't need to have a perfect finish either – imperfections in the wood such as knots or cracks are all part of adding the character and charm to the look.

Wall-to-wall carpet contributes a more contemporary and luxurious approach, but does add a warm, soft touch underfoot and can be introduced into a scheme, then topped with soft rugs to add layers of pattern and texture. Rugs are key to the romantic style and tapestry, woven or homespun designs with faded vintage looks can be used in every room in the house – incorporate two or three into a room scheme by positioning them around the room or by making a patchwork effect with them.

Soft furnishings and fabrics

Fabrics and soft furnishings are a highly recognizable element of romantic style and give the look its enduring appeal. As with wallpaper, light pretty florals and faded stripes are the patterns of choice. Original vintage fabrics are now hard to find, but luckily there are many modern yet authentic-looking alternatives available – from beautiful faded linens and eiderdowns to well-worn aged ticking upholstery and quilted silk cushions. Pretty voiles, organzas and sheer fabrics in plain muted colours or printed florals also suit the romantic style.

Layering is an important part of achieving the romantic look. In the bedroom, team pretty embroidered linens with plump pillows, feather-filled duvets (comforters) or eiderdowns and a stack of cushions. In the living room, cover a linen sofa with a pretty silk throw and plump cushions made from scraps of vintage fabric teamed with velvets, wool, linen and tapestry. Fading and patching are all part of the appeal and give the style its unique look.

Bed or table linens made from a mix of vintage floral remnants and pretty trims instantly add a romantic and personal touch to a furnishing scheme.

Natural style

Using nature as its starting point and inspiration to make the most of natural organic materials, their soft colours and timelessly classic shapes, the natural-style home is unfussy, tactile, unique and one that simply cannot go out of fashion.

The look

Mellowed or rough-hewn wood, cool stone, fresh cotton, natural wicker and heavy textured fabrics are some of the key elements essential to the natural look and, with its soft neutral colour palette and wonderfully eclectic mix of texture, there is an enduring appeal to the natural look that works equally well in both urban and country homes. Using natural colours and tones allows plenty of freedom of choice in the look and style of your home. In its purely natural form with a mix of rough textures, chunky shapes and heavy textiles, it can take on a simple rustic

With a sophisticated neutral palette and classic enduring finishes, this natural bathroom combines tactile textures such as smooth and rough with glossy and matt surfaces.

appearance; while teaming natural materials with sharper, more modern and streamlined forms and finishes sees a contemporary take on the natural look.

Whichever style you like, the fundamental rules are the same when it comes to creating the basis of the style. Start with a simple canvas for walls, floors and surfaces and use a few dominant textures from nature, such as smooth stone or reclaimed wood flooring, mellow timber cladding or exposed beams for these surfaces, keeping the canvas tonal and avoiding hard contrasts to create a harmonious and cohesive look. By incorporating your chosen furniture and accessories within a neutral shell, the end result will be a calm and restful environment.

Why it appeals

With its soft colours drawn from nature and tactile organic elements, natural style cannot fail to inspire and appeal, as it is simple, understated and creates a relaxed and comfortable look. As well as being a good choice of style for those who like the easy-on-the-eye

Soft white walls teamed with furniture and furnishings in an interesting mixture of textures and tones creates a restful living room with a sense of space and light.

Key characteristics

• **Natural materials** Wood is the recurring element of natural style and indispensable for adding warmth. It can be untreated timber with its rough-hewn rustic looks, smooth and bleached driftwood, antique and mellowed or sleek contemporary veneer. Use this versatile material on floors, walls, ceilings, panelling, furniture and accessories. Stone is also a practical surface material – use rough flagstones, sandstone and limestone or rustic bricks and terracotta.

• **Texture** Aim for interesting surfaces around the home and create a mix and match of texture on fittings, furniture, furnishings and accessories. Juxtapose soft with hard and rough with smooth – a basin or bath made from stone or wood will create a spectacular and streamlined look for a bathroom, while a dining table made from simple blocks of rustic wood adds individuality and character to a scheme, and a linen-covered sofa adds a cosy yet informal touch to a room.

• **Focal fireplace** A fireplace is a bold focal point and practical addition to the look – choose a rustic log burner for a more traditional look or a sleek, contemporary hole-in-the-wall-type fireplace in smooth limestone for a more modern approach.

• **Simple window dressings** Window dressings should provide an unstructured simple outline and be made from natural materials. Floor-length flowing curtains in calico, linen or wool, or simple streamlined wood or hessian (burlap) blinds are ideal for an unfussy scheme.

• **Natural light** Forget atmospheric lighting – the key to creating a natural look is to use as much daylight as possible. Pale colours, simple window styles and uncluttered rooms all help achieve a light, airy feel in a home.

colour palette and informality, the natural look gives plenty of scope for those who want to explore and utilize the increasing range of natural materials on offer. With the rising trend for more ecofriendly interiors, the natural look continues to gain popularity, with the widespread availability of specialist paints, furniture and furnishings that don't compromise on style, comfort, efficiency or functionality.

The natural look with all its variations of style – from country to restrained modern – is extremely versatile for all types of lifestyles. Even the mix of pale colours can be practical, as natural fibres and textures tend to be extremely durable and hardwearing – ideal for a busy family home.

Key colours

The natural colour palette is made up of colours derived from natural objects in their purest form. Think of surfaces from the landscape for key colours, then work up a palette of tone-on-tone colourings for a harmonious scheme or contrasting shades for a more dramatic look. Take inspiration from soft white chalk, mottled stone in greys and taupes, mellow wood tones, sand and earth – colour combinations that will work in almost any interior. Although the palette extends to deep nut browns and dark slate greys, unless these are introduced into a scheme cleverly and with balance, they can look overpowering. For example, a dark slate floor used with pale wood may make a room appear smaller and rather cold, while a softer-coloured stone floor would blend harmoniously.

Choose elegant accessories in pared-down shapes and with interesting surface textures to coordinate with the natural look – these rattan coasters perfectly complement the neutral table.

Kitchens

Planning the use

Traditionally regarded as the heart of the home, the kitchen is certainly the most practical space, so before you start thinking about decorating and the designs and colours you would like, you need to work out exactly what you want from your kitchen. If you are able to start from scratch with a newly rebuilt or extended kitchen, so much the better.

Wear and tear

Consider how much time you will spend in your kitchen, whether it's a room just for cooking or whether it needs to provide an eating space, too, and what appliances and storage are essential to the way

A breakfast bar with stools provides a compact eating area in the kitchen and encourages social interaction while the cook is hard at work.

Laundry location

Count yourself lucky if you already have a separate utility room for washing, drying and ironing clothes, and for storing recycling bins, outside coats and muddy shoes and other unsightly sundries. If you don't have a utility room but you have enough space in a large kitchen to consider partitioning it off to create a utility area, it is well worth doing so.

For those without utility space, consider whether you need to do your laundry in the kitchen, or whether it is feasible to install your washing machine in a separate room, such as the bathroom. If space is limited and you are also reliant on a tumble dryer for getting washing dry, especially in winter, consider getting a single washer/drier model rather than having two separate appliances each taking up valuable room.

you like to cook. The classic image of a spacious kitchen with a huge range and a table large enough to seat the extended family for Sunday lunch does not accurately portray many people's homes today. In fact, in a small home, the chances are that the kitchen is more of a pitstop than a centre of operations, with traditional 'family kitchen' activities like meals and homework redistributed to living areas and bedrooms. Whatever its size, the kitchen still needs to function efficiently. And some will have to put up with more knocks than others, so also take into account how many people will regularly be using the room.

Be honest about your cooking capabilities. Don't plan for a mass of appliances you will never use, and think about what your cupboards really need to hold. Will you want a permanently stocked larder (pantry) of essential ingredients and exotic extras ready to produce meals at a moment's notice, or do you buy just what you need for one meal at a time?

Then there is the issue of worksurface capacity to consider – do you need enough space to prepare three courses at once and still have room to roll out the pastry, or are you realistically more likely to cook one-pot meals that can be chopped and assembled in very little space?

Besides preparing and cooking food and perhaps dining in the kitchen, you may also want to use the room as a utility area in which to carry out household chores such as washing, drying and ironing clothes (see box, opposite).

Rethinking and redesigning

If you have the opportunity to make some structural alterations, decide whether you want to keep your kitchen as an entirely separate entity or if it might actually work better as part of a larger open-plan room. The great advantage about open-plan living is that you can choose where you want your boundaries to fall. This means that if your current kitchen is somewhat on the small side, it's a brilliant way to steal back a bit of space and make your kitchen bigger if you feel you need it.

If you are planning to redesign the entire kitchen layout, this is the ideal opportunity to think about making any lighting and plumbing alterations that

A table and chairs in the middle of an open-plan kitchen/living room create a dining area, while folding seats and light-transmitting perspex chairs enhance the small space.

would improve its working effectiveness and, if necessary, reposition or add extra power points for your kitchen appliances.

Creating a dining area

Some large kitchens give you the option of using them as full-scale dining rooms, too. Even in small kitchens, you can often still carve out an eating area big enough for breakfast or coffee. A breakfast bar is useful in a fitted kitchen, but you may be able to accommodate a small table. Circular tables take up less space than square designs – little garden tables are ideal – and a couple of folding chairs will help keep floor space free.

If the kitchen adjoins an open-plan living room, you can extend into the latter and create a more spacious dining area in the border territory between the two. A kitchen unit or storage cabinet can act as a boundary if you want a clear divide between the two areas (curved-end island units would soften the divide and avoid jutting corners), or rely on a shift in decorating style to define the territory. Changing the floorcovering – from linoleum or aluminium to warm wood, for example – will make the dining area feel more 'furnished', or simply paint the walls a different colour.

An island unit is useful in a large kitchen, providing extra worksurface and storage space, and possibly marking the boundary of the kitchen in an open-plan area.

Planning the layout

Take your time deciding on the design of your kitchen, thinking through how it will function rather than concentrating purely on the visual effect. It is worth talking to kitchen planning experts to get their advice – many companies offer this as a no-obligation service.

The work triangle

The key thing to remember when planning a kitchen is what kitchen designers usually refer to as the 'work triangle' of cooker, sink and fridge. These need to be positioned so that moving between them is easy and unobstructed. Site the sink area first because, once you include the draining boards, it is probably going to be the longest unit. (If you want a dishwasher and washing machine, position them near the sink to keep plumbing costs down.) Now plan your food preparation and cooking areas close by. Make sure that the hob and oven are only a few steps from the sink so that you do not need to wander around with hot pans. Position the fridge near the food-preparation area, but keep it away from the busy traffic around the sink. Try to have worksurfaces between all key areas, but keep the main food preparation area between hob and sink.

Tailor-made treatment

For small kitchens that will be used by only one person, a U-shape or horseshoe layout is ideal. If there is going to be more than one of you in the kitchen, a run of cupboards and fittings creating an open L-shape provides a better layout, so that you have room to get past each other. Long, thin rooms suit a galley-style kitchen with appliances and worktops running the length of the long sides of the room. Larger kitchens obviously allow more flexibility and you may be able to incorporate an island unit, breakfast bar or dining table into the scheme.

A practical work triangle of cooker, sink and fridge is the basis for most kitchen layouts, allowing you to access all three points equally and work efficiently in your kitchen.

Unit size

Modern units come in a range of widths so that you can make the best of the space that you have. As a general guide, however, standard kitchen base units are 60 cm (24 in) wide and around 60 cm (24 in) deep (and therefore accommodate most standard kitchen appliances within them), although there are specialist slimline ones available, which are 50 cm (20 in) deep. Wall cabinets at eye level are usually about 32–35 cm (12½–13¾ in) deep.

A small kitchen can easily be swamped by furniture and fittings. To keep it feeling open and light, find out whether you can have the depth of units reduced (it is easy to cut a slice off the back of cabinets without affecting the look of the front) or have only floor-standing cabinets and leave the upper walls clear.

Circular walls in a kitchen present a challenge and require precision fitting of units and worktops, so you would do well to consult a bespoke fitted kitchens company.

Think, too, about individual preferences for height and reach. Decide whether you find drawers or cupboards easier to access. Plan for ovens, fridges and freezers to be at a convenient height – it is usually best to have the oven and fridge at eye level, with the freezer lower down, as it is used less often. Make sure that worktop and sink levels can be adjusted to suit your height – it is very uncomfortable standing and working in the kitchen if they are the wrong level for you; well-made units will have legs that can be individually adjusted to suit.

Measuring up for cabinets

Before you start visiting kitchen showrooms and choosing the units you would like to install, you need to know exactly how much room you have available for them. Even if you have decided that your kitchen will be planned by a professional, taking your own measurements and creating an accurate floor plan on graph paper will give you and the expert planner a much better feel for the project and for what is practically possible:

- Measure the overall dimensions of the room, including the height of the ceiling – many manufacturers produce wall cabinets in a range of different heights.

- Mark fixed features such as radiators or other heaters, doors, windows, alcoves and skirting boards (baseboards), including their dimensions, and note the positions of gas, water and electricity supplies.
- Measure any awkward projections such as pipes that will need to be avoided. Additionally, you might find it useful to take a photograph for reference, so that you can use this as guide if and when you visit a kitchen planning company.
- Measure any existing appliances you want to keep, so that you can work them into your plan.

Worktop lighting

Plan for good lighting to illuminate your worktops. Choose well-positioned ceiling lights if there are no wall-hung units above your worktops. If you do have eye-level wall cabinets, lights can be fixed beneath them so that the fittings do not show, or you can position individual halogen spotlights in corners for a more sparkling effect. Lights fitted inside glass wall cupboards will provide a gentle background glow, and you can also fit lighting panels to the wall behind a worktop, creating a sort of illuminated splashback.

A narrow galley-style kitchen demands streamlined furniture and surfaces to complement the room's shape. The absence of eye-level units means ceiling spotlights can directly light the worktops.

These fitted kitchen units have a light-coloured metallic plinth (kick board), which gives the illusion of space and light below the units, more typical of free-standing furniture with legs.

Practical issues

The practical design of the cabinets is an important consideration, too – including the mechanics of how they open and close – especially in small kitchens. Normal hinged doors can take up masses of space when fully open, which is a big issue in narrow galleys or U-shaped layouts where an open door can block the whole kitchen. If you need space-saving solutions, look for doors that fold or slide, either sideways on runners, or upwards on a sash-window principle. Half-height or split wall cupboards are useful, with doors that swing upwards or fold back into the cupboard recess rather than swinging out into the room. Bear in mind that normal hinged doors on narrow units will not swing too far out into the room when open.

Off-the-peg or made-to-measure?

Off-the-peg units come flat-packed for home assembly with all the necessary hardware (like wall fixings, screws and hinges) or as ready-assembled carcasses. You will be restricted to standard unit sizes, which may mean some wasted space in corners. If precision fit and use of space is a priority for you, or you have a particularly unusual-shaped room, you might prefer to seek out a bespoke or custom-build company, who will design exactly what you want, made to measure. Weigh up the costs carefully. Flat-pack units are much cheaper, but you will have to pay for fitting on top, and you might decide it is worth spending more on a bespoke design with fitting included.

Free-standing furniture

Free-standing kitchen furniture allows you to be flexible with your kitchen layout and has the advantage that you can take it with you if you move house. It is not as dominant as fitted cabinets, it can be used throughout a large kitchen or it can be interesting to include the occasional free-standing piece to give a fitted kitchen a more individual look and added character, particularly if the kitchen adjoins a living or dining room that naturally has a more 'furnished' feel.

A big plus point for choosing free-standing units over fully fitted modular units is that the time and cost of installation is much reduced – just put the pieces of furniture in place and connect services (gas, electricity or water) where appropriate. In addition, you can buy a single piece at a time when convenient rather than footing the bill for an entire fitted kitchen in one go.

Ideas to consider

A wooden dresser (hutch) or a tall cupboard in the traditional French armoire style – like a sort of wardrobe (closet) for the kitchen, but fitted with shelves – is ideal for storing tableware and linens, and

A wooden dresser (hutch) is a popular choice of free-standing furniture, especially in a country- or farmhouse-style kitchen, and offers both hidden and display storage options.

An island unit in a fitted kitchen has the appearance of free-standing furniture, but cannot be moved and may even house the cooker or be plumbed in for a sink.

makes an intriguing contrast with modern fitted units. Other wooden free-standing items worth considering for your kitchen include a sideboard (buffet), butcher's block, pastry table with a stone insert (see page 188) and a traditional-style settle bench. All of these offer possibilities for storage and are available from specialist kitchen cabinet-makers or may even be found as individual pieces in second-hand and antique shops.

For a more industrial feel, an alternative to a wooden dresser (hutch) is a steel shelf rack or mesh-front metal cupboard – rather like an updated version of the traditional meat safe. For up-to-the-minute free-standing pieces specifically designed to help make limited space work harder, look out for steel laboratory-style units incorporating a sink and worktop – and sometimes even a hob, too – which can take the place of a run of units, or provide an island workstation dividing the kitchen from the adjoining space. Some of these units are fitted with cupboards underneath. Other types are constructed on legs, which creates a more open, spacious effect at the same time as providing room for individual storage crates, trolleys or spare folding seating underneath.

Space-saving tips

In a kitchen with limited space, you need to keep the shapes clean and streamlined, avoiding unnecessary decoration.

• Consider installing half-height wall cupboards where ceilings are low, or to help prevent units from appearing to dominate the whole room.

• Fit a run of cabinets with an end piece of curved shelves in order to cut off the corner and give a more streamlined look.

• Fit doors and drawers with hidden handgrips so that the sleek run of the unit fronts is not interrupted by knobs or handles.

• Hide appliances behind fascia doors to keep the cabinet lines clean.

• Have the cupboard depth reduced by cutting a bit off the back so that they do not protrude so far into the room.

• Look for doors that slide or fold up and swing back into the cupboard cavity instead of opening outwards.

Walls, worktops and floors

You cannot choose fitted cabinets in isolation from worktops, splashbacks and other surfaces, so include these in your plans when thinking about colours and materials. There is plenty of choice, so do your research thoroughly for materials that suit your budget and lifestyle.

Hard-working backdrops

Painted walls will provide a backdrop for whatever kitchen furniture you choose, but you will need to add some sort of protective covering behind worktops and sinks. Ceramic tiles are the traditional splashback material – tough, washable and decoratively versatile. Painted tongue-and-groove wood panelling is a slightly mellower option – a good way of covering up a less-than-perfect wall surface and still washable as long as you use an oil-based paint such as gloss or eggshell. Slate and granite will look smart and dramatic; and newer, more unusual materials include stainless steel, aluminium and heat-resistant glass (available in sheet or tile form) for an urban, industrial look.

Choosing worktops

Worktops can make all the difference to the look of a kitchen – cheap surfaces can downgrade a smart kitchen, while smart choices can make basic kitchen furniture look like an expensive designer fit. But you need to think of practicality as well as style – different materials have different merits.

Laminates Cheapest – and probably the most common choice – are easy-to-clean laminates, available in a huge choice of colours and patterns, including imitation marble, slate and granite. However, laminates are not very tough, and cutting directly on to the surface can cause lasting damage.

Stone Genuine granite or marble are much more expensive, but practical for hob areas as they are hardwearing, resistant to heat and low on maintenance. In a small kitchen, where you are looking at fairly short runs of worktop, it is more feasible to pay for exactly what you want. Granite, slate and marble are also good for pastry making, so you might want to consider insetting a slab as part of a longer surface.

Wood When it comes to choosing appropriate surfaces for chopping, you are better off with softer, cushioned wood. Look for end-grain blocks, where wood has been turned on its end and glued together, then sliced through into cross sections. These surfaces are more resistant to knife damage and are easy to restore if marked, although they can be susceptible to heat and must be regularly oiled to keep them conditioned. Again, wood can be inset as a cutting slab if you don't want a full worktop.

Water-resistant oiled hardwood is useful as a soft landing for glass and china around the sink.

Modern alternatives Expensive but increasingly popular options, stainless steel creates a seamless effect where it adjoins a steel sink, and manmade composites such as Corian or Surell can be moulded to provide a single-piece sink and worktop.

Ceramic tiles are the traditional choice for kitchen walls and splashbacks, providing a durable and easily wiped surface as well as plenty of design and colour options.

Mix and match different surfaces for a successful modern kitchen – here, laminate worktops work with brightly coloured base units, white and chrome accessories and a cushioned vinyl floor.

Forgiving flooring

Kitchen flooring (see pages 70–85) needs to be easy to clean, anti-slip and preferably soft enough for dropped crockery to survive the impact. Wood fulfils these criteria and always looks good. Existing boards that are in good condition can be sealed with varnish, or new wood can be laid on top. Vinyl or linoleum (sometimes referred to as Marmoleum) will be softer and (in the case of vinyl) cheaper. Linoleum is harder-wearing and more resilient because it is made from a blend of natural ingredients including cork and linseed oil, which actually gets tougher with age. It also offers you more design choices, including customized patterns cut to match your room.

Surface detail

Use your kitchen appliances (see page 190) to add another element of style and colour. There is now a much wider choice of finish than ever before. As well as the traditional plain whites, you will find lacquered finishes in bright colours making much more of a statement. Steel, aluminium and chrome, in brushed or polished finishes, give a professional look, and heat-resistant glass can be used for surprising items like extractor hoods, to match glass splashbacks and worktops.

Material effects

For a contemporary style, try mixing different surfaces – for instance, industrial steel with mellow wood, or smooth glass with slate.

• Wood – for worktops, chopping blocks and floors. Make sure the surface is sealed and/or oiled to protect the finish.

• Tiles – for floors, walls and worktops. They are colourful, but unforgiving if you drop something, and the grouting may get dirty.

• Glass – for shelves, splashbacks, worktops and even extractor hoods.

• Metal – for appliances, flooring, cupboard fronts, splashbacks and worktops, plus power points, handles and hanging rails, or try it along the front of a wooden worktop to give it an industrial edge.

• Granite – for sinks, splashbacks, worktops.

• Slate – for worktops, splashbacks, floors.

• Marble – for floors, worktops and pastry boards. Dark marble will be marked by acidic substances such as lemon juice.

• Rubber – for floors. Available in sheet or tile form and polyurethane, it combines a modern, functional look with a warmer texture underfoot.

Kitchen accessories and appliances in brushed or polished metallic finishes are very popular. They provide a light-reflective surface and often come in stunning contemporary designs.

Living
rooms

Planning the space

However tempting it is to launch straight into designing the ultimate comfort zone in your living room, your first move is to think about what other roles the room may need to play. How many people are going to use it, what for and at what times of day?

Multi-purpose living

Few homes today can afford to have a room that is kept for visitors only – most living rooms have to work hard all day and every day. It may well have to act as dining room, study and playroom as well as sitting room, so it needs to adapt easily from practical to comfortable, day to night, casual to formal. If your living space is destined to be a multi-purpose room, it needs careful planning to get the best out of it. Flexibility is everything. There are two ways to approach this:

In this very large open-plan room, rugs on the wooden floor help zone the space by marking the boundaries of the distinct seating areas in the room.

- To allocate different parts of the room for different activities.
- To keep the whole area as a unified space, but furnish it with dual-purpose furniture.

Zoning your space

Décor To some extent, the furniture itself will define the purpose of the area – seating, dining table or computer desk, for example. But the distinction between each one can be reinforced by a shift in decorating mood, so that it feels right for the purpose as well as being practically equipped. Painting a couple of walls at one end of the room a richer colour will create a dining area with a sense of drama for evening use, while a corner or alcove could be allocated a different shade for use as a study space.

Flooring Flooring can signal a change, too, with carpet or natural matting in the sitting area, for example, giving way to wooden boards in the dining space.

If your living room also has to function as a dining room, careful positioning of the furniture can help visually separate the two areas – a sofa often provides the perfect division.

Textures Furnishing textures have their own subliminal effect, with fabrics and other soft surfaces automatically denoting comfort, and hard surfaces spelling practicality. But fabrics themselves can divide into simple, robust linens and cottons for daytime and working areas, with more luxurious materials restricted to the comfort zone.

Lighting Lighting is crucial in establishing and altering atmosphere – make sure that you can adjust your lighting levels to create different effects in different areas and at different times of day.

Furniture as boundaries Some rooms lend themselves to being divided up more obviously, with furniture used to form the boundaries of different areas. For example, a sofa positioned at right angles halfway across a long living room will effectively divide the room into sitting and dining areas, while two small sofas at right angles to each other will create a corner with a protected, enclosed feel. A tall shelving or storage unit standing against the wall at right angles can also create the beginning of a partition – just enough to mark the start of a separate area without actually cutting the room in half. For a more emphatic division, you can extend this further, or use folding screens and Japanese-style sliding partitions – these are a neat solution for providing instant privacy for a study area.

Adapting your furniture

If you can, choose your furniture to match the different roles that your living room needs to play. A decent-sized table can act as a desk and turn into a dining table when needed. A sofa bed is slightly more expensive than a sofa, but will give you the option of turning the living room into a temporary guest room if necessary. Futons are a good alternative if you are on a budget. The mattress spreads out on the floor or on a low-level wooden slatted base, but is rolled up into useful seating at other times.

Footstools are useful surfaces for trays and magazines as well as providing extra seating, and ottomans and blanket boxes go one better by supplying handy storage inside for toys, papers and other items. For more organized storage, use the multiple small drawers of classic pharmacy chests, perfect for filing paperwork and stationery, and providing an extra display surface, too.

Remember that the furniture can be rearranged to demarcate different areas at different times of day or year. A work table can stand under a window during the day, then be moved into the centre of the room for a dinner party. Sofas and chairs arranged round a fireplace during winter months can be regrouped to face the room in the summer. Furniture on wheels is always helpful in a multi-purpose room, as it makes rearranging the layout an easier option.

If you don't have a dedicated guest room, invest in a good-quality, comfortable sofa bed for the living room so that you can accommodate overnight visitors.

Comfort and colour

Comfort means different things to different people. Some need a living space that feels calm and restful; others need it to be lively and stimulating. The important thing is that the effect can be felt as well as seen – furnishings need to establish the right mood as well as style.

Choosing your colours

Calm and elegant, neutral colours (see page 102) are a classic option for a living room that has to please everyone. Add plenty of warm honeys and browns if the room is dark and receives little daylight, or introduce more grey and stone shades if you want to cool it down.

If you fancy some colour without being overwhelmed by it, pastel shades are restful and understated. Their effect is cool but not cold, providing a good sense of space without the starkness of plain white. Soft pastel shades mix effectively together because of their similarity of tone, so try

mixing a pastel palette of accents and contrasts – pale blues, greens and mauves, with touches of pink and creamy yellow to add warmth.

Avoid anything too deep or dark, as it will feel gloomy and oppressive during the daytime. Instead, if you want stronger colour, go for bright, clean shades that feel fresh and stimulating, or for muted colours with a 'natural' character, for example denim blue, earthy ochre and soft moss or leaf green.

Layering texture

Where the colour scheme is restrained, you need plenty of texture to provide warmth, contrast, variety and a degree of pattern. In living rooms, build up layers of texture with soft furnishings – rugs, cushions, throws and upholstery all contributing their own element of comfort. Try a streamlined square-cut sofa covered in heavy slubby linen and piled with cushions in different textures – suede, flannel and chunky cord – then complete the look with neat textural detail such as buttoned cushions and blanket-stitch throws.

All-day rooms need to provide a relaxing background, so opt for a classic colour scheme like neutrals and warm browns rather than hard-to-live-with strong colours and busy patterns.

Muted tones of violet, blue and green are easy on the eye and, combined with light-coloured walls and floor, help create a calm and restful living space.

Look for plain-coloured fabrics that provide their own self-patterning – like herringbone woollens, jacquard weaves and corduroy – and adjust the quality and quantity of texture to suit different times of day and year. Add extra layers of warmth and richness in winter, or to boost comfort for evening use, and use fewer layers, in cooler, crisper fabrics during summer.

Carefree furnishings

Since comfort is largely to do with peace of mind, think about practicality as well as aesthetics. If your living space has to suit a busy working or family life, there is no point in going for cream sofas. You need to keep the furnishing fabrics tough and washable – an upholstery fabric with some texture or a small repeating pattern is the best choice, as plain, light colours show marks easily and can become shabby on furniture that is used daily. Try traditional denim or drill, or thick linen in mid-tone colours – these will create an understated, contemporary look without showing every mark, and will also be easy to clean when necessary.

Window dressings

Curtains are expensive, but a living room can look bare without some sort of window dressing (see page 142). Even a small amount of fabric used in a roman blind is a great improvement. If you live on your own or on the ground floor, some form of window dressing is essential for giving a sense of security.

Curtains are often the focal point of a room, so they need to harmonize with the rest of the furnishings. Choose neutral colours for curtains (and sofas) to make them last, then bring colour to the scheme with accessories such as cushions, rugs, lamps and flowers.

Wide picture windows are harder to dress successfully than tall windows. Try to divide them into separate sections using two or three roman blinds side by side, for example, rather than one large one across the whole window. If you have an ugly radiator or heater below a window, floor-length sheer fabrics are useful for covering it without trapping the heat.

Flooring solutions

Flooring contributes its own layer of colour and texture (see pages 70–89). Natural wood – the original boards or new-laid wood (solid planking, parquet or laminate veneers) – always looks smart and creates a feeling of space, while white-painted boards have a modern simplicity. But these may not provide enough comfort, especially if you have children to consider.

Neutral carpet is the soft option, with natural floorcoverings (see page 86) falling somewhere between the two – although quieter than wood, the hardwearing, rough naturals like coir can be even less comfortable, while some of the softer jute and cotton weaves are not tough enough for an all-day area. If you want comfort, look for a weave with a high proportion of wool, which will also provide medium resilience.

A room with a restrained colour scheme needs plenty of texture – like deep-pile rugs and throws and cushions in assorted fabrics – to keep the scheme lively.

Planning your seating

Instead of the traditional layout of the three-piece suite arranged in a semi-circle around the fireplace or television, today's living areas are more likely to be fitted out with individual pieces of furniture in a mixture of sizes and designs, letting you tailor your seating plan to the shape of the room and establish a more personal sense of style.

Flexible options

The furnishings need to be flexible – a small sofa, several armchairs and some spare folding chairs with cushions are more useful than a single row of seating. Some people like to sit down, and some like to sit up, so you should have both options available, if possible.

Arrange corner sofas to suit your room. Use them against two walls to save space or in the middle of the room to create a cosy seating area.

Sofa shapes

Sofas are still the most luxurious idea, and you will probably want to include one if you have room for it, but measure up carefully. Look at different sizes, and don't just go for the biggest one you can fit in the room. You might find that two small sofas, creating an L-shape or arranged opposite one another, suit the proportions of the room better than a single large one. Supplement sofas with footstools, pouffes or outsized floor cushions if there is not enough space for individual chairs.

Particularly compatible with open-plan living, corner sofas are widely available. Low backed and square cut, they squeeze extra comfort out of wasted corners and, best of all, come as modular units, which can build up to whatever shape and size you want.

Look out for variations on the basic sofa shape, too. Antique couches, many of them with decorative wooden frames and neat, slightly padded arms, are often smaller and more graceful in design, and create interesting contrasts in modern settings. The classic

This modern version of a traditional antique couch is right up to date with its simple clean lines and use of contemporary natural materials.

Handy hints

• Don't try to cram too much furniture into your living space – if it doesn't look right, it won't feel comfortable. Keep to a minimal seating layout and supplement it when necessary with chairs from other rooms.

• Don't keep adding new pieces just for the sake of change. If you are bored with the furniture you have, introduce variety by creatively adapting what is there – rearrange the layout or change covers and cushions to suit different seasons or occasions.

• Don't let the television dominate your seating plan. Consider the positions of windows, doorway and fireplace, if there is one, then put the television in the space that's left.

• In a small living room, choose low-level seating that will make the ceiling feel higher and the room generally more spacious, but bear in mind that low seating may not be convenient for elderly or disabled visitors.

• Large pieces of upholstered furniture are expensive. Budget sofas and chairs are available, but you get what you pay for and they will not last for ever. Quality second-hand upholstered furniture can be an excellent investment.

Keep your seating options flexible by incorporating some individual pieces of furniture that can be moved around the living room as necessary.

chaise-longue, with a head-rest at one end and only a rudimentary back support, looks trimmer and lighter than a full-scale sofa (although its position will be limited by which end has the head-rest, so think carefully when you are buying). And elegant daybeds, with frames that sweep up into a curved arm at either end, can stand flat against the wall and be piled with cushions to create a comfortable back.

Chairs

Don't feel obliged to have a sofa if your living room is simply not the right shape. Individual chairs, chosen for their own style and outline, will look more comfortable and have more impact than a sofa that has

been forced into a space that is too small for it. Leather chairs are particularly good at holding their own, with a smart, classic look that works well in neutral modern settings. Don't let the effect get too functional and office-like, however, and avoid the Regency study look at the other extreme. The most comfortable and attractive style is squashy and slightly worn – suggesting a practical, hardwearing quality without actually looking antique.

Measuring up

After the kitchen, the living room is the room that benefits most from careful measuring and planning. Don't be tempted to buy major pieces of furniture without double-checking the measurements first. Tape together some large pieces of old wallpaper or newspaper to the size of the piece of furniture you are considering, and try it out at home. You may be surprised at how much space it will eat up. Can you walk around it comfortably if necessary? Is there room for people to get past with drinks, or loaded plates? Will it block the source of heat, or perhaps be too visually overpowering? Most rooms end up with some gentle compromises, and it may be impossible to make your own living room absolutely perfect, but good interior design is about understanding and working with all the options.

Eating and entertaining

Some homes have a dedicated dining room; others have a dining area within the kitchen. Alternatively, if the main dining area in your home is within your living room, you need to think carefully about the style of furniture you choose.

Adaptable dining furniture

If the dining area gets constant use for family meals or entertaining, you can make a clear shift in style to distinguish it from the sitting area (see page 196), but there is no point in maintaining a dining area if it is needed only occasionally. You would do better to incorporate it into the main space and create a streamlined link between the areas, then use lighting and furnishing tricks to turn it into a dining room when you need one.

Dining tables

Where space is short, dining tables either need to be compact and unobtrusive or must earn their keep by serving another purpose when not being used for meals. Circular and oval designs often make more efficient use of space than square or rectangular, because they can seat the same number of people without wasting space with jutting-out corners. Tables that can be enlarged when necessary are always useful. A drop-leaf design can be folded to half its size and placed against a wall to act as a desk, then opened out for meals. Some tables have an 'envelope' design that works the opposite way, with flaps that reduce the size by folding up instead of down, so that the tabletop folds inwards on itself and the corners meet in the middle. Others have extra panels that can simply be slotted in to extend the tabletop when the need arises.

Alternatively, you could replace a single oblong table with a pair of smaller square ones, using just the one for everyday eating, with the second acting as a

A folding table and chairs set, ideally with castors, that can be pushed to one side when not in use is ideal in a living room/diner where space is limited.

side table in the living room, then pushing the two together to accommodate more people as necessary. This is also a useful option if you have a narrow or awkward doorway that makes access hard for larger pieces of furniture.

Temporary tables

A basic worksurface such as the classic trestle table is incredibly useful, as it can switch from worktop during the day to dining table at night, and will also fold away for quick storage if you want to reclaim the space. Another option is to keep a separate tabletop that can be laid on top of an existing smaller table to enlarge it for when you are entertaining. For this purpose, you could use a sheet of 2 cm (³/₄ in) MDF (medium-density fibreboard) or plywood, appropriately painted or covered with a cloth.

Folding bistro-style chairs are stylish options for occasional seating and can be stored flat when not required. The Mediterranean blue of these chairs is perfectly suited to alfresco dining.

Seating

For chairs, you need to decide between upright dining chairs and softer tub or basket shapes. Some of the uprights are a little less straight-up-and-down than they used to be, with streamlined curves, providing a modern functional look in wood or moulded plastic. However, the overall effect is still hard rather than soft. If you want more comfort, add some cushions or opt for the tub shapes with arms and sprung or padded seats, which are more conducive to leisurely meals and relaxed entertaining.

If you need extra seating, use folding chairs that can be stored out of sight – slatted garden or metal bistro designs are stylish and handy, while canvas-slung director's chairs contribute a rather more relaxed feel.

Mood lighting

Atmospheric lighting is essential for entertaining, as a badly lit table can have a profound effect on enjoyment of the meal. A low-hanging chandelier or candelabra will shed diffused light over the table and create extra shine. Other ceiling-hung lights can be fixed on rise-and-fall mechanisms, and wall sconces will provide gentle background illumination. This combination provides a good level of ambient light and enhances the decorative style of the room. Don't underestimate the importance of candlelight, which adds a sense of life and movement as well as a beautifully intimate soft glow. Keep your china plain and pale, and set the table with plenty of glassware (including coloured glass) to reflect the light and increase sparkle.

Painted wooden furniture

Be aware that too much wood can have a deadening effect on a room. To relieve this, you can include other surfaces in a dining area such as metal-framed furniture, glass-topped tables and perspex chairs. Alternatively, choose painted wood, which has a slightly fresher, lighter feel.

White-painted wood gives a clean, Shaker-simple look, or you could opt for mixed shades for a more casual effect. This is a good way of recycling second-hand chairs, which can be turned into a dining set with paint in toning shades. Try ice-cream pastels or bright schoolroom colours.

Carefully prepare the chairs before painting. If they are already painted, sand them down thoroughly to create a key to help the new paint bond to the old. If they are varnished or waxed, use a proprietary stripper to remove the coating (see page 18), then scrub with wire wool dipped in white spirit. When the chairs have been prepared, apply a primer and undercoat, then add a topcoat in gloss or eggshell (see page 12) for easy cleaning – gloss gives a shinier, more reflective finish; eggshell a subtler semi-matt sheen.

Candlelight is ideal for entertaining. In addition, the candles can be scented and there are hundreds of styles and colours of holders to add interest to the room.

Creating a work space

Like a dining area within a living room, a work space within the same room will probably not need to be in use all the time, so you will want to make the most of dual-purpose ideas that let you adapt the room as necessary.

Ergonomic seating

When creating a space to work, the one thing that should really be purpose-built is the seating. If you are going to be sitting at a desk for any length of time, you need a proper office chair, ergonomically designed for comfort and support, rather than something borrowed from the kitchen or dining room.

Practical pointers

• Make sure that you have enough power and telephone points to fulfil all your needs within easy reach of your work area.

• Look for compact desktop equipment when it comes to equipping your home workstation – choose laptops instead of full-sized computers and combined systems offering, for example, printer/fax/photocopier/scanner all in one, to save on desk space and eliminate trailing wires.

• Never lose sight of the fact that this is still part of the living space – don't let your work clutter spill over and spoil the relaxed atmosphere of the room.

Well-designed cupboards that neatly store everything out of sight help create a home office that doesn't overdominate the rest of this open-plan living area.

Working surfaces

After the chair, the most important items are a desktop that is a comfortable height and good lighting to see what you are doing. You will need a good directional lamp that can focus on your work, and ideally some natural light from a window, too. For a fixed desk, you could consider building a worktop into a corner of the room as part of a run of fitted cupboards. This is a good way of making use of low or sloping ceilings and awkward corners – spaces that are too cramped for most other purposes, but fine for one person sitting down. To give yourself the deepest desk space while taking up the least room, fit a worktop right into the corner with a curved front edge to sit at, making it less obtrusive.

Dining tables and trestles are another option, especially folding designs. A neat alternative is to fit a tailor-made desktop that folds back against the wall, hinged to a wall-fixed batten so that it can be pulled up (or down) when needed and supported by fold-away legs or by brackets that swing out from the wall. If you don't want to improvise like this, there are various free-standing purpose-built furniture designs that provide organized workspace. Like an updated version of the traditional bureau, these incorporate pull-out shelves to take paperwork or a computer keyboard, and the most impressive of them include filing racks and have fabric-backed pinboard space on the inside of the doors, so that in effect you open up a complete miniature office when you open the cabinet doors. And, of course, when it is not in use, you can neatly close the doors on it as well.

If you have a handy alcove or recess, you can build a series of chunky shelves into it and make the lowest one deeper to create a desktop. This will keep your work space confined in a neat area, and when not needed for work, it can be used as an extra shelf or a display surface.

Filing systems

To keep paperwork and other office clutter in order, the neatest solutions are all 'closed' storage. Filing cabinets and trolleys can be slotted under your desk, and there is no shortage of handy-sized boxes and files, readily available in wood, cardboard, plastic, metal and other materials, that can be neatly shelved to present a smart face to the world, while hiding all sorts of horrors. Label or colour-code them so that you don't need to ransack the whole lot to find your latest bank statement. Domestic, work and personal papers should be kept separately, and make sure that you have easy-access places for essentials like passports, licences and emergency-repair contact numbers.

As well as ready-made storage systems, you can improvise your own 'files' from everyday containers such as baskets, bins and buckets. Wicker picnic hampers are a useful size for document files and stationery; kitchen cutlery trays and bottle carriers will create neat pen holders; and gleaming galvanized buckets can be used either as smart wastepaper bins or to hold rolled-up plans, sheets of wrapping paper and other larger papers that you do not want to fold.

Make use of wall space, too, with hanging storage on hooks and pegs. A row of square-cut aluminium cans – the sort sold as kitchen containers – fixed to the wall above your desk will hold pens, paperclips and other bits and pieces. Maps do double duty as decoration and information, and a pinboard fitted into an alcove will provide practical memory prompts for invitations and messages to be answered.

If your work space is incorporated within the living room, take care to keep the area tidy so as not to detract from the rest of the room.

You'll find a huge range of filing boxes available in all sizes, colours and materials, designed to provide inconspicuous storage or to coordinate with your colour scheme.

Storage and home entertainment

Since most living rooms are multi-purpose spaces, you need enough storage for all the different items associated with each activity – books, files, toys, tableware and entertainment equipment. It's easy for things to accumulate, so be strict about clutter control and don't allow the room to become a general dumping ground.

Conceal or display?

There are two options for finding room for your possessions in your living space: hiding things away or putting them on display. There is no simple rule that dictates which is best, or which of the many forms of storage suits which type of object. Some people enjoy keeping books, CDs and tableware out on show, while others prefer everything behind a slick, streamlined

Since you can never have too much storage space, do consider useful dual-purpose furniture like this carousel-style revolving coffee table that also offers storage for books and CDs.

façade of closed doors. The only general guideline is that the less tidy you are by nature, the more hidden storage you will need.

Behind closed doors

Closed storage is the neatest option. Cupboards, cabinets and drawer units provide storage slots of different shapes and dimensions, so work out what you need them to hold before you commit yourself to any particular design. You may find individual free-standing pieces that fit the bill, but built-in furniture will make the best use of your space, by taking advantage of the wall height so that it can afford to stay slim-fitting and use up less of the floor area.

Remember that things like books, CDs, DVDs and videos don't need deep storage. The average paperback is only 11 cm (4½ in) wide, so you can store masses of

Warehouse-style shelves

If you don't mind a few rough edges, you can create your own flexible storage system by stacking up wooden crates. Make sure that they are sturdy and clean, and 'anchor' each one by giving it something heavy to hold (books are ideal). The effect is functional and slightly industrial, with the printed contents or freight details on the timber adding an individual touch. More shallow storage along the same lines can be achieved with planks stacked on bricks, giving a basic, warehouse look that suits simple modern furnishings.

Drawer units allow you to store possessions out of sight. Seagrass storage solutions are particularly popular, being part of the current vogue for natural materials in the home.

Selective display
Be disciplined. Use photographs to establish a personal feel, but don't let them cover every surface. Instead, select your favourite pictures and turn them into a wall display rather than scattering them over shelves and tables.

these items against a single wall without shaving more than 15 cm (6 in) off the size of the room. Where you do need deeper storage – for things like vinyl record collections or stacks of tableware – make use of existing recesses by fitting them with shelves and then fronting them with doors to sit level with the adjacent wall or with other built-in cupboards.

If the idea of a whole wall of doors feels too oppressive, but you are not tidy enough to trust yourself with open storage, consider partly glazed or frosted glass doors instead, or line the panels with wire mesh to screen off the contents without obscuring them completely.

Supplement your wall cupboards with floor-level furniture containing hidden storage space. Look for chests and coffee tables that have lift-up lids or integral drawers, or buy individual under-sofa drawers – huge shallow trays on wheels that cleverly slot underneath the sofa, so all that is visible is a wooden panel that appears to be part of the sofa base.

On display

The same principle can be applied to open shelving, with deep shelves built into alcoves (useful for things like televisions and video recorders) and shallow ones fitted against walls. Plan the shelf heights carefully, or use a flexible system that allows you to shift them up and down to different levels, otherwise you will waste a

lot of space unnecessarily. Do not assume, for instance, that books all need the same amount of space. Create shelves for different book heights and you will find you can fit two shelves of paperbacks in the same space needed for a single row of illustrated hardbacks.

Cube units provide very effective open storage, with the neat, fitted benefits of built-in furniture yet the flexibility of free-standing pieces, letting you move them where and when you want. They make excellent room dividers and you can also arrange your cubes to suit the shape and height of the room – particularly useful under sloping ceilings. Perfect for hi-fi equipment, they will also effectively house photographs, display items, tableware and glasses.

With no visible fixings, so-called 'floating' shelves are a neat, streamlined, open storage solution – the perfect place for a mini sound system, pictures and ornaments.

Bathrooms

Bathroom layouts

Unlike living and sleeping areas equipped with free-standing furniture, bathrooms do not let you rearrange the layout when you feel like a change, so you need to try to get it right first time. You can always introduce new accessories if you want to update the look.

Planning on paper

Using graph paper and a scale of 2.5 cm (1 in) to represent 1 m (3 ft), draw up a floor plan of your bathroom. Mark all the fixed features such as doors, windows and alcoves, and try out the fittings you want to include in different positions until you are happy with the effect. Take a look at the bathroom furniture designs on the following pages to get an idea of the different shapes and sizes available, including pieces specifically and often ingeniously designed to fit into corners or other small spaces.

A partition wall provides extra wall space against which to install this washbasin. The bath is ideally placed below the sloping ceiling where head room is not essential.

Don't automatically position the bath along the longest wall. A slightly shorter design may well fit across a shorter wall instead and leave you with more free floor space. Other useful positions for baths are against walls with windows (the bath will fit beneath the window, whereas a basin or lavatory cistern here might be a problem) and beneath sloping ceilings (where there is insufficient space to stand at full height, but plenty of room to lie down).

Bear in mind that doors and windows can be moved if necessary to improve the layout. A doorway repositioned across the corner of a bathroom will leave you with more usable wall space for essential fittings. And if the bathroom is under the roof, you might consider installing skylight windows and blocking up an existing window to provide extra wall space.

If you are short of wall space on which to hang heating fixtures such as radiators, consider installing plinth (kick board) heaters instead in the base panels of fitted furniture. Keep the floor and other surfaces clear wherever possible. Look for wall-hung furniture and fittings, and take advantage of any space-saving accessories such as wall-mounted soap dispensers.

Fitting in furniture

Bathrooms can be fitted with cabinet furniture on the same basis as kitchens, with streamlined units housing the sanitary ware and offering useful storage in the spaces between them. Tailor-made floor-to-ceiling units, incorporating just one or two panels of open shelving, provide a particularly sleek and stylish solution, although at a relatively high cost. Alternatively, you could build cabinets around individual items. Washbasins in particular benefit from cupboard housings that readily turn them into vanity units, providing a handy surface for toiletries around the basin as well as additional storage space below, behind closed doors.

If the room has any useful alcoves or recesses, make the most of these, too. A lavatory or basin can be positioned in a recess and the space above it fitted with a series of shelves to hold towels or toiletries.

Louvre doors and partition walls give the bath its own little room – a luxurious retreat from the rest of the family, who can still access the lavatory and washbasin.

Using screens

If you are planning to divide the room with screens for privacy, their positions need to be taken into account at the planning stage. A shower or lavatory, for instance, could be sited at the far end of the bathroom, with a partition built out at right angles to screen it.

If there is enough space in your bathroom, you could install both fittings like this, on opposite walls, each of them screened off from the rest of the room. This also has the advantage of giving the main room additional wall space, against which you can site the bath or washbasin.

Handy hints

• Check how much space is taken up by the door when it opens into the room – you might be able to free up space by rehanging it so that it opens outwards instead, or by replacing it with a sliding design.

• Positioning your bathroom suite around the edges of the room or against internal partition walls is generally the most sensible option for making the most of the floor space and for best dealing with all the necessary pipework.

Plumbing and pipework

Installing a new bathroom suite is a relatively uncomplicated and inexpensive process if you are happy to stick with the position of the existing fittings. The pipework is already in place, so all you need to do is plumb in a new bath, basin and lavatory.

To improve a badly designed room, though, especially if it is small, you may well find the best solution is to rearrange things, and this is when extra pipework becomes necessary, so be prepared for extra expense and upheaval:

- Moving a lavatory involves running an internal pipe to the external position of the waste pipe or repositioning the waste pipe, which is very costly. Consider simply rotating the lavatory by 90 degrees instead, which may be a good enough compromise for your requirements.
- Adding a shower, either over the bath or in a cubicle, means new pipework, plus connection to the electricity supply if the shower is a powered model.
- To save on running costs, keep the pipes for hot water as short as possible.
- Bear in mind that if you are going to end up with ugly internal pipework, it could be a good idea to install fitted furniture that will hide it all away behind neat panels and inside cupboard units, or to employ a good carpenter to box in the pipes for you.

An alcove makes a useful recess in which to site a washbasin or lavatory, while recessed shelves within a wall provide handy storage space for towels and toiletries.

Surface interest

It is the materials, though, that have made the biggest difference to modern basin design, and opened the way for sleek new shapes. Glass (chunky and textured like ice, or frosted and opaque) and stainless steel both create shallow, elegant 'dishes' that can be wall-mounted on brackets or can sit on countertops. These are the best for saving space, as the curves are streamlined, with no intrusive corners. If you have more room, robust materials like stone and marble create a more down-to-earth look, but still conjure up imaginative effects such as circular 'tub' basins or deep butler's-style troughs, like a smaller version of the traditional kitchen scullery sink.

Lavatories

Like basins, these are neatest when they are wall-hung, so that there is no pedestal cluttering up the floor, and the cistern is concealed behind the wall. Some are designed to fit right into corners; others to protrude as little as 50 cm (20 in) into the room. If you fit a floor-standing lavatory, it might be a good idea to look for a two-piece circular pan (originally a Victorian design, but still made by some specialist companies), so that the waste pipe link can be swivelled to the best position

before the two parts are cemented together – really useful if you are trying to fit the lavatory into an awkward space. Another space-saving option, but in complete contrast to the sleek modern look, is to fit a high-level cistern with a pull chain if you fancy an Edwardian-style bathroom.

Buying and fitting a new lavatory seat – there are plenty of modern designs to choose from – is the easiest way to improve the look of an existing lavatory that you don't want to have to replace.

Shower spaces

Now that the technology is so advanced, showers are increasingly desirable as the most refreshing, water-efficient and space-saving way of washing. Powerful pump-driven showers can be fitted over a bath (so as not to take up any extra room), in very restricted spaces such as under the stairs, in the corner of a

By freeing up the floor space beneath them, wall-hung lavatories and basins give the impression of space and are an ideal solution in small bathrooms.

Tap talk

There are dozens of different tap (faucet) styles and finishes – you need to make appropriate choices for your style of bathroom and decide whether you want individual hot and cold pillar taps or mixer taps (faucets). For example, a streamlined contemporary suite needs minimalist lever-style taps (faucets) with similarly clean lines, whereas traditional cross-head taps (faucets) suit a classic-style bathroom suite. Use the same design for your bath, basin and shower fittings for consistency.

Taps (faucets) can be fitted directly to the basin or bath, to the horizontal surface around it or wall mounted for a really neat look. Wall-mounted taps (faucets) help save space and keeps the surround free for toiletries. The mechanism will take up about 8 cm (3¼ in) of concealed depth behind the wall. Wall-mounted taps (faucets) are easier to keep clean, as there are fewer visible fittings to trap dirt. They are particularly recommended in the shower.

You can install a shower above an ordinary bath, but a specially designed shower bath is slightly wider at the taps (faucets) end and gives you more room in which to stand.

Open options

In many ways, it is better not to enclose the shower at all. If the bathroom is already small, why build an even smaller room inside it? Instead, leave the cubicle open, so that it has sides to contain the water jet (best tiled floor to ceiling for maximum water resistance), but no door boxing it off. If you do want to create a screen for privacy, the best curtains to use are double-sided, with a waterproof liner on the inside and a fabric outer layer in towelling or smart textured waffle, to soften the clinical edge.

Most luxurious of all is the 'wet room', where the whole space is tiled, and water simply flows away through a drain in the floor. This is costly to install because of the expanse of tiles required and the need to devise safe water drainage, especially in timber-framed houses. However, designers are already coming up with ways to solve the problem, for instance by setting an aluminium tray into the floor beneath the tiling, so it is worth thinking about. If you do opt for an open shower, remember to install effective storage for towels and bathroom tissue, otherwise these will get drenched along with the rest of the room!

The semi-enclosed shower in this completely tiled 'wet room' has no door and no shower tray – the water simply runs away via a drain in the room's tiled floor.

bedroom or as part of a small en-suite. They can even, if necessary, take the place of the bath altogether in the main bathroom.

If you don't want to give up your bath, you have the option of fitting it with an overbath shower or installing a specially designed shower bath, which is shaped with an integral shower tray at one end to create a larger area in which to stand. If you can find the space, though, a separate cubicle is more satisfactory – partly because it allows the bath and shower to be used at the same time, and also because the latest cubicles are now being designed with luxury extras such as body and steam jets.

You need to allow a square metre or yard for the average-sized floor tray, and make sure that there is enough space around it to allow you to get in and out of the enclosure easily – if you're short of room, a shower curtain or bi-folding door takes up less room than a normal door. Alternatively, corner cubicles are a neat way of maximizing the floor space.

reflective finish, which acts almost like a mirror in bouncing light back into the room and helping to open up a small space, but the effect can be fairly easily overdone, creating too prominent a glare. So matt tiles are an effective compromise, combining practical water resistance and plain colour with a more subtle, comfortable finish.

Mosaic effects, created from hundreds of tiny tile chips, are always intriguing and can sometimes work in small rooms, where their 'miniature' nature seems appropriate for the proportions. But again, be careful not to overdo and in consequence spoil the effect – it will be accentuated by a confined space, and therefore may be better limited to a single wall or shower lining rather than covering the whole room.

Other types of tile worthy of consideration in the bathroom are modern glass and lustred tiles. These are not to be confused with the chunky glass bricks used for partitions and interior walls, but are delicate wall tiles with either a clear, jewel-like glow or an iridescent mother-of-pearl finish. These provide a glorious range of colours and subtly shifting shades, with a gently reflective surface that gives a light-enhancing gleam rather than a bright shine.

Most often used on walls and floors, tiles are in fact useful for sealing any surface exposed to water. Here, they make a very effective bath surround.

Underfoot issues

The bathroom floor needs to balance aesthetics with practicality – you can bend some of the rules if the room will be used by just one or two adults, whereas a busy family bathroom will need a more rigorous approach (see pages 70–89).

Tiles Ceramic tiles, along with stone and marble, look smart and are waterproof and easy to clean, but they may feel cold and hard underfoot, and are inclined to get slippery when wet. If you want a little more comfort, or have children to worry about, you will need to add washable cotton rugs or go for a softer option altogether.

Wood Wood has a practical appeal that looks good and suits many different styles of décor, but floorboards will need to be well sealed (or painted) to make them waterproof, and you need to be careful with laminates, which have a tendency to warp and lift if subjected to constant damp.

Vinyl and rubber You may find sheet vinyl a better option; it is easy to lay (see page 84) and can imitate

Wooden flooring in the bathroom is unusual, but very stylish, as these decking-style tiles demonstrate. The wood must be well sealed to prevent water damage.

almost any look you want, from wood to stone or marble. Alternatively, you could go for something uncompromisingly modern such as rubber, which is warm, waterproof and incredibly hardwearing.

Bear in mind that you can always add wooden slatted duckboards to give a vinyl floor a more utilitarian look, or to provide a safe, non-slip surface on top of marble or ceramic tiles.

Turning up the heat

Bathroom heaters are part of the furniture these days, with sleek designs and contemporary colours contributing as much to the look of the room as the suite and tiles you choose. If you have wall space to spare, fit a ladder design that will double as a towel rail.
Radiators If your home's central heating system relies on radiators and you have limited wall space, extra-narrow radiators can be fitted into narrow strips, or look for floor-level cylindrical designs that run along a skirting board (baseboard) and pump out ample heat from a compact shape. The chunky scale of old school-style pillar radiators would be too big for a small bathroom, but you will find new versions with a reduced number of pillars and sometimes an integral towel rail, too, bringing this classic design right up to date for the modern home.

A chrome ladder-style heater/towel rail is an effective and stylish way to heat the room and to store and dry damp towels all year round.

Adding warmth and texture

Glass, mirror, chrome and ceramics are the key elements of most modern bathrooms – hard, cold, reflective surfaces that are practical, but not terribly inviting or comfortable for a room in which you want to be able to linger and relax. You therefore need to add softer materials that balance the chill and contribute warmth.

The most obvious of these ameliorating elements are bathroom linens – piles of towels in thick fluffy cotton or textured waffle, a deep-pile bath mat for wet bare feet, a bath pillow for relaxing against in the bath and an attractive fabric shower curtain. Classic white linens are always the simplest and soundest, but you can select other shades to coordinate with the room's colour scheme or add contrasting accents.

You might also want to incorporate areas of plain wood, for bath surrounds and fitted furniture. Antique pine or stained tongue-and-groove panelling suits a country-style bathroom, while the mellow finish of beech and cedar adds a smart European-style simplicity to contemporary bathrooms.

Piles of fluffy towels add a sense of comfort to the bathroom, while the texture provides a good contrast to the tiles, glass, mirrors and metallic surfaces typical of bathrooms.

Bedrooms

Planning your sleeping space

The bedroom is the one place where you can afford to indulge your creative decorating instincts without restraint. This is private space that does not need to keep the rest of the world happy 24 hours a day, so you can choose colours purely to please yourself and design tailor-made storage to suit your clothing.

Creating calm

For bedrooms to provide the restful space you need for sleep, they should be free of clutter. That means clothes put away, the floor clear of books and shoes and not letting the casual layering of bed linen degenerate into a tousled muddle.

Feng shui principles go so far as to dictate that bedrooms should be used for sleep and nothing else, that you shouldn't keep books here or have a study area, or have clothes out on show. In particular, outdoor clothes should not be visible, as these distract you from the inward calm of rest and sleep.

This approach may be a little extreme for some, but you can see the point. Especially in small bedrooms where space is limited, it is all the more important that the 'sleep centre' takes precedence.

To get your sleep centre right, you need to make good storage a priority so that you keep clothes, footwear and accessories out of the way and leave the floor clear for the bed. Do not automatically buy a massive wardrobe (closet) if most of your clothes do not need full-length hanging room – it will be wasted on separates and shirts that take up only half the space. Work out whether your clothes need shelf, drawer or hanging storage, allocate space accordingly (see page 234) and stick to the system – don't let things stray.

If possible, position the bed where you can walk freely all the way round it and don't have to climb over it to reach essential items. If space is really tight, for example in attic bedrooms or beneath awkward

Make your bedroom a calm haven to which to retreat at the end of the day – decorate the room in restful colours and keep it free of clutter.

ceilings, fit the bed where the height is lowest. This space will not be much use for storage, but is fine where you will only be lying down, and can create a cosier place in which to lay your head.

Dual-purpose bedrooms

In reality, contrary to the feng shui ideals above, many people's bedrooms are increasingly used not just for sleeping and dressing, but also function as a work or sitting room, especially if the room is a particularly large one. Indeed, bedrooms can successfully double as work spaces as long as the decoration is kept neat and simple and the space is well organized between the two areas. If the room is multi-functional, you might want to disguise the bed with tailored covers during the day.

Adaptable furniture allows your bedroom to be used for more than just sleeping. Here, an attractive daybed doubles up as a comfortable place to sit during the daytime.

Clutter control

• Always put clothes away when you take them off, or put them in a linen basket or laundry bag ready to wash.

• Throw away clothes that no longer fit you, that you regret having bought, that are marked or damaged beyond practical use or that are no longer in fashion (you can keep the occasional favourite for fancy-dress parties). Don't hold on to them on the basis that you paid for them and want to get your money's worth – better to cut your losses and get rid of the guilty irritation you feel every time you see them.

• Don't allow magazines and books to accumulate beside the bed. Recycle, file or return them to the bookshelves.

Screens and partitions

One way of keeping the bed area separate and free of clutter is to create a floating wall at the head of the bed, which divides the bed from storage. This need not take up much space – it simply gives the bed a sort of extra-high, free-standing headboard, pulling it forwards into the room rather than standing it against a structural wall. The extra space behind the bed can then be used for storage or a dressing room. Although it reduces the floor space that the bed stands in, it makes the area look bigger by freeing it up from other furniture. If you don't want to construct a fixed wall, you could achieve the same effect with a free-standing screen.

Proportion and scale

A large bedroom gives you plenty of scope for some grand-scale furniture like a four-poster and a double wardrobe (closet). If the bedroom is small, you need to create a light and airy impression by making sure that any peripheral furniture is small and neat. Slim console tables – the sort designed to stand against a hall wall – can take the place of full-sized dressing tables, and bedside tables only really need to be big enough to take a lamp, clock and perhaps a separate radio.

The one item that is worth having as big as possible in the bedroom is a mirror. As well as providing that all-important clothing check, it will reflect the available light and make the whole room feel twice the size. You don't even need to fix it in place – a huge mirror leaning casually against the wall can look wonderful.

A large bedroom can provide a quiet and private work space, but keep the décor simple and the desk area tidy so that it doesn't encroach on the room.

Colour and texture

A bedroom should be a haven – comforting, relaxing and in your own taste and style. A good night's sleep is essential, so bedroom furnishings should be as comfortable as possible, and window dressings designed carefully.

Creating atmosphere

The atmosphere in your bedroom is up to you – go for luxury and decadence, with rich fabrics and sumptuous textures; or for calm sophistication, defined by elegant tailored lines and cool, neutral colours; or perhaps the refreshing simplicity of painted floors and furniture and crisp gingham covers. The trick is to devise a bedroom that is a retreat from the rest of the world, where you can switch off and totally relax.

Since you see the fabrics in your bedroom first thing in the morning and last thing at night, it is worth taking time to find something you really like.

Handy hints

• Cream-coloured curtain linings, and cream or off-white sheers and voiles, are often better than pure white ones in urban homes, where the city grime seems to creep in at every window.

• A small stool or armchair in the bedroom that is rarely sat on could be covered in a lightweight silk or a pretty cotton rather than a heavier furnishing fabric.

• A bed that doubles as a seating area during the day needs a hardwearing cover made in a durable upholstery fabric. Light, curtain-weight fabrics are unsuitable for this purpose, as they crease and look worn if sat on regularly.

• A simple cream or white cotton bedspread is an excellent way of brightening a dark bedroom, as it will reflect light back into the room. It can easily be livened up with patterned cushions.

Bedroom colours

Colour is a crucial element here. Choose shades that will be gentle and relaxing at night and look good by artificial light – and if used during the day, refreshing and comfortable in natural light, too.

Soft blues, greens and mauves are cool and restful; pinks and yellows will add more warmth if you want it; creams and whites are unfailingly calming and elegant. However, you might want bright, rich or deep shades that surround you with more definite colour. In a small bedroom that already feels warm and protective, such colours will accentuate the womb-like effect.

Layered linens

Bedrooms automatically provide masses of fabrics with which to build up your colour scheme and add texture. Most people now opt for a duvet (comforter) on the bed, which is practical and comfortable, but does swamp the bed in a layer of single colour or pattern. The contemporary revival of classic sheets and blankets offers more elegant lines. Alternatively, for sheer luxury, indulge in an extravagant collection of contrasting fabrics and varied textures. For instance, combine the elegant drape of heavy woven blankets with silk eiderdowns, then add deep piles of different-sized cushions and pillows on top.

The sparkle factor

With so much light-absorbent texture around, a bedroom can feel slightly muffled by fabric, so try to incorporate a few reflective surfaces to catch the light and bring the space to life. Mixing silks and satins among your bed linens will help, as will using mirrors and metal-framed furniture, lamps and frames, while the glass droplets of chandelier light fittings can add a more decorative sense of sparkle.

Flat wall colour can be enlivened by painting any woodwork in eggshell paint, which has a slight sheen to it. You could even cover one wall or a chimney breast with painted wood panelling to extend this effect over a larger area of the room.

Barefoot comfort

A neutral-coloured carpet, or a natural floorcovering in cotton, linen, wool or jute, will be soft and warming beneath bare feet. Wood – especially white-painted floorboards – looks good in bedrooms, too, and is

White-painted floorboards look good in a bedroom, although you may want a soft rug beside the bed for your bare feet when you first get up in the morning.

useful in small rooms for opening up the space. If you have a hard floor surface, consider using soft rugs or carpet pieces to both soften the room a little and provide more sound insulation.

Low lighting

Lighting needs to be kept subtle and atmospheric; aim for mood rather than practicality – unless, of course, your bedroom is a multi-purpose room. Fit dimmer switches to keep levels soft and adaptable, and use lamps to diffuse the light source and accent individual areas rather than flooding the whole room, providing illumination where required for reading or applying make-up. If you have built-in cupboards, consider fitting interior lights (in heat-resistant casings), which make it easier to find things.

Restful windows

In bedrooms, curtains and blinds need to exclude light, provide privacy and also be heavy enough to help soundproof the room against outside noise.

Keep the colour of your curtains or blinds similar to the walls to avoid jarring blocks of sudden colour or pattern. Curtains will be drawn closed more often here than in daytime rooms, and a dominant fabric that provides an attractive frame for the window when drawn back or rolled up may look overpowering when opened out in larger panels. It is usually more restful to surround yourself with a single sweep of colour.

Instead of heavy curtains, you could use a light-resistant or roller blind, and combine it with a drape in a sheer fabric to soften the look.

A large mirror in this run of cupboards (which slides up to reveal a concealed television) helps reflect light and make the room seem larger.

Bedroom storage

Most of your bedroom storage will be for clothes and accessories, but you may also need to house toiletries, make-up, books, personal papers and mementos. As with storage in living rooms (see page 206), you need to decide how many of your possessions can stay out and how much room you need to hide things away.

Bespoke systems

As in living rooms, closed storage will help keep the room free of clutter and, once again, the most efficient system is to build it in to make full use of the area available and to create customized spaces for different kinds of item – in effect, taking the principle of the fitted kitchen and applying it to the bedroom. Fitted-

bedroom companies have perfected this as an art form, and it is worth looking at the gadgets and tricks that they employ to find a place for everything, even if you don't want to buy their cabinets. The key is to create several individual compartments behind your closed doors rather than one long hanging space, giving you the flexibility to treat each section independently and devise different storage systems for each one.

First, work out how many of your clothes need to be hung rather than folded, then divide hanging clothes into full length and half length. Dresses, coats and long skirts need full-length space, so one section of your cupboards will need to be fitted with a high rail. Shirts, trousers, skirts and jackets need only half the length, so you can save space in another section either by fixing two rails, one above the other to create double-decker storage, or by fixing one high rail and then fitting cupboards or shelves beneath them.

Built-in storage makes full use of the space available, while glass-fronted rather than solid doors help avoid the potentially austere effect of a run of matching cupboards.

Headboard shelves

By building a slimline cupboard out of MDF (medium-density fibreboard) and securing it firmly to the wall behind a plain divan bed, you can make a custom-made headboard and bedside shelf at the same time, thereby creating space for books, photographs, radio and a reading light, and eliminating the need for bedside tables if floor space is in short supply.

Make the cupboard about 15 cm (6 in) deep, 1 m (3 ft) high and a little wider than your bed. If you design it with end panels that open to let you slot flat items inside, you can create extra storage space for things such as pictures that you have no room to hang, artwork and document portfolios.

The rest of the space can be taken up with shelves and drawers in different sizes and depths. Fit extra-deep drawers for jumpers, with smaller, shallower ones for underwear and socks. Look out for transparent drawers in wire or acrylic, which let you see their contents without too much rummaging. Make use of high-level shelves to store items not needed so often, then take advantage of additional ideas that are designed to be slotted into larger systems to help keep things neat.

Racks fixed to the inside of the door, for example, will hold items like scarves and ties; honeycomb organizers can be fitted inside drawers to 'file' small items individually; shoe racks will stand in the bottom of the cupboard to keep pairs together without scuffing; and shoe hangers comprise a strip of open-sided canvas 'cubes', each one taking a single pair of shoes. Incorporating a few deep shelves will also give you space to stack boxes and baskets commandeered as improvised 'drawers'.

Space-saving doors

If space is tight, you don't want to obstruct it with cupboard doors. Consider fitting sliding doors rather than hinged ones, so that they don't swing out into the room. In a very small space, it helps if the doors are also mirrored, to increase the amount of light and make the room appear twice as large.

Double-depth cupboards

If you have two bedrooms linked by a partition wall, you can 'pool' their storage space by replacing the wall with a deep layer of cupboards, in effect creating a double wall with a storage cavity between them. This lets you give each room a double-depth cupboard, each one using up the full depth of the wall space yet only half its length, so that the two cupboards sit side by side, but face in opposite directions, like the storage equivalent of a loveseat.

To create a 'shared' closet with access from both rooms, leave one end of the wall free of cupboard fittings, but fix a door on both sides so that you can walk from one room to the other. This area can then be equipped with a rail running the full depth of the double wall and will provide ample hanging space.

Creative hanging

Additional hanging storage can be supplied by free-standing clothes rails on wheels. As an alternative to the full-sized wardrobe (closet), canvas-sided designs that close with zips (zippers) or fabric ties are a good temporary measure, as they can be collapsed and folded away when not in use.

Portable clothes rails are useful for temporary storage, for example for overnight visitors, and for odd corners where there is not enough room for a full-sized wardrobe (closet).

Rigid wicker baskets are widely available in all sizes and make attractive storage containers for all manner of items. Baskets lined with fabric are ideal for storing spare bed linen.

Hanging storage is particularly useful if it keeps floor space clear, so look for sets of fabric pockets for hanging inside a cupboard or on the back of a door to hold shoes, socks or underwear. A peg rail or row of hooks fixed to the wall in traditional school-cloakroom style will keep belts, scarves and hats neatly lined up, or can be used for drawstring bags to hold dirty laundry and clothing accessories.

Extra storage ideas

There are masses of additional storage devices to help keep bedroom paraphernalia hidden and turn your room into a stress-free space for relaxation and sleep.
Chests, trolleys and cabinets Look for free-standing chests fitted with extra-deep drawers designed to hold jumpers, dressing tables with drawers or deep-drawer trolleys that can be pushed into place wherever you need them. Or improvise the same effect with an office filing cabinet painted in bright enamel paint.
Baskets and boxes Useful for items that don't need everyday access, big laundry baskets – reminiscent of out-size picnic hampers – and wooden blanket boxes can hold spare bed linen and bulky sweaters, and additionally provide a useful table surface or window

seat on top. Similarly, a collection of wicker hampers or old leather suitcases stacked stagecoach-style in descending order can be used to hold out-of-season clothes or other items not used regularly, creating a smart display in an alcove or corner.

Smaller boxes are perfect for storing scarves, gloves, underwear, belts and jewellery. Look for decorative hat boxes that will sit on a chest of drawers (dresser) or in a disused fireplace, and Shaker boxes, traditionally made from thin layers of cherry wood curved round into an oval shape. On a more prosaic level, salvage and cover shoe and boot boxes with fabric or paper, or choose from the numerous designs of ready-decorated storage boxes in assorted shapes and styles. Label them and store under the bed, on top of cupboards and in shelf units.
Under-bed storage Rigid plastic 'under-bed' boxes are another option for storing things away. These large, shallow, lidded boxes on castors are specially designed to fit under beds (even low ones) and can be easily pulled out when required. Alternatives are heavy-duty plastic or canvas zip-up box-shaped 'bags'.

Use an ordinary vacuum hose to suck out all the air from plastic vacuum-storage bags and the bags will then take up less than half the space that they would otherwise occupy. This makes them ideal for storing bulky items like spare duvets (comforters) and pillows.

These attractive seagrass storage baskets with lids can be filled and then stacked up on one another to create a display feature in themselves.

Indulge in some creative thinking and you may come up with some unusual storage solutions, like this idea of using a wooden stepladder to hold shoes.

Shelving Open shelves keep things more colourful, so you could stack T-shirts and jumpers in colour-coordinated ranks like a shop display, but you will need to be scrupulous about folding them, cleaning them regularly and guarding against moths, which like nothing better than a soft pile of woollens.

Shelves are great, too, for books and ornaments, but are obvious dust traps and need regular cleaning and tidying to prevent accumulation of clutter. If you don't trust yourself to keep them looking neat, consider hanging fabric in front to curtain them off from view.

Lateral thinking

Make use of storage not specifically designed for clothes or bedrooms. Pharmacy chests and kitchen spice chests provide useful little drawers for items that would easily get lost in a larger space. Modern stationery shops stock all sorts of miniature filing cabinets, originally intended for desk essentials, but perfect for things like jewellery and hair accessories. Vegetable racks can slot inside cupboards to hold foldable clothes, and old-fashioned bicycle baskets, with one usefully flat side, can be strapped on to a tie rack on the inside of a cupboard door.

Tailor-made storage

Coats really need to be hung on well-padded hangers to keep their shape. Hanging them by a skimpy collar loop drags the fabric and gradually spoils the line of the tailoring. An even better proposition is a proper dressmaker's dummy, complete with life-sized curves so that it provides coat storage and an innovative sculptural display all in one. Unfortunately, the classic solid wood models are rarely found nowadays, but simpler versions with cotton stretched over a basic frame are more commonly available. Alternatively, you could use a shop-window dummy made from moulded plastic or clear perspex.

Family heirlooms or antique finds like wooden sea chests or old leather trunks are great for storing out-of-season clothes or other items that you don't need to access every day.

Children's rooms

Creating room to grow

Children's rooms are self-contained worlds quite unlike anywhere else – they can easily become 24-hour hideouts, where children live by their own timescale, and where work, rest and play merge seamlessly into one another. Decorating needs to be practical – plan for durable surfaces, effective storage and children's preferences for bright colours and peer-group fads.

Easy as ABC

You could say that the best approach to furnishing children's rooms is to make it as easy as ABC – A being for adaptable, B for bright and C for cleanable:

- **Adaptability** is essential if the room is to cope with growing children (and possibly extra children), multiple uses, all-day activities and sleepovers.
- **Brightness** is stimulating and space making, and steers a neat path between soft baby pastels and teenage grunge.
- **Cleanability** is a must for all ages, however sophisticated the children regard themselves.

In addition there are issues like safety to consider (see box, opposite), and good lighting for bedtime reading and desk work, plus a tabletop for homework and enough floor space to spread out toys, games and puzzles. You also want to create an environment where the child's imagination will flourish and develop.

Work, rest and play

Children's rooms must provide areas for private work and play, as well as a safe space for sleeping. So the room must either have a separate section for each, or be flexible enough to adapt. In most homes it has to be the latter, so think laterally and be inventive.

Allow an older child some input into the decisions regarding the décor of their room. They are more likely to look after it and may even help paint the walls!

Children's tastes and needs change much faster than those of adults. You don't want to have to redecorate at every stage of their development or every time they acquire a new cartoon hero or pop idol. The trick, therefore, is to stick to a background of plain colour that can be updated with accessories when a trend is outgrown or a new look called for.

Safety check

• Bunk beds are not advisable for children under six years old.

• Avoid trailing flexes and fit unused power points with safety covers.

• Be careful with furniture that can trap fingers – use stoppers on doors and drawers to stop them slamming shut, and replace heavy-lidded wooden blanket boxes with lighter baskets.

• Fit childproof window locks if window access is at all possible.

It is also a good idea to get the children involved in the initial decorating decisions, because if they feel that the room is really theirs, they are more likely to look after it and keep it tidy. That doesn't mean wallpapering in the latest 'brand' pattern that will quickly outgrow its useful life – just let them help choose colours and fabrics.

Defining with colour

Children need colour for stimulation, and although pale, space-making pastels are soothing for nurseries, once children become aware of their surroundings, they tend to be more inspired by bold, bright colours with a sense of fun and contemporary style. If you are letting them choose for themselves, you may need to steer them in the right direction to prevent their choice of colour resulting in an enclosed, oppressive feel, especially if the room is small. Bear in mind that yellows (being naturally lighter) and blues and greens (being receding colours) will feel less oppressive than bright reds and deep purples.

The best way to satisfy the demand for current crazes is to have painted walls as your base and accessorize them with peel-off borders and cut-outs that can be replaced and updated. Look out, too, for luminous stick-on effects such as planets and stars to create a night sky on the ceiling, or metallic or glow-in-the-dark paint. Areas of blackboard paint will encourage creative games and writing, and magnetic paint can be used with letter and number magnets, or with fridge magnets to clip pictures to the wall. You can repaint these as often as necessary.

Give children floor space for their games and areas for creativity like this boarded-up fireplace painted with blackboard paint to encourage drawing.

Decorate children's bedrooms in bold, bright colours – this pink and yellow combination makes a stimulating colour scheme for a little girl's room.

Carpet is always comfortable underfoot in bedrooms, but a neutral-coloured one is probably best kept for older children who are beyond the messiest age.

Ground rules

Children's floors need to be tough, washable and – again – adaptable to different themes and colour schemes. If your children are past the messiest stage and have reached an age where they demand a few luxuries, you could go for a comfortable carpet in a mid-tone neutral colour that will not show too many grubby marks, can be cleaned when necessary and does not restrict your decorating plans.

For younger children, cushioned vinyl or plain wood both fit the bill, and can be covered with washable rugs and playmats for colour and comfort. It is important, however, that you add a non-slip backing so that they don't slide on polished surfaces.

The advantage of wooden floorboards is that you can repaint them as often as you like (see page 76), sticking to a single all-over colour to match the furnishings, or adding fun patterns and games for a more inventive effect. Either use specialist floor paints for a tough, hardwearing finish, or add several coats of clear varnish to seal the colour after you have finished painting.

An alternative – if you like the idea of colour on the floor, but want to keep it adaptable – is to use multi-coloured carpet tiles that can be relaid in different patterns as often as you want, creating stripes or chequers, or a block of central colour contrasting with the outer border, to mimic a rug. Carpet tiles also have the practical advantage of allowing you to move worn or marked areas to a less-obvious position, such as under the bed, if the flooring is past its best.

Flexible furniture

If you are thinking about adaptability, keep the furniture as simple as possible. A few specialist buys are invaluable. Beds need careful thought (see page 246) and it is useful to choose a cot that will 'grow' into a bed. A changing unit is essential for a small baby, but look for one with plenty of practical cupboard space so that it will still have a useful life when the baby develops beyond the nappy (diaper) stage.

Over and above that, think flexibly and plan for furniture that will adapt to different uses. Instead of miniature chairs and tables, opt for squashy beanbags and basket chests that can be used for seating and storage when the child is older, too. Modular furniture that can be rearranged into different layouts is always an advantage, and anything on wheels is useful.

You can easily attach add-on castors to furniture and storage boxes – this gives you plenty of options for keeping room layouts flexible.

A changing unit is useful for a baby, but choose one that has the potential to be adapted into an equally useful piece of furniture for an older child.

This is also the place to make use of unwanted furniture from other rooms, as well as jazzed-up junk shop or flea market finds. Keep a lookout for any useful, adaptable cupboards and chests that can be transformed with a coat of bright, washable paint. Again, you can change the colour as often as you like or select a range of tones or contrasting colours for doors and drawer fronts.

Versatile themes

Avoid blue-for-a-boy and pink-for-a-girl colour schemes that will limit your furnishing ideas and make it much more difficult to arrange for room swaps and sharing at a later stage. The same goes for gender-specific patterns such as flowers and footballers – you are better off with generic designs such as checks, stripes, spots and plain bands or blocks of colour.

Accessory accents

Fabrics and bed linens give children a bit more scope to indulge in their current crazes, although it is still best to avoid strongly branded products that will be out of date or outgrown by next season. Look for bright colours and stimulating patterns – reversible, if possible, to provide extra variety in a single duvet (comforter) cover or pillowcase.

Mixed textures make children's furnishings more fun, and help to stimulate younger children's imagination, so try to incorporate different fabrics such as furs, felt, velvet and wool. Furry cushions and soft rugs are always popular with children.

If your children can't live without their favourite character in their bedroom décor and you are reluctant to introduce it in fabrics, look for lampshades, wastepaper bins, cardboard storage boxes, coat hooks and of course posters which are relatively inexpensive items that can be replaced more often and may be a suitable compromise against a fairly plain background.

For older children, introduce intriguing accessories such as lava lamps to bridge the gap between child and adult – they will enjoy the combination of 'fun' and 'vaguely scientific', and these items also create a very soothing effect to help them sleep.

Children love bold colours, patterns and textures like shiny metallic fabric, animal prints and faux fur – these shocking pink and orange furry cushions will delight most little girls.

Shared rooms

Children need their own territory to help them establish their personality and take responsibility for their possessions. If they have to share a room, find ways to create room divisions so that each child feels they have their own private space.

Division by colour

Colour-coding different areas is one of the simplest solutions with room sharing. Let each child choose their own wall colour – or, if that seems too risky, pick two distinctive colours yourself that will work together yet create a definite contrast. You could then give each child two adjacent walls in their colour, so that the corner formed by each pair becomes their own territory. Alternatively, divide the room down the middle so that the colour changes halfway across two facing walls. You can have decorative fun with this border line by painting a wavy line or jagged zigzags so that the two colours fit together like a jigsaw.

   Provide different styles of bed linen to maintain the division – where possible, look for duvet (comforter) designs that contain both colours, so that they

You may not be able to literally divide a shared room down the middle, but different-coloured bed linen and accessories help mark 'territory'.

Individual style

To emphasize the idea that each child has their own separate territory in their shared environment, make sure that both sides of the room have plenty of display space to allow them to get creative in their own individual way. Fix pinboards to the walls to display pictures and poems, and include areas of magnetic paint (brilliant for attaching things without the need for drawing pins/thumb tacks or sticky pads) and blackboard paint, so that they can use coloured chalks to scribble on the walls to their heart's content.

coordinate well, but provide contrasting patterns – and make sure that the floor changes colour from one area to the other, too. Wooden floorboards can be painted – and repainted – as often as the need arises. Loose rugs (with non-slip backing) in contrasting shades will effect a quick change and add a soft surface. Alternatively, you could create two reverse-image patterns with carpet tiles, designing a layout of, for example, blue tiles scattered across a green background on one side of the room, with green tiles against a blue background on the other.

An open shelf unit can divide a room very effectively, either partitioning a shared bedroom or, as here, separating a work area from a play area.

Dividing lines

For a more definite divide, furniture can be used to great effect, forming an actual partition. Open shelf units are the most practical because they provide useful storage for both sides – for books, toys, CDs and school stuff – but their open structure still lets light through, so that the room does not feel blocked off.

Look for wide-based, solid structures to make sure that they can stand securely without support, and children should be made clearly aware that they are not for climbing on. Modular cube systems work particularly well, because you can devise a partition in exactly the design you want. Stack the cubes to different heights to build a battlement-like structure, or arrange them in an ascending slope with full height at one side of the room, reducing to ground level.

Trellis screens

If there is not enough space for a chunky storage system to divide the room, create a trellis screen instead. Again, this will maintain plenty of light in the room, but will also provide useful hanging space and a pinboard background for displaying pictures and posters or sticking reminders in place. You can use standard garden trellis, either fixing it in place as a permanent divide, or joining two or three sections together with hinges so that they form a free-standing zigzag screen that can be moved into different positions or folded away when not needed.

Temporary partitions

Children are very capricious, and the clamouring for a partition could easily change to demands for a return to the open-plan room, so a flexible screen that can be supplied and removed again at short notice is a sensible solution. If you are short of floor-standing space, you could achieve a screen effect simply by fixing a row of ceiling-hung blinds to form a removable fabric divider, which can then be lowered or raised in just a few seconds.

Children's beds

Adults spend about one-third of their lives in bed, while small children with early bedtimes and older children with a reluctance to get up in the mornings spend closer to half. It's therefore important to get the bed right, although for young children it's not just for sleeping – it's also a play area for a thousand imaginary games.

Beds with storage

Children tend to have so many games and toys nowadays that the usual rules of clutter control have to be relaxed a little when it comes to planning a child's domain. It is therefore worth looking for a bed that incorporates or allows for some form of storage. At its most basic, this could simply be a bed with a frame that stands clear of the floor so that there is plenty of space to push large robust boxes and crates underneath (see page 236). Alternatively, choose a wooden divan-style bed with storage built into the base. This can take the form of full-length storage drawers or a mixture of drawers and cupboards.

More imaginative options are the mid-height play beds or 'cabin' storage beds designed along bunk bed principles, with the mattress on a platform reached by a ladder, and the lower level providing cupboards, drawers, shelves and pull-out desks. Some styles even incorporate a full-scale playhouse underneath, with doors and windows cut into the side of the raised bed; others have fabric to create a tent. Alternatively, you could always make a play area or private den in the space under a simple platform bed yourself by hanging colourful curtains that can simply be drawn across and placing a low level table and chair inside.

Fantasy beds

Young children will devise their own games around any structure in the home – hence the enduring popularity of the cupboard under the stairs, and the tent improvised from blankets and a clothes airer – and this includes their bed, however basic it is. If you want to give their imagination a prompt, however, you can find children's beds designed as racing cars, fire engines, boats, buses, cottages, castles and spaceships. It may seem indulgent, but it may be worthwhile if it provides a complete play centre in one piece of furniture and keeps the rest of the house clear of toys and chaos.

Bunk beds

Full-scale bunk beds with double-decker sleeping spaces are always a favourite with children, so exploit their enthusiasm – bunks are a godsend if you have two children sharing a single room or if their friends need to be accommodated overnight.

The frames can be made of traditional wood (which can be painted if you want to introduce extra colour in the room) or contemporary steel, and both types come flat-packed for home assembly. Some bunk bed designs

A recess with a sloping ceiling is the ideal location for a snug built-in child's bed like this one, which has handy storage space incorporated within the base.

Handy hints

• Instead of a cot that becomes redundant once the child has outgrown it, opt for one with removable sides that converts into a proper bed, so that he or she can carry on sleeping in the same place. The most elegant ones are designed like French lits-bateaux, and can even be used with cushions along the back as a sofa or daybed when the child reaches teenage years.

• Bunk beds make good use of vertical space, but standard sizes are often wider and longer than they need to be. For very small rooms, consider commissioning a reliable carpenter to build tailor-made bunks.

• Headboards provide extra colour and stimulation for children, and you can always create your own by painting a headboard-sized panel on to the wall. Or fix a piece of corkboard to the wall and cover it with fabric, creating a few deep pockets for books, soft toys, torches (flashlights) and various other bedtime essentials.

can be detached into two single beds – a useful feature if your sleeping requirements are likely to change, for example if you intend to move house or as the children get older and require separate bedrooms. Some of the most convenient designs are made with a lower-level sofa bed with a deep padded cushion along the back, so that it provides cosy seating during the daytime or when not needed as a conventional bed.

Make sure, if you do end up with a high-level bed of some sort, that you make the most of the room's height by fitting high-level shelving alongside it to match. A set of slimline shelves next to the upper mattress will provide a useful bedside bookcase as well as a surface for an alarm clock or radio, although some bunk beds come with their own clip-on tables, which is useful.

Lamps should be clipped on to the bedframe for increased stability rather than free-standing on a high shelf, so that there is no risk of either trailing flexes or hot bulbs being knocked over on to the bedclothes.

This standard child's bunkbed has been given a creative makeover with a fairytale theme. It comes complete with a turret and slide and would capture the imagination of any little girl.

This high-level bed has a sofa bed beneath, providing comfortable daytime seating and extra sleeping space for an overnight guest, as well as space for a desk.

Clever storage

The key to kids' storage is to make it fun, so that they will actually deign to use it. Exploit every possible corner and encourage your children to organize their possessions by keeping clothes, school things and play things separate.

Low-level storage

Low-level storage is the most practical for young children, as they can easily reach and put things away for themselves. Baskets and crates that slide under a bed can also be stacked against a wall so that they don't take up so much floor space. Picnic hampers and laundry baskets make good toy boxes, as the lids are light and will not trap small fingers. Plastic crates that fit together for neat stacking create a colourful display rather like a tower of building blocks, or you can buy or make frames to hold them. The most useful of these

come on wheels to form a sort of movable trolley; if you choose semi-transparent boxes, you'll be able see what is inside.

Cube systems

Cube storage is ideal, as the arrangement can be changed as children grow, starting with a single row of units along the wall to provide low-level cupboards or open slots with a deep shelf or seat on top, and then building them up to higher levels as the child gets older and wants space for books, CDs and a music system. You can also slot individual containers such as shoe boxes and filing crates into the cubes to hold smaller items. Cubes are particularly useful in attic rooms, as you can create triangular systems under sloping ceilings and fit a run just one or two cubes high where the walls are lowest.

Choose storage options suitable for the age and height of the child – you cannot expect a toddler to put his own toys away if the boxes are beyond his reach.

Unfinished wooden cubes can be painted to suit and provide plenty of options for storage, which can be moved around and adapted as your child's needs change.

It is worth painting the cubes to make them more interesting, as colour is important if you want the idea of storage to appeal to messy children. Similarly, basic wardrobes (closets) and chests of drawers (dressers) can be painted in harlequin shades to brighten them up, each drawer a different colour, or doors and handles contrasting with the main structure.

Fabric covers

If you despair of children being able to keep open shelves tidy, you can easily turn them into cupboards by fitting curtains or blinds to cover them. Fabric covers are simple to make and an economical and space-saving alternative to solid woodwork. You could also opt for canvas or plastic clothes rails rather than full-scale wooden wardrobes (closets); these are much less bulky and provide plenty of adaptable storage to keep children organized. Bear in mind that children do not in general need as much hanging clothes space as adults (most clothes can be folded and laid flat), and what hanging space they do need will be shorter.

Fix storage tubes (open cylinders of cardboard or perspex) to the walls to hold school artwork and posters without crushing or folding them. Neatest of all, fix up a length of wall-mounted trellis to which you can attach as many hooks as you want without damaging the wall behind.

Don't forget overhead space either. Lengths of webbing (like fishing net) and trellis panels can be suspended from the ceiling to provide extra storage, especially for soft toys that won't hurt themselves or anyone else if they fall.

Creative cupboards

Use your creative talents to devise colours, effects and trompe l'oeil façades that incorporate storage into the overall style of the room, so that the cupboards become part of the decoration and open up imaginative new worlds rather than seeming a waste of good play space.

- Paint a lighthouse on a tall cupboard to create the background for a seaside-style room. Do the same with a helter-skelter design to conjure up a fairground setting.
- Turn a little girl's cupboard into a doll's house façade. You could even divide up the shelves inside so that they make different 'rooms'.
- Fit tongue-and-groove panelled doors across alcoves to turn them into cupboards, with rope pull handles for a nautical, beach-house effect.
- Paint a narrow cupboard door in bright stripes reminiscent of a Punch and Judy theatre, or hang a panel of striped fabric across the front for an even more authentic effect.
- Create a jousting-tent cupboard, with a painted pelmet fixed over an alcove to form the turret top, and a length of striped fabric that can be rolled up like a blind for access to the shelves behind.
- Paint a cupboard in alternating blocks of colour for a harlequin pattern. Mark out your design in pencil first, and use masking tape to give each square a neat edge, peeling the tape off after the paint is dry and the pattern complete.

With some creative thinking, pots of paint and novelty cupboard or drawer handles, you can easily transform plain or ugly items of furniture into child-attractive pieces.

Index

Acknowledgements

Executive Editor Katy Denny
Editor Charlotte Wilson
Executive Art Editor Joanna MacGregor
Designer Colin Goody
Senior Production Controller Manjit Sihra
Picture Library Manager Jennifer Veall

Picture acknowledgements

Graeme Ainscough 8, 14 top left, 29 top left, 70, 77, 82, 86, 88, 89, 90, 93 bottom right, 97 top left, 99 top left, 102 top right, 104, 105 bottom right, 116, 117 bottom right, 124, 126 bottom left, 130, 131 bottom right, 134, 136, 138, 140, 150, 152, 153, 154 top right, 154 bottom left, 155 bottom left, 157 bottom, 158 top right, 159 top left, 161, 162, 163 top left, 163 bottom right, 164, 165 top left, 165 bottom right, 166 top left, 166 bottom right, 167 top left, 168, 169, 170 top right, 171 top left, 172 top right, 172 bottom left, 173, 174 top left, 174 bottom right, 175 top left, 175 bottom right, 176, 178, 179 top right, 182 bottom left, 183 top right, 186, 187 top left, 189 bottom right, 191 bottom left, 192, 193 top left, 197, 198 left, 198 right, 200 top right, 200 bottom left, 201, 203 bottom right, 205 top right, 205 bottom right, 207 top left, 208 top left, 212, 213 top left, 216 top right, 216 bottom left, 217, 220, 221 top left, 222 bottom left, 223 top right, 225 bottom right, 226, 230, 231 top right, 232, 242 top left. **Alamy/Nick Baylis** 72 bottom right; /Nick Hufton/VIEW Pictures Ltd. 210; /Tim Street-Porter/Beateworkls Inc. 245. **Alma Home** (12-14 Greatorex Street, London, E1 5NF; Tel: +44 (0)20 7377 0762; Fax: +44 (0) 20 7375 2171.) Black leather embossed crocodile 81 bottom left. **The Alternative Flooring Company** (Tel +44 (0)1264 335111; www.alternativeflooring.com) 87 bottom right, 87 bottom left, 87 centre left top, 125 centre left top. **Amtico** (www.amtico.com; Tel: +44 (0)800 667766) 125 centre left bottom; /AH838/Colour-Flash Blue to Purple 81 centre left bottom; /GL014/Champagne Glass 81 top right; /W685R/Drift Wood 81 centre right. **Anaglypta**/Wallwhite by Anaglypta 111 bottom right. **Armitage Shanks**/Sandringham 215 bottom right. **Armstrong** (www.armstrong.com) Timberwork dark silver vinyl flooring 83. **Laura Ashley** (Enquiries/stockists +44 (0)870 5622116; www.lauraashley.com) Small Pink Roses/Pink 42 bottom left; /Candy Stripe/Sapphire-White 43 top right; /Harlescott/Pale Gold 125 bottom right. **B&Q plc** (www.diy.com) Lemon Shaker style kitchen 182 bottom right; /Mondo Carousel185 bottom right. **Brintons** (www.brintons.net) 125 top left. **Ceramic Dolomite** 219 top left. **Jane Churchill** (Tel: +44 (0)20 8877 6400) 118 bottom right. **Clarence House** (3/10 Chelsea Harbour Design Centre, London, SW10 0XE; Tel: +44 (0)20 7351 1200; Fax: +44 (0)20 7351 6300; www.clarencehouse.com) 119 centre left top. **Cole & Son** (Wallpapers) Ltd. (Tel: +44 (0)20 8442 8844; www.cole-and-son.com) 61-3039/Buckingham Stripe/Green 43 centre left; /62-1003/Humming Birds/Lavender 42 centre left. **Neisha Crosland**/Hawthorn/Gold Leaf 43 bottom left; /Speckle Dot/Pretty Blue 43 bottom right. **Crown Paint** (www.crownpaint.co.uk) 17 bottom, 73 bottom right, 95 top left, 111 top left, 113 bottom right. **Crown Wallcoverings**/Aquino Wallpaper by Crown 40; /Palazzo by Shand Kydd 103. **Crowson Fabrics** (www.crowsonfabrics.com) 14 bottom right; 127 top left, 143 top right. **Crucial Trading Ltd** (Tel: +44 (0)1675 433505; www.crucial-trading.com) 87 centre left bottom. **Dalsouple** (Tel: +44 (0)1278 727733; www.dalsouple.com) Hortensia 81 centre left top. **DIY Photo Library** 16, 50, 72 bottom left. **Dulux** (www.dulux.co.uk) 13 top right, 27, 95 bottom right, 108, 109 top left, 112, 209 top. **Fired Earth Interiors** (Tel: +44 (0)1295 814300; www.firedearth.com) 87 top left; /Moustier 54, 60 top left; /Early English Delft 125 bottom left; /Planet metallic tiles 59; /Antique reclaimed terracotta 75 bottom left; /Flagstones Slate 75 centre right; /Jerusalem Stone/Jericho 75 centre left bottom; /Kobe Mosaic 60 centre right; /Marble Mosaic/Crema 75 bottom right; /Retro Metro 60 bottom left; /Roman Mosaic 60 centre left;/Space 60 bottom right; /Turin Stone/Garda Blue 60 top right; /Red Terracotta 75 centre left top; /White Oak 75 top left. **Focus** (www.focusdiy.co.uk) Juniper bathroom 214. **Forbo** (Marmoleum by Forbo; Tel: +44 (0)800 731 2367; www.marmoleum.co.uk) 621 Dove Grey 81 top left. **Furniture 123** (www.furniture123.co.uk) 247 bottom left. **Getty Images** 119 bottom right; /Bob Elsdale 6

Oliver Gordon 7 top right, 13 bottom right, 15, 63, 71 top left, 76, 78, 81 bottom right, 93 top right, 94 top left, 98, 99 bottom right, 101 bottom right, 102 bottom left, 105 top left, 109 bottom right, 114, 117 top left, 131 top left, 133, 139, 156, 157 top, 158 bottom left, 159 bottom right, 160 left, 160 right, 167 bottom right, 170 bottom left, 171 bottom right, 179 bottom left, 180, 181 top left, 187 bottom right, 189 top left, 190, 193 bottom right, 194, 196 top right, 196 bottom left, 199, 202, 203 top left, 206, 207 bottom right, 223 bottom right, 224, 225 top left, 228, 229 top left, 236 bottom right, 238, 248 bottom left. **Graham & Brown Limited** (Tel: +44(0)800 3288 452; www.grahambrown.com) Arc wallpaper by Hemmingway design 48; /Cuba Super Fresco Texture 41. **H&R Johnson Tiles** (Tel: +44 (0)1782 575575; www.johnson-tiles.com) Coast 96; /Decadence 36; /Takara 122, 125 centre right top, 126 top right. **Heritage Bathrooms plc** (www.heritagebathrooms.com) Unity wetroom 219 bottom right. **The Holding Company** (Tel: +44 (0)20 8445 2888; www.theholdingcompany.co.uk) 235, 237 bottom right. **Homebase** (www.homebase.co.uk) 222 top right, 240, 241 bottom right, 241 bottom left, 242 bottom right, 243 top left, 244, 248 top right, 249 bottom, 250 bottom right, 250 bottom left, 251. **Ideal Standard Ltd** (www.ideal-standard.co.uk) Revue 215 top left; /Sophie 7 bottom left. **International Paints** (Tel: +44 (0)1480 484284; www.international-paints.co.uk) Applying anti-damp paint 17 top right; /Applying primer paint 23 bottom left. **KA International** (www.ka-international.com) 42 top right, 42 bottom right, 43 centre right, 143 bottom right. **Cath Kidston Ltd** (Information and order line: +44 (0)20 229 8000) 118 centre right, 119 top right, 119 centre right bottom. **Magnet** (www.magnet.co.uk) Harvard Maple bedroom 229 bottom right; /Minimalist bedroom 234; /Seville Kitchen 184 top left; /Stainless steel kitchen 183 bottom left. **Ray Main/Mainstream** 39 bottom right, 247 bottom right. **Marimekko** (www.marimekko.com) 119 top left. **MFI** (www.mfi.co.uk) 185 top left, 204. **Octopus Publishing Group Limited** 10 bottom right, 11 centre right top, 11 centre right bottom, 12, 18, 30 top right, 30 bottom right, 31 centre left, 31 bottom left, 33 top left, 33 centre left, 33 centre right, 33 bottom right, 107, 121 bottom left, 125 centre right bottom, 208 bottom right, 209 bottom left; /Graham Atkins-Hughes 135 top left, 137, 141; /Dominic Blackmore 97 bottom right; /Paul Forrester 11 left, 32, 35 top right, 35 bottom right, 35 centre right top, 35 centre right bottom, 58 bottom left; /Sebastian Hedgecoe 181 bottom right, 231 bottom left; /Tom Mannion 67, 71 bottom right, 237; /David Parmitter 142; /Paul Ryan 123 bottom left; /Russell Sadur 128; /Debi Treloar 144, 155 top right; /Shona Wood 10 bottom left, 10 bottom centre left, 10 bottom centre right, 29 top right, 29 centre right, 29 bottom right, 30 top left, 30 bottom left, 38, 39 bottom left, 56, 57; /Polly Wreford 73 top left, 94 bottom right, 221 bottom right, 236 top left. **Morris & Co. by Sanderson** (Tel: +44 (0)1895 830 044; www.sanderson-uk.com) 119 centre right top. **Mulberry Home** 119 bottom left. **Narratives Jan Baldwin** 100; /Kate Gadsby 246. **New House Textiles** (Tel: 01989 740684; www.newhousetextiles.co.uk) Carousel 149. **Roger Oates Design** (www.rogeroates.com) 87 centre right. **David Oliver** (Distributor: Paint and Paper Library, 5 Elystan Street, London, SE3 3NT) Basketweave/Olive 43 top left. **Original Style Ltd** (www.originalstyle.com) 58 top right, 65. **Osborne & Little** (Tel: +44 (0)20 7352 1456; www.osborneandlittle.com) F5052-03/Modello/Piolo 125 top right. **Red Cover**/Johnny Bouchier 243 bottom right; /Jake Fitzjones 191 top right; /Huntley Hedworth 113 top left; /Paul Massey 53; /Niall McDiarmid 213 bottom right; /Chris Tubbs 233. **Reed Harris** (www.reedharris.co.uk) Atlantic Glass Mosaic tile 62. **Romo** (www.romofabrics.com) Leoni 101 top left; /Arianne 121 top right; /Lorenzo 145; /Mazara 118 top right; /Paros 132, 133 bottom right; /Simonii 120 top right, 120 bottom left. **Sanderson** (Tel: +44 (0)1895 830 044; www.sanderson-uk.com) Reminiscence 118 centre left; /Joie de Vivre 127 bottom right; /Willow Leaf/Terracotta 42 centre right. **Sottini** (Tel: +44 (0)1482 449513, www.sottini.co.uk) 218. **Titley & Marr** (Tel: +44 (0)20 351 2913) 119 centre left bottom. **Wickes Building Supplies Ltd** (www.wickes.co.uk) Evesham dresser 184 bottom right. **Wilman Interiors** (www.wilman.co.uk) Elysees Collection Limoges Opal 110. **Zimmer & Rohde** (15 Chelsea Harbour Design Centre, London,SW10 0XE; Tel: +44 (0)20 7351 7115; www.zimmer-rohde.com) 118 bottom left.